Volkhard Helms

Principles of Computational Cell Biology

Related Titles

Emmert-Streib, F., Dehmer, M. (eds.)

Analysis of Microarray Data

A Network-Based Approach

2008

ISBN: 978-3-527-31822-3

Höltje, H.-D., Sippl, W., Rognan, D., Folkers, G.

Molecular Modeling

Basic Principles and Applications

2008

ISBN: 978-3-527-31568-0

Klipp, E., Herwig, R., Kowald, A., Wierling, C., Lehrach, H.

Systems Biology in Practice

Concepts, Implementation and Application

2005

ISBN: 978-3-527-31078-4

Baxevanis, A. D., Ouellette, B. F. F. (eds.)

Bioinformatics

A Practical Guide to the Analysis of Genes and Proteins

2004

ISBN: 978-0-471-47878-2

Flaig, R.-M.

Bioinformatics Programming in Python

A Practical Course for Beginners

2008

ISBN: 978-3-527-32094-3

Volkhard Helms

Principles of Computational Cell Biology

From Protein Complexes to Cellular Networks

WILEY-VCH Verlag GmbH & Co. KGaA

The Author

Prof. Dr. Volkhard Helms
Saarland University
Centre for Bioinformatics
P.O. Box 15 11 50
66041 Saarbrücken
Germany

Library of Congress Card No.: applied for

British Library Cataloguing-in-Publication Data
A catalogue record for this book is available from the British Library.

Bibliographic information published by the Deutsche Nationalbibliothek
Die Deutsche Nationalbibliothek lists this publication in the Deutsche Nationalbibliografie; detailed bibliographic data are available in the Internet at http://dnb.d-nb.de.

Typesetting Thomson Digital, Noida, India
Printing Strauss GmbH, Mörlenbach
Binding Litges & Dopf Buchbinderei GmbH, Heppenheim

Printed in the Federal Republic of Germany
Printed on acid-free paper

ISBN: 978-3-527-31555-0

Contents

Principles of Computational Cell Biology – From Protein Complexes to Cellular Networks. Volkhard Helms

ISBN: 978-3-527-31555-0

Preface

This book grew out of a course for graduate students in the first year of the MSc bioinformatics program that the author teaches every year at Saarland University. Also included is some material from a special lecture on cell simulations. The book is designed as a textbook, placing emphasis on transmitting the main ideas of a problem, outlining algorithmic strategies for solving it and describing possible complications or connections to other parts of the book. The main challenge during the writing of the book was the concentration on conceptual points that may be of general educative value rather than including the latest research results of this fascinating fast-moving field. It is considered more important for a textbook to give a cohesive picture rather than mentioning all possible drawbacks and special cases where particular general guidelines may not apply. We apologize to those whose work could not be mentioned due to space constraints.

The intended audience includes students of bioinformatics and from life sciences disciplines. Consequently, some basic knowledge in molecular biology is taken for granted. The language used is not very formal. Previous knowledge of computer science is not required, but a certain adeptness in basic mathematics is necessary. The book introduces all of the mathematical concepts needed to understand the material covered. In particular, Chapter 2 introduces mathematical graphs and algorithms on graphs used in classifying protein–protein interaction networks. Chapter 6 introduces linear and convex algebra typically used in the description of metabolic networks. Chapter 7 discusses ordinary and stochastic differential equations used in the kinetic modeling of signal transduction pathways. Chapter 8 introduces the method of Fourier transformation used for protein–protein docking and pattern matching. Also introduced are Bayesian networks in Chapter 4 as a way to judge the reliability of protein–protein interactions and inference techniques to model gene regulatory networks. We note, however, that the emphasis of this book is placed on discrete mathematics rather than on statistical methods. Not included yet are classical network flow algorithms such as Menger's theorem or the max-flow min-cut theorem as they are currently rarely used in cellular modeling. The book focuses on proteins and the genes coding for them, as well as on metabolites. Less room is given to DNA, RNA or lipid membranes that would, of course, also deserve a

Principles of Computational Cell Biology – From Protein Complexes to Cellular Networks. Volkhard Helms

ISBN: 978-3-527-31555-0

great deal of attention. The main reason for this was to provide a homogenous background for discussing algorithmic concepts.

I am very grateful to Dr Tihamér Geyer who coordinated the assignments for the lectures for valuable comments on the manuscript, and for many solved examples and problems. The following co-workers from Saarbrücken and elsewhere have provided valuable suggestions on different portions of the text: Kerstin Kunz, Jan Christoph and Florian Lauck, I thank Dr. Hawoong Jeong, Dr. Julio Collado-Vides and Dr. Agustino Martínez-Antonio, Dr. Ruth Sperling, Dr. James R. Williamson, Dr. Joanna Trylska, Dr. Claude Antony and Dr. Nicholas Luscombe for sending me high-resolution versions of their graphics. I thank Dr. Andreas Sendtko and the publishing staff at Wiley-VCH for their generous support of this book project, for their seemingly endless patience during the revision stage, and for the excellent type-setting.

I also thank the Center of Theoretical Biophysics at the University of California, San Diego for their hospitality during a sabbatical visit in summer 2007 that finally allowed me to complete this work. Finally, this book would not have been possible without the support and patience of my wife Regina and our two daughters.

March 2008

Volkhard Helms
Center for Bioinformatics
University of Saarland

Problems

To really absorb the content of this textbook it is advisable to also try to solve some of the problems enclosed.

1
Networks in Biological Cells

Modern molecular and cell biology has worked out many important cellular processes in great detail, although some other areas are known to a lesser extent. What often remains is to understand how the individual parts are connected. One may wonder whether mathematical modeling can make a contribution to this field before the missing details are known. Figure 1.1 displays a cartoon of a cell as a highly viscous soup containing a complicated mixture of many particles. Certainly, several important details are left out here that introduce a partial order, such as the cytoskeleton and organelles of eukaryotic cells. The point of Figure 1.1 is to remind us that there is a myriad of biomolecular interactions taking place in biological cells at all times and that it is pretty amazing how considerable order is achieved in many cellular processes that are all based on pairwise molecular interactions.

The focus of this textbook is placed on presenting mathematical descriptions developed in recent years to describe various levels of cellular networks that are mostly based on molecular interactions. We will learn that many biological processes are tightly interconnected and this is exactly where many links still need to be discovered in further experimental studies. It is the belief of many researchers in the field of molecular biology that only combined efforts of modern experimental techniques and mathematical modeling and bioinformatics analysis will be able to arrive at a sufficient understanding of the biological network of cells and organisms.

In this first chapter we will start with some principles of mathematical networks and their relationship to biological networks. Then we will briefly look at several of the key biological players to be used in the remainder of this textbook (cells, compartments, proteins, pathways). Without going into any further detail, we will now jump right into the fast-growing field of network theory with the amazing "small-world phenomenon".

1.1
Some Basics about Networks

Network theory is a branch of applied mathematics and more of physics that uses the concepts of graph theory. Its developments are led by application to real-world

Principles of Computational Cell Biology – From Protein Complexes to Cellular Networks. Volkhard Helms

ISBN: 978-3-527-31555-0

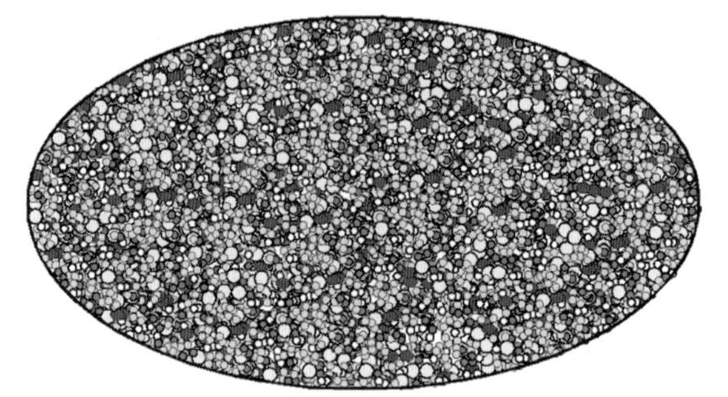

Figure 1.1 Is this how we should view a biological cell? This schematic picture makes an important point: about 30% of the volume of a biological cell is taken up by millions of individual proteins. Thus, biological cells are really "full". However, such pictures do not tell us much about the organization of biological processes and, as we will see later in this textbook, there are many different hierarchies of order in such a cell.

examples such as acquaintance networks and collaboration networks (which fall under the class of social networks), technological networks (such as the Internet, the World Wide Web and power grids), and biological networks (such as neural networks, food webs and metabolic networks).

1.1.1
Random Networks

In a random network, every possible link between two "vertices" (or nodes) A and B is established according to a given probability distribution irrespective of the nature and the connectivities of the two vertices A and B. This is what is "random" about these networks. If the network contains n vertices in total, the maximal number of undirected edges (links) between them is $n \times (n-1)/2$. This is because we can pick each of the n vertices as the first vertex of an edge and there are $(n-1)$ other vertices that this vertex can be connected to. In this way, we will actually consider each edge twice, using each end point as the first vertex. Therefore, we need to divide the number of edges by 2.

If every edge is established with a probability p, the total number of edges in the graph is $p \times n \times (n-1)/2$. The mathematics of random graphs was developed and elucidated by the two Hungarian mathematicians Erdös and Renyi. However, the analysis of real networks showed that they often differ significantly from the characteristics of random graphs.

1.1.2
Small-World Phenomenon

The term **small-world phenomenon** was created after the observation that everyone in the world can be reached by some other person through a short chain of social acquaintances. A 1967 **small-world experiment** by psychologist Stanley Milgram found that any two random US citizens were connected by an average of six acquaintances and this gave rise to the famous phrase "six degrees of separation". However, after more than 40 years, its status as a description of heterogeneous social networks still remains an open question. The average distance between vertices in a network is short, usually scaling logarithmically with the total number n of vertices.

In a paper published in the journal *Nature* in 1998, the two mathematicians Duncan J. Watts and Steven H. Strogatz (Watts and Strogatz, 1998) reported that small-world networks are common in a variety of different realms ranging from neuronal connections of the worm *Caenorhabditis elegans* to power grids. Watts and Strogatz also showed that the addition of a handful of random edges can turn a disconnected network into a highly connected one. For example, the addition of a few judicious routers makes a vast communication network (such as the Internet) no more than six hops wide.

1.1.3
Scale-Free Network Model

Only 1 year after the discovery of Watts and Strogatz, Albert-László Barabási from the Physics Department at the University of Notre Dame introduced an even simpler model for the emergence of the small-world phenomenon. While Watts and Strogatz's model was able to explain the high clustering coefficient and the short average path length (these terms will all be introduced in Chapter 3) of a *small world*, it lacked an explanation for another property found in real-world networks such as the Internet: these networks are **scale-free**. In simple terms, this means that while the vast majority of vertices are weakly connected, there also exist some highly interconnected super-vertices or **hubs**. The term scale-free expresses that the ratio of highly to weakly connected vertices remains the same irrespective of the total number of links in the network. We will see in Chapter 4 that the connectivity of scale-free networks follows a power law. If a network is scale-free, it is also a small world.

Barabási's scale-free model is strikingly simple, elegant and intuitive. To produce an artificial scale-free network possessing the small-world properties, only two basic rules must be followed:

- **Growth.** The network is seeded with a small number of initial vertices. At every time step, a new vertex is added that forms connections to m existing vertices.
- **Preferential attachment.** The probability of a newly added vertex connecting to an existing vertex n is assumed to depend on the degree of n (the number of connections already formed between vertex n to other vertices). The more connections n has, the more likely new vertices will connect to n. This behavior is also described by the saying "the rich become richer".

The same mechanism is at work, for example, in the World Wide Web. Obviously, this network is in a constant state of growth where new pages are added every second. Also, we know from our own experience that once a user creates a new webpage, they will most likely include links to other well-known pages (hubs) on this page. In the early exciting days when the study of large-scale networks took off like a storm, it was even suggested that the scale-free network model may be the foundation for a law of nature which governs the formation of natural small-world networks.

However, recent work on integrated biological networks showed that the concept of scale-free networks may rather be of theoretical value and that it may not be directly applicable to certain biological networks. We will return to this issue at the end of the textbook (Chapter 10) when looking at integrated networks. For the moment, we will consider the idea of network topology (scale-free networks, small-world phenomenon) as an enormously powerful concept, and useful for understanding the mechanism of network growth and vulnerability.

1.2 Biological Background

Until recently, the paradigm of molecular biology was that genetic information is read from the genomic DNA by the RNA polymerase complex and is **transcribed** into corresponding RNA. Ribosomes then bind to messenger RNA (mRNA) snippets and produce amino acid strands. This process is called **translation**. Importantly, the paradigm involved the notion that this entire process is unidirectional (Figure 1.2).

It is now well established that feedback loops are provided in this system, e.g. by the proteins known as transcription factors that bind to sequence motifs on the genomic DNA and mediate (activate or repress) transcription of certain genomic segments. An

Central Paradigm of Molecular Biology

DNA → RNA → Protein → Phenotype (Symptoms)

Central Paradigm of Structural Biology

Genetic Information → Molecular Structure → Biochemical Function → Phenotype (Symptoms)

Central Paradigm of Molecular Systems Biology

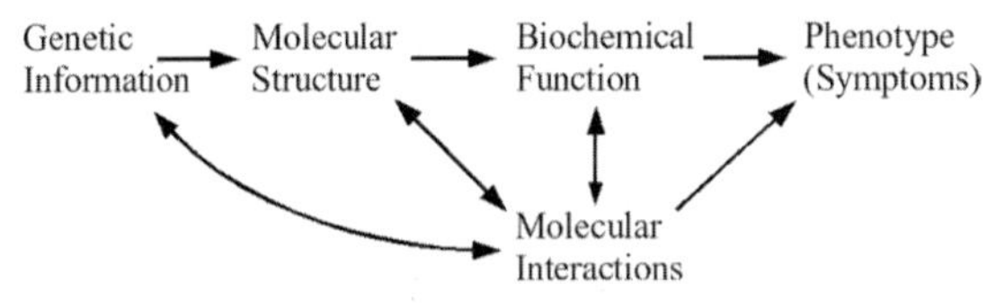

Figure 1.2 (Top) Since the 1950s, a paradigm became established that information flows from DNA over RNA to protein synthesis, which then gives rise to particular phenotypes. (Middle) The emergence of structural biology – the first crystal structure of the protein myoglobin was determined in 1960 – emphasized the importance of the three-dimensional structures of proteins determining their function. (Bottom) Today, we have realized the central role played by molecular interactions that influence all other elements.

important discovery of the last 10 years showed that small RNA snippets may also mediate gene expression. The cellular network therefore certainly appears much more complicated today than it did 50 years ago.

This brings us to the world of **gene regulatory networks**. To discover which gene is activated or repressed by a particular transcription factor, one could create a knock-out mouse lacking the gene coding for this transcription factor and see which genes are no longer expressed or are now expressed in excess. However, in this way, we can only discover those combinations that are not lethal for the organism. Also, pairs or larger assemblies of transcription factors often need to bind simultaneously. It simply appears impossible to discover the full connectivity of this regulatory network by a traditional one-by-one approach. Modern microarray experiments, however, probe the expression levels of large numbers of genes simultaneously. Yet, it quickly turns out that the analysis of this large-scale data is complicated by the noisy nature of the data and by the fact that genes do not interact directly with each other.

Here, we will be mostly concerned with the following four types of biological cellular networks: protein–protein interaction networks, gene regulatory networks, signal transduction networks and metabolic networks. We will discuss them at different hierarchical levels as shown in Figure 1.3 using the example of regulatory networks.

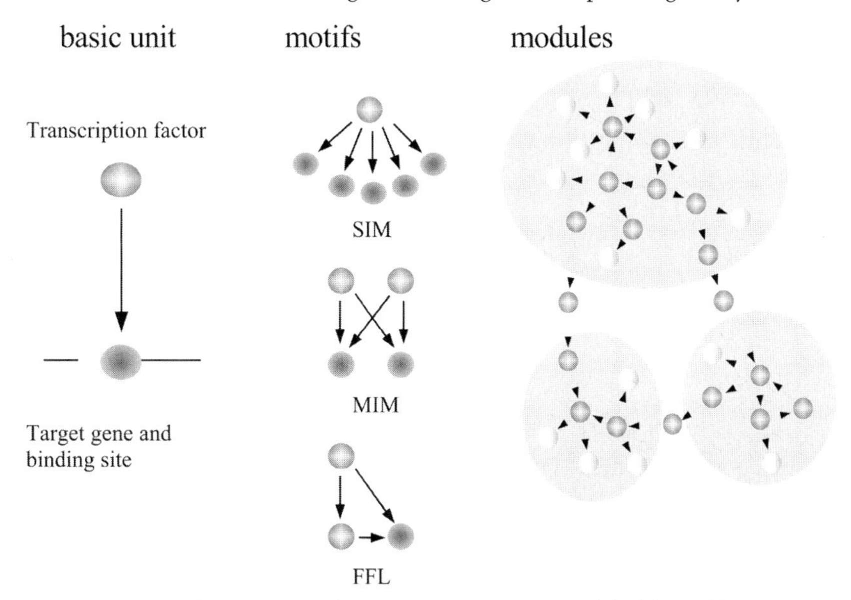

Figure 1.3 Structural organization of transcriptional regulatory networks. (Left) The "basic unit" comprises the transcription factor, its target gene with DNA recognition site and the regulatory interaction between them. (Middle) Units are often organized into network "motifs" that comprise specific patterns of inter-regulation that are over-represented in networks. Examples of motifs include single-input/multiple-output (SIM), multiple-input/multiple-output (MIM) and feed-forward loop (FFL) motifs. (Right) Network motifs can be interconnected to form semi-independent "modules", many of which have been identified by integrating regulatory interaction data with gene expression data and imposing evolutionary conservation. The next level consists of the entire network (not shown). Figure drawn after Babu *et al.* (2004).

1.2.1 Cellular Components

Cells can be described at various levels of detail. Here, we will mostly use three different levels of description:

(a) Inventory lists and lists of processes:
 –Proteins in particular compartments
 –Proteins forming macromolecular complexes
 –Biomolecular interactions
 –Metabolic reactions

(b) Structural descriptions:
 –Single protein structures
 –Protein complexes
 –Subcellular compartments

(c) Dynamic descriptions:

– Cellular processes ranging from nanosecond dynamics for the association of two biomolecules up to processes occurring in seconds and minutes such as the cell division of yeast cells

We will assume that the reader has a basic knowledge about the organic molecules commonly found within living cells and refer those who do not to basic textbooks on biochemistry or molecular biology. These biomolecules in a cell can be divided into several categories based on their role in metabolism.

(1) **Macromolecules** including proteins, nucleic acids, polysaccharides and certain lipids.

(2) The **building blocks** of macromolecules including sugars as the precursors of polysaccharides, amino acids as the building blocks of proteins, nucleotides as the precursors of nucleic acids (and therefore of DNA and RNA) and fatty acids which are incorporated into lipids. Interestingly, in biological cells, only a small number of the theoretically synthesizable macromolecules exist at a given point in time. At any moment during a normal cell cycle, many new macromolecules need to be synthesized from their building blocks and this is meticulously controlled by the complex gene expression machinery. Even during a steady-state of the cell, there exists a constant turnover of macromolecules.

(3) Metabolic intermediates (**metabolites**). The molecules in a cell have complex chemical structures and must be synthesized step-by-step beginning with specific starting materials that may be taken up as energy source. In the cell, connected chemical reactions are often grouped into metabolic pathways (Section 1.3).

(4) Molecules of **miscellaneous function** including vitamins, steroid or amino acid hormones, molecules involved in energy storage (e.g. ATP), regulatory molecules (e.g. cyclic AMP) and metabolic waste products such as urea.

Almost all biological material needed to construct a biological cell is either synthesized by its RNA polymerase and ribosome machinery or is taken up from the outside via the cell membrane. Therefore, as a minimum inventory every cell needs to contain the construction plan (DNA), a processing unit to transcribe this information into mRNA (polymerase), a processing unit to translate these mRNA pieces into protein (ribosome) and transporter proteins inside the cell membrane that transport material through the cell membrane.

1.2.2
Spatial Organization of Eukaryotic Cells – Compartments

Organization into various compartments greatly simplifies the temporal and spatial process flow in eukaryotic cells. As mentioned above, at each time point during a cell cycle only a small subfraction of all potential proteins are being synthesized (and not yet degraded). Also, many proteins are only available in very small concentrations, possibly with only a few copies per cell. However, due to localizing these proteins to particular spots in the cell, e.g. by attaching them to the cytoskeleton or by partitioning them into lipid rafts, their local concentrations may be much higher. We assume that the reader is vaguely familiar with the compartmentalization of eukaryotic cells involving the lysosome, plasma membrane, cell membrane, Golgi complex, nucleus, smooth endoplasmic reticulum, mitochondrion, nucleolus, chromatin, rough endoplasmic reticulum and cytoskeleton.

1.2.3
Cellular Organisms

Table 1.1 presents some statistics of the organisms considered in this textbook. Determination of RNA-coding genes is still in its infancy.

1.3
Cellular Pathways

1.3.1
Biochemical Pathways

Metabolism denotes the entirety of biochemical reactions that occur within a cell (Figure 1.4). In the past century, most of these reactions have been grouped into **metabolic pathways** that each contain a sequence of chemical reactions in which each reaction is catalyzed by a specific enzyme and the product of one reaction is the substrate for the next one. Unraveling the individual enzymatic reactions was one of the big successes of applying biochemical methods to cellular processes. Metabolic pathways can be divided into two broad types. **Catabolic pathways** lead to the disassembly of complex molecules to form simpler products. They provide the raw materials for the synthesis of other molecules and they provide chemical energy

Table 1.1 Data on the genome length and on the number of protein and RNA genes are taken from the Kyoto Encyclopedia of Genes and Genomes (KEGG) database (July 2007); data on the number of putative transporter proteins are taken from www.membranetransport.org.

Organism	Length of genome (Mb)	Number of protein genes	Number of RNA genes	Number of transporter proteins
Prokaryotes				
Methanococcus jannaschii	1.7	1786	43	49
Bacillus subtilis	4.2	4106	119	297
Escherichia coli	4.6	4131	168	354
Eukaryotes				
Saccharomyces cerevisiae	12.2	5879	410	327
Drosophila melanogaster	180	14081	24	615
Caenorhabditis elegans	97	20077	24	656
Homo sapiens	2880	25307	25	784

required for many cellular activities. Energy released by catabolic pathways is stored temporarily either as high-energy phosphates (primarily ATP) or as high-energy electrons (primarily in NADPH). **Anabolic pathways** lead to the synthesis of more complex compounds from simpler starting materials. Anabolic pathways are energy requiring and utilize chemical energy released by the exergonic catabolic pathways.

The traditional biochemical pathways were often derived from studying simple organisms where these pathways constitute a dominating part of the metabolic activity. For example, the **glycolysis** pathway was discovered in yeast (and in muscle) in the 1930s. It describes the disassembly of the nutrient glucose that is taken up by many microorganisms from the exterior. Figure 1.5 shows the glycolysis pathway in *Homo sapiens* as represented in the KEGG database (Kanehisa *et al.*, 2006).

1.3.2
Enzymatic Reactions

Enzymes are proteins that catalyze biochemical reactions. Like all catalysts, enzymes work by lowering the **activation energy** of a reaction, thus allowing the reaction to proceed much faster than in aqueous solution. Remarkably, enzymes may speed up reactions by factors of many thousands to billions of times. An enzyme, like any catalyst, remains unaltered by the completed reaction and can therefore continue to function. As enzymes do not affect the relative free energy difference between the products and reagents, they do not affect the equilibrium of a reaction. The great advantage of enzymes compared to most other catalysts is their stereo-, regio- and chemoselectivity and -specificity.

For the binding reaction $A + B \leftrightarrow AB$, the **binding constant** k_d:

$$k_d = \frac{[A] \cdot [B]}{[AB]},$$

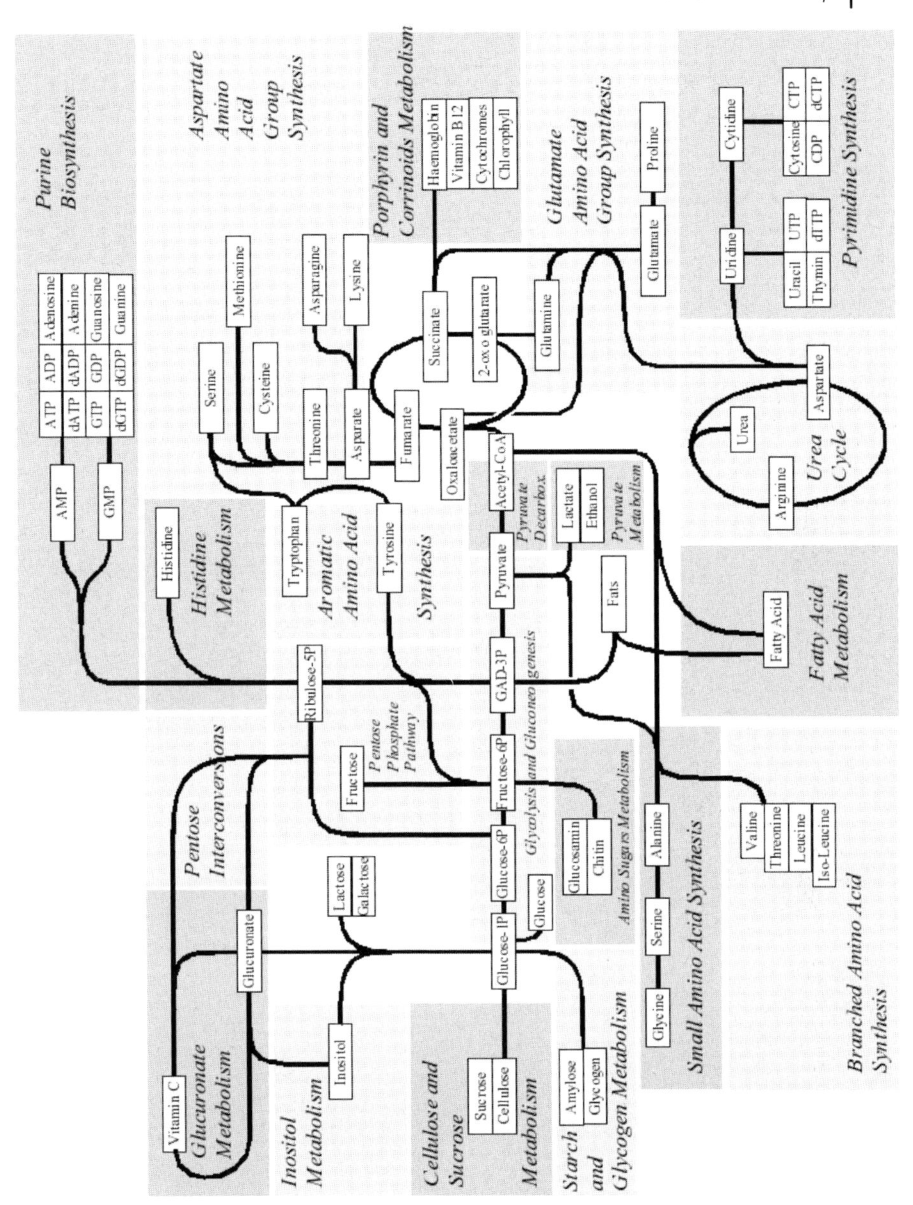

Figure 1.4 Major metabolic pathways.

determines how much of the substrate concentration is bound to the enzyme under equilibrium conditions. The binding constant has the unit M. In the case of a "nanomolar inhibitor", for example, the equilibrium is very strongly on the complexed form and only a few free molecules exist. The binding constant k_d is also the ratio of the kinetic rates for the forward and backward reactions, k_{on} and

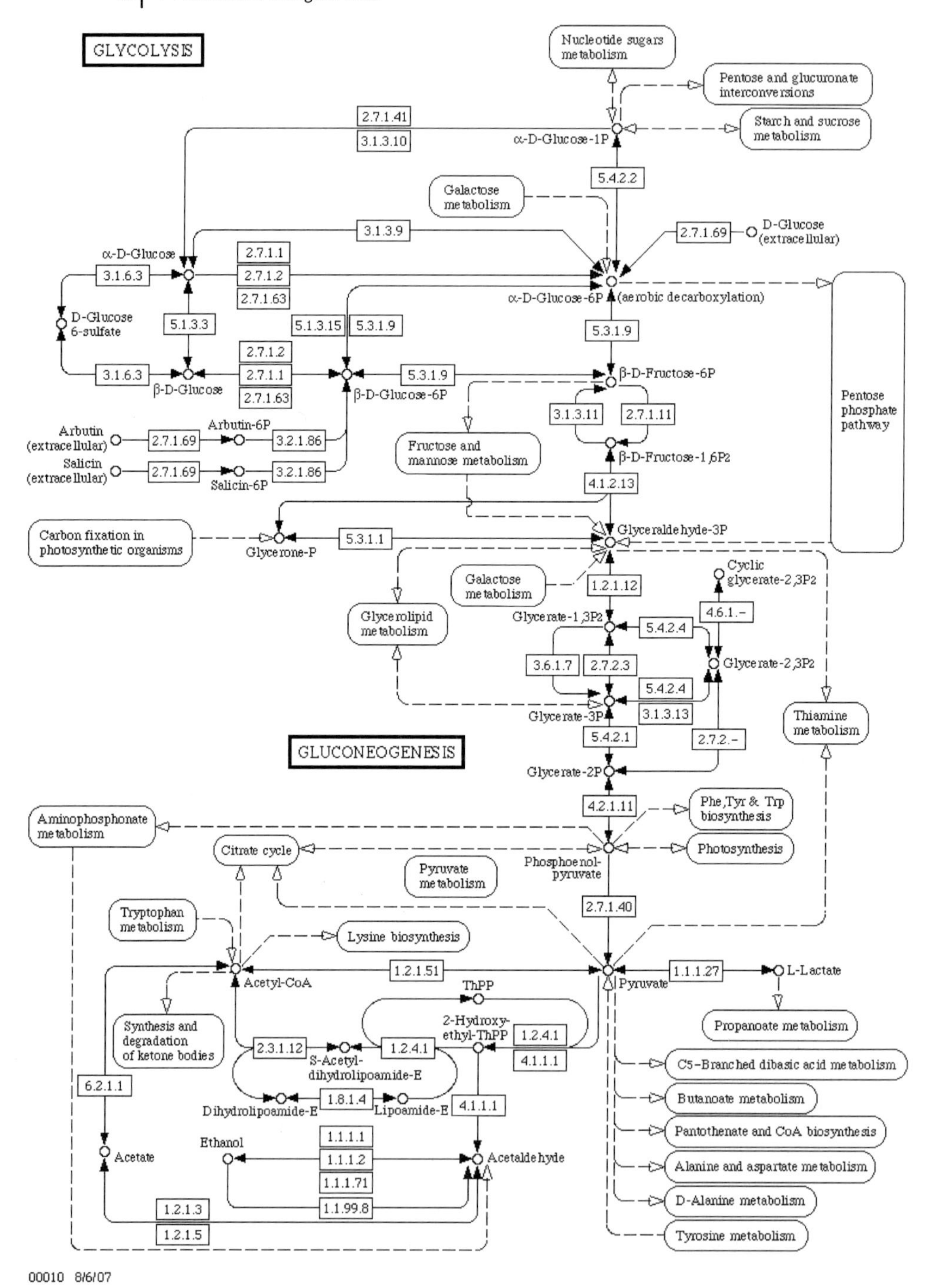

Figure 1.5 The glycolysis pathway as visualized in the KEGG database is connected to many other cellular pathways (picture taken with permission from http://www.genome.ad.jp/kegg/).

k_{off}. The units of the two kinetic rates are $M^{-1} s^{-1}$ for the forward reaction and M^{-1} for the backward reaction.

Understanding the fine details of enzymatic reactions is one of the main branches of biochemistry. Fortunately, in the context of cellular simulations, we need not be interested in the enzymatic mechanisms themselves. Here, instead, it is important to characterize the chemical diversity of the substrates a particular enzyme can turn over, and to collect the thermodynamic and kinetic constants of all relevant catalytic and binding reactions. A rigorous system to classify enzymatic function is the Enzyme Classification (EC) scheme. It contains four major categories, each divided into three hierarchies of subclassifications.

1.3.3 Signal Transduction

Here, we denote by **signal transduction** the transmission of a chemical signal such as phosphorylation of a target amino acid, and separate this from energy transduction cycles that will be described separately (see, e.g. Section 7.7). Signal transduction is a very important subdiscipline of cell biology. Hundreds of working groups are looking at separate aspects of signal transduction and large research consortia such as the Alliance of Cell Signaling have been formed in the past.

1.3.4 Cell Cycle

The **cell cycle**, or **cell division cycle**, is the cycle of events in a eukaryotic cell from one cell division to the next. It consists of interphase, mitosis and, usually, cell division. The cell cycle is regulated by cyclins and cyclin-dependent kinases. In 2001, the Nobel Prize in Physiology or Medicine was awarded to Leland H. Hartwell, R. Timothy Hunt and Paul M. Nurse for their discovery of these central molecules in the regulation of the cell cycle. The phases of the cell cycle are:

- The **G_0 phase** that is a period in the **cell cycle** where cells exist in a quiescent state.
- The **G_1 phase** that is the first growth phase.
- The **S phase**, during which the DNA is replicated, where S stands for the synthesis of DNA.
- The **G_2 phase** that is the second growth phase, also the preparation phase for the cell.
- The **M phase** or mitosis and cytokinesis that covers the actual division of the cell into two daughter cells.

A surveillance system, so-called "**checkpoints**", monitors the cell for DNA damage and failure to perform critical processes. Checkpoints can block progression through the different stages of the cell cycle if certain conditions are not met. For instance, one checkpoint monitors DNA replication and prevents cells from proceeding to mitosis before DNA replication is completed. Similarly, the spindle checkpoint blocks the transition from metaphase to anaphase within mitosis if not all chromosomes are attached to the mitotic spindle. We will see in Section 7.2 how cellular processes may dynamically regulate each other.

1.4 Ontologies and Databases

1.4.1 Ontologies

"Ontology" is a term from philosophy and describes a structured controlled vocabulary. Why have ontologies nowadays become of particular importance in biological and medical sciences? The main reason is that, historically, biologists worked in separate camps, each on a particular organism, and each camp discovered gene after gene, protein after protein. Due to this separation, every subfield started using its own terminology. These early researchers did not know that, at a later stage, biologists wished to compare different organisms to transfer useful information from one to the other in a process termed **annotation**. Thus, proteins deriving from the same ancestor may have been given completely different names.

It would require many years of intensive study for any one of us to learn these associations. Instead, researchers realized quite early that it would be extremely useful to generate general repositories for classification schemes that connect corresponding genes and proteins belonging to different organisms or which provide access to functional annotations. One important project in this area is the **Gene Ontology** (GO) (www.geneontology.org). This collaborative project started in 1998 as a collaboration of three model organism databases, FlyBase (*Drosophila*), the *Saccharomyces* Genome Database (SGD) and the Mouse Genome Database (MGD).

In the GO project, gene products are described in terms of their associated biological processes, cellular components and molecular functions in a species-dependent manner. A gene product might be associated with or located in one or more cellular components; it is active in one or more biological processes, during which it performs one or more molecular functions.

1.4.2 Systems Biology Markup Language

The **systems biology markup language (SBML)** has been formulated to allow the well-defined construction of cellular reaction systems and allow exchange of simulation models between different simulation packages. The idea is to be able to interface models of different resolution and detail. Cell simulation methods usually import and export (sub)cellular models in SMBL language.

1.4.3 KEGG

Initiated in 1995, KEGG is an integrated bioinformatics resource consisting of three types of databases for genomic, chemical and network information. KEGG consists of three graph objects called the gene universe (GENES, SSDB and KEGG Orthology databases that contain more than 1.2 million genes from 30 eukaryotic, 250 bacterial and 25 archeal genomes), the chemical universe (COMPOUND, GLYCAN and REACTION databases that contain more than 13 000 chemical compounds and more than 6000 reactions) and the protein network (PATHWAY database) (Table 1.2). The gene universe is a conceptual graph object representing ortholog/paralog relations, operon information and other relationships between genes in all the completely sequenced genomes. The chemical universe is another conceptual graph object representing chemical reactions and structural/functional relations among metabolites and other biochemical compounds. The protein network is based on biological phenomena, representing known molecular interaction networks in various cellular processes.

1.4.4 Brenda

Since 1987, the Brenda resource has been developed at the German National Research Center for Biotechnology and later at the University of Cologne. It is a comprehensive information system on enzymatic reactions. Data is stored in a relational database containing all data in 46 tables, enabling different queries (Table 1.3). Data on enzyme function are extracted directly from the primary literature by scientists holding a degree in Biology or Chemistry. Formal and consistency

Table 1.2 The three graph objects in KEGG (after Kanehisa *et al.*, 2004).

Graph	Vertex	Edge	Main databases
Gene universe	gene	any association of genes (ortholog/paralog relation, sequence/structural similarity, adjacency on chromosome, expression similarity)	GENES, SSDB, KO
Chemical universe	chemical compound (including carbohydrate)	any association of compounds (chemical reactivity, structural similarity, etc.)	COMPOUNDS, GLYCAN, REACTION
Protein network	protein (including other gene products)	known interaction/relation of proteins (direct protein–protein interaction, gene expression relation, enzyme–enzyme relation)	PATHWAY

Table 1.3 Overview on Information stored in the Brenda system on each particular biochemical reaction.

Nomenclature	Enzyme names, EC number, common/recommended name, systematic name, synonyms, CAS registry number
Reaction and specificity	pathway, catalyzed reaction, reaction type, natural and unnatural substrates and products, inhibitors, cofactors, metals/ions, activating compounds, ligands
Functional parameters	K_m value, K_i value, pI value, turnover number, specific activity, pH optimum, pH range, temperature optimum, temperature range
Isolation and preparation	purification, cloned, renatured, crystallization
Organism-related information	organism, source tissue, localization
Stability	stability with respect to pH, temperature, oxidation, and storage; stability in organic solvent
Enzyme structure	links to sequence/SwissProt entry, three-dimensional structure/Protein Data Bank entry, molecular weight, subunits, posttranslational modification
Disease	disease

checks are performed by computer programs; each dataset on a classified enzyme is checked manually by at least one biologist and one chemist.

One may wonder whether all this detail is required by a computational cell biologist analyzing the network capacities of a particular organism. In some ways no, in other ways yes. No, if you only want to analyze the pathway space (Chapter 6). Yes, if you are interested in particular reaction rates or in modeling time-dependent processes. Computer scientists among the readers of this text should become aware that the rates of biochemical reactions vary significantly with temperature and pH, and may even change their directions.

1.5
Methods in Cellular Modeling

Table 1.4 presents an overview of the methods in cellular modeling that are covered in this textbook.

Summary

This introductory chapter gives a first look at the cellular components that will be the objects of computational and mathematical analysis in the rest of the textbook. Obviously, it was not intended to provide a rigorous introduction, but rather to whet the appetite of the reader without spending too much time on subjects that many readers will be very familiar with.

We have seen that the central paradigms of molecular biology (a linear information flow from DNA → RNA → proteins) and of cellular biochemistry (grouping of

Table 1.4 Mathematical techniques used in computational cell biology that are covered in this text.

Mathematical concept	Object of investigation	Analysis of complexity	Time dependent	Chapter(s)
Mathematical graphs	protein–protein networks and protein complexes	yes	no	2–4, 9
Stoichiometric analysis; matrix algebra	metabolic networks[a]	yes (count number of possible paths that connect two metabolites)	no	6
Differential equations	signal transduction, energy transduction	no	yes	7
Equations of motion	individual proteins, protein complexes	no	yes	9
Correlation functions, Fourier transformation	reconstruction of two- and three-dimensional structures of cellular structures and individual molecules	no	yes, when applied on time-dependent data	8

[a]May also be applied to gene regulatory networks and signal transduction networks.

biochemical reactions into major pathways) are being challenged by new discoveries on the roles of small RNA snippets, and by the discovery of highly interconnected hub proteins and metabolites that seem to connect almost "everything to everything". This is certainly why mathematical and computational analysis is now needed to keep the overview over all of the data being generated and to deepen our understanding about cellular processes.

Further Reading

Small-World Networks

Watts DJ, Strogatz SH (1998) Collective dynamics of 'small-world' networks. *Nature* **393**, 409–410.

Gene Regulatory Networks

Babu MM, Luscombe NM, Aravind L, Gerstein M, Teichmann SA (2004) Structure and evolution of transcriptional regulatory networks, *Current Opinion in Structural Biology* **14**, 283–291.

The KEGG Database

Kanehisa M, Goto S, Kawashima S, Okuno Y, Hattori M (2004) The KEGG resource for deciphering the genome, *Nucleic Acids Research* **32**, D277–D280.

Kanehisa M, Goto S, Hattori M, Aoki-Kinoshita KF, Itoh M, Kawashima S, Katayama T, Araki M, Hirakawa M (2006) From genomics to chemical genomics: new developments in KEGG. *Nucleic Acids Research* **34**, D354–D357.

2
Algorithms on Mathematical Graphs

In this chapter, we introduce the mathematical object of a graph and some basic algorithms that operate on graph structures. These concepts are essential for analyzing the topologies of protein–protein interaction networks in Chapter 4.

2.1
Primer on Mathematical Graphs

A **graph** G is an ordered pair (V,E) of a set V of **vertices** and of a set E of **edges**, see (Figure 2.1). E is always a subset of the set $V^{(2)}$ of unordered pairs of V which consists of all possible connections between all vertices. If $E = V^{(2)}$ we say the graph is fully connected. Below we will consider fully connected subsets of the full graph that are also called cliques. A **weighted graph** has a real or integer valued weight assigned to each edge. A **subgraph** of a graph G is a graph whose vertex and edge sets are subsets of those of G.

A **path** in a graph is a sequence of vertices such that from each of its vertices there is an edge to the successor vertex. The first vertex is called the **start vertex** and the last vertex is called the **end vertex**. Both of them are called **end or terminal vertices** of the path. The other vertices in the path are **internal vertices**. Two paths are **independent** (alternatively called **internally vertex-disjoint**) if they do not have any internal vertex in common (Figure 2.2). Given an undirected graph, two vertices u and v are called **connected** if there exists a path from u to v. Otherwise they are called disconnected. The graph is called a **connected graph** if every pair of vertices in the graph is connected. A **connected component** is a maximal connected subgraph. Maximal means here that it can only be enlarged by rearranging edges. The **giant component** is a term from network theory referring to a connected subgraph that contains a majority of the entire graph's vertices.

A **walk** is an alternating sequence of vertices and edges, beginning and ending with a vertex. The length l of a walk is the number of edges that it uses. A **trail** is a walk in which all the edges are distinct. A **cycle** denotes here a closed path with no repeated vertices other than the starting and ending vertices.

The **shortest path problem** is the problem of finding a path between two vertices such that the sum of the weights w_i of its constituent edges is minimized. More

Principles of Computational Cell Biology – From Protein Complexes to Cellular Networks. Volkhard Helms

ISBN: 978-3-527-31555-0

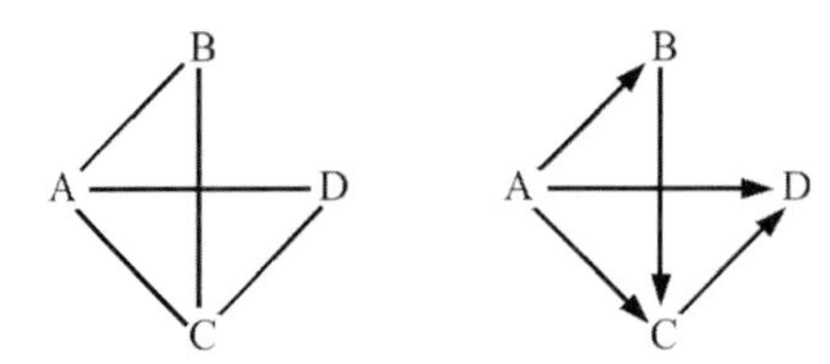

Figure 2.1 A mathematical graph consists of vertices and edges. (Left) An undirected graph is shown consisting of four vertices (A, B, C and D) and five edges (connections). This example could represent the results from a yeast two-hybrid experiment probing binary protein–protein interactions that gave positive results for five interactions A–B, A–D, A–C, B–C and C–D. (Right) Almost the same system is shown, but this time as a directed graph with arrows (arcs) instead of edges. This example could, for example, visualize a gene regulatory network where a transcription factor A controls the expression of genes B, C, D, etc. Here, A, B, C and D are the four vertices of the graph, and the five arcs are the directed edges of the graph.

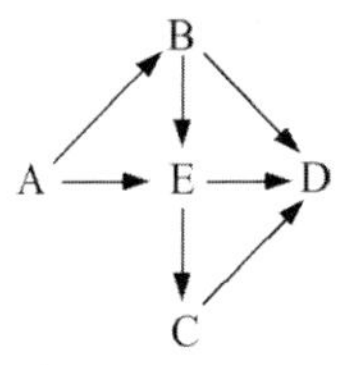

Figure 2.2 Vertices A and D are connected by five paths (A → B → D, A → B → E → D, A → B → E → C → D, A → E → D, A → E → C → D). Only two of these paths are independent: A → B → D and either A → E → D or A → E → C → D.

formally, given a weighted graph, and given further two elements u, v of V, find a path P from u to v so that:

$$\sum_{i \in P} w_i,$$

is minimal among all paths connecting u to v. The **all-pairs shortest path problem** is a similar problem, in which we have to find such paths for every two vertices n to n'. A solution to this problem is presented in Section 2.4.

A **tree** finally is a graph in which any two vertices are connected by exactly one path. Alternatively, a tree may be defined as a connected graph with no cycles. A **labeled tree** is a tree in which each vertex is given a unique label (Figure 2.3).

2.2
A Few Words about Algorithms and Computer Programs

In mathematics and computer science, an **algorithm** is a finite set of well-defined instructions for accomplishing some task. Given an initial state, it will terminate in a corresponding recognizable end-state.

The concept of an algorithm is often illustrated by the example of a recipe, although many algorithms are much more complex. Algorithms often have steps that repeat

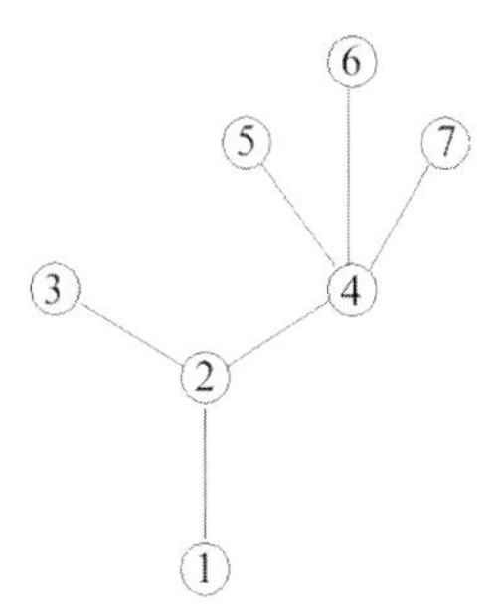

Figure 2.3 A labeled tree with seven vertices and six edges connecting them.

(iterate) or require decisions (such as logic or comparison) until the task is completed. Correctly performing an algorithm will not solve a problem if the algorithm is flawed or not appropriate to the problem. Different algorithms may complete the same task with a different set of instructions in more or less time, space or effort than others.

2.2.1 Implementation of Algorithms

The analysis and study of algorithms is one core discipline of computer science and is often done in an abstract way without the use of a specific programming language or software implementation. In this regard, it resembles other mathematical disciplines in that the analysis focuses on the underlying principles of the algorithm, and not on any particular implementation. One way to explain the principle structure of an algorithm is formulating it as **pseudocode**.

- Example
- One of the simplest algorithmic problems is finding the largest number in an unsorted list of numbers. The solution necessarily requires looking at every number in the list, but ideally only once at each. From this follows a simple algorithm:
  (1) Look at each item in the list. If it is larger than any item that has been seen so far, make a note of it.
  (2) The latest noted item is the largest item in the list.

  Here is a more formal coding of this algorithm in pseudocode:

```
Algorithm LargestNumber
  Input: A non-empty list of numbers L.
  Output: The largest number in the list L.
  Largest ← -∞
  for each item in the list L, do
        if the item > largest, then
              largest ← the item
  return largest
```

Some of the notation used here may be unfamiliar to you. "←" is a loose shorthand for "changes to". For instance, `largest ← the item` means that the `largest` number found so far changes to this `item`. The last **`return`** instruction terminates the algorithm and outputs the value listed behind it.

It is often of interest knowing how much of a particular resource (such as time or storage) a given algorithm requires. For example, the algorithm above has a time requirement of $O(n)$, using the big O notation with n as the length of the list. This means that the total number of operations required is linearly proportional to n or "requires on the order of n" operations. The actual number of iterations could be $2 \times n$, $3 \times n$ or something equivalent depending on how clever the implementation is done and what programming language is used. However, it will certainly not be proportional to $3 \times n \times n$. At all times, the algorithm only needs to remember a single value, the largest number found so far. Therefore this algorithm has a space requirement of $O(1)$.

Developing efficient algorithms scaling with $O(n)$ or $O(n \log n)$ is particularly important when dealing with biological networks where vertex sets may contain thousands up to hundreds of thousands of vertices. For certain problems, no algorithm exists that can solve the problem in polynomial time (where the running time scales as $O(n^x)$ with $x \in N$). These problems may therefore require very long computations on large networks or may not be computable at all in polynomial time. They are then said to be **NP-complete**.

2.2.2 Classes of Algorithms

Out of the many ways to classify algorithms, one way is classifying them by their design methodology or paradigm. Some commonly found paradigms include:

- **Divide and conquer.** A divide and conquer algorithm repeatedly reduces an instance of a problem to one or more smaller instances of the same problem (usually recursively), until the instances are small enough to solve easily.
- **Dynamic programming.** When the optimal solution to a problem can be constructed from optimal solutions to subproblems, and overlapping subproblems, we can often solve the problem quickly using *dynamic programming*. This approach avoids recomputing solutions that have already been computed. For example, the shortest path to a goal from a vertex in a weighted graph can be found by using the shortest path to the goal from all adjacent vertices. A shortest path algorithm is used in Chapter 4 to compute the betweenness of vertices in a protein–protein interaction graph (these are vertices with few connections that are located "between" more densely connected regions of other vertices).
- **The greedy method.** A greedy algorithm is similar to a dynamic programming algorithm. The difference is that solutions to the subproblems do not have to be known at each stage. Instead a "greedy" choice can be made of what looks best for the moment. We will encounter greedy algorithms in Chapter 2 (Kruskal's algorithm) and in Chapter 8 (CombDock).
- **Linear programming.** When solving a problem using linear programming, the program is formulated as a number of linear inequalities and then an attempt is

made to maximize (or minimize) the inputs. Many problems (such as the maximum flow for directed graphs) can be stated in a linear programming way and then be solved by a "generic" algorithm such as the Simplex algorithm. Finding a particular solution maximizing the biomass production by linear programming will be addressed in Chapter 6 in flux balance analysis.

- **Search and enumeration.** Many problems (such as playing chess) can be modeled as problems on graphs. A graph exploration algorithm specifies rules for moving around a graph and is useful for such problems. This category also includes the search algorithms and backtracking.
- **The probabilistic and heuristic paradigm.** Algorithms belonging to this class fit the definition of an algorithm more loosely.
  (1) **Probabilistic algorithms** are those that make some choices randomly.
  (2) **Genetic algorithms** attempt to find solutions to problems by mimicking biological evolutionary processes, with a cycle of random mutations yielding successive generations of "solutions". Thus, they emulate reproduction and "survival of the fittest". We will see an example of this type at the end of Chapter 7. In genetic programming, this approach is extended to algorithms by regarding the algorithm itself as a "solution" to a problem.
  (3) **Heuristic algorithms**, whose general purpose is not to find an optimal solution, but an approximate solution when the time or resources to find a perfect solution are not practical.

2.3 Data Structures for Graphs

As introduced in Section 2.1, a **graph** is an abstract data type that consists of a set of vertices and a set of edges that establish relationships (connections) between the vertices. In typical graph implementations, vertices are implemented as structures or objects. There are several ways to represent edges, each with advantages and disadvantages. One of these ways is to associate each vertex as an **adjacency list** in an **array** of incident edges (Figure 2.4). If no information is required to be stored in edges, only in vertices, these arrays can simply be pointers to other vertices and thus

(2,5,8,9)
(3,4,8)
(4,5,9)
...

$$\begin{pmatrix} \ldots & 1 & 0 & 0 & 1 & 0 & 0 & 1 & 1 & \ldots \\ & \ldots & 1 & 1 & 0 & 0 & 0 & 1 & 0 & \ldots \\ & & \ldots & 1 & 1 & 0 & 0 & 0 & 1 & \ldots \\ \ldots & & & & & & & & & \end{pmatrix}$$

Figure 2.4 Two popular ways to store connectivity information. (Left) The first array contains pointers to the proteins interacting with protein 1, the second array contains the interaction partners of protein 2, etc. (Right) An $n \times n$ matrix contains values of "1" for interacting proteins and values of "0" for those where no interactions were recorded.

represent edges with little memory requirement. An advantage of this approach is that new vertices can be added to the graph easily and they can be connected with existing vertices simply by adding elements to the appropriate arrays. A disadvantage is that determining whether an edge exists between two vertices requires $O(n)$ time, where n is the average number of incident edges per vertex.

- Example
- If we would like to find out whether vertex 17 is connected to vertex 53 in the array of pointers, this requires a search whether 53 is contained in the list of edges of vertex 17.

An alternative way is to keep an **adjacency matrix** (a two-dimensional array) M of Boolean values (or integer values, if the edges also have weights or costs associated with them). The entry $M_{i,j}$ then specifies whether an edge exists that goes from vertex i to vertex j. An advantage of this approach is that finding out whether an edge exists between two vertices becomes a trivial look-up in the memory for the value of the matrix element belonging to the two vertices which requires a constant amount of CPU time. Similarly, adding or removing an edge is a constant-time memory access. A disadvantage is that adding or removing vertices from the graph requires re-arranging the matrix accordingly, which may be costly depending on its size. Also, this representation requires $O(n^2)$ of storage, which is quite wasteful for sparse networks.

Another, less often used, way of representing a graph with n vertices v_i and m edges e_j in a computer is the $n \times m$ **incidence matrix**. Here, the vertices are labeled from 1 to n and the edges from 1 to m. The matrix entries b_{ij} describe the relation of vertices and edges. In an undirected graph:

$$b_{ij} = \begin{cases} 1 & \text{if } v_i \in e_j \\ 0 & \text{else} \end{cases} .$$

Figure 2.5 shows an example of a graph and its corresponding incidence matrix. We will re-encounter this form of representing connectivities in Chapter 6. The stoichiometric matrix used there is a close relative of the incidence matrix although, there, one reaction may sometimes connect more than two vertices.

In the general case, a graph may consist of many edges between many vertices and, unless the matrix representation for the edges is chosen, there may even be more than one edge connecting the same pair of vertices. Edges can be bidirectional or unidirectional. In most cases, the only information given by an edge is that there is a relationship between the two vertices connected and the information is stored in the vertex itself. However, some graphs have numerical values associated with each edge. These graphs can be used for different problems such as the traveling salesman problem.

There exist two general strategies to search graphs. **Breadth-first search** is a graph search algorithm that begins at the root vertex and explores all its neighboring vertices. Then, for each of those nearest vertices, it explores their unexplored neighbor vertices, and so on, until it finds the goal. In contrast, **depth-first search**

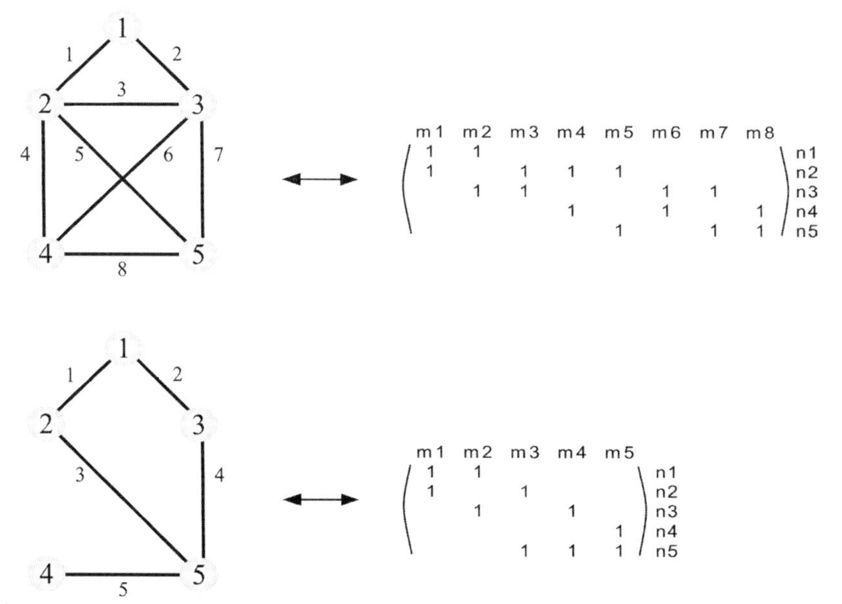

Figure 2.5 (Top) A densely connected graph and its corresponding incidence matrix. Each edge connects exactly two vertices and that is why every column contains two "1" entries. Apparently, the upper example generates an even larger matrix than an adjacency matrix. This form of representing connectivities is most space saving if the number of vertices exceeds the number of edges.

starts at the root and explores as far as possible along each branch before backtracking to follow up on the next branch.

Graph search algorithms are a significant field of interest for computer scientists. One of the most well-known algorithms is **Dijkstra's algorithm** which efficiently finds the shortest path between two vertices in a graph.

2.4 Dijkstra's Algorithm

Dijkstra's algorithm, named after the Dutch computer scientist Edsger Dijkstra, is an algorithm that solves the single-source shortest path problem for a directed graph with nonnegative edge weights. For example, if the vertices of the graph represent cities and edge weights represent driving distances between pairs of cities connected by a direct road, Dijkstra's algorithm can be used to find the shortest route between two cities (Figure 2.6).

The input of the algorithm consists of a weighted directed graph $G(V,E)$ and a source vertex s in G. Each edge of the graph is an ordered pair of vertices (u,v)

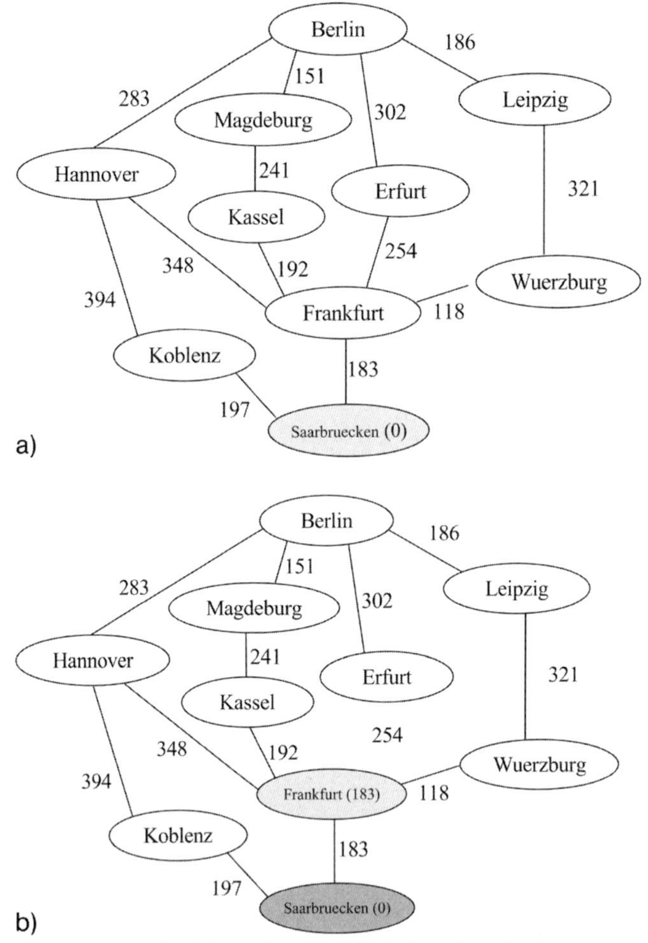

Figure 2.6 Example of the application of the Dijkstra algorithm to compute the shortest connection between the cities of Saarbrücken and Berlin. The numbers on the arrows denote the distance in kilometers. The vertex considered in each step is shaded in light grey. Vertices dealt with in previous steps are shaded in dark grey.

representing a connection from vertex u to vertex v. Weights of edges are given by a weight function w: $E \rightarrow [0,\infty]$. Therefore $w(u,v)$ is the nonnegative cost of moving from vertex u to vertex v. In the case of a road network, the cost of an edge can be thought of as the distance between those two vertices. The cost of a path between two vertices is the sum of costs of the edges in that path. For a given pair of vertices s and t in V, the algorithm finds the path from s to t with lowest cost (i.e. the shortest path). As a simple extension, it can also be used for finding the costs of shortest paths from a single vertex s to all other vertices in the graph.

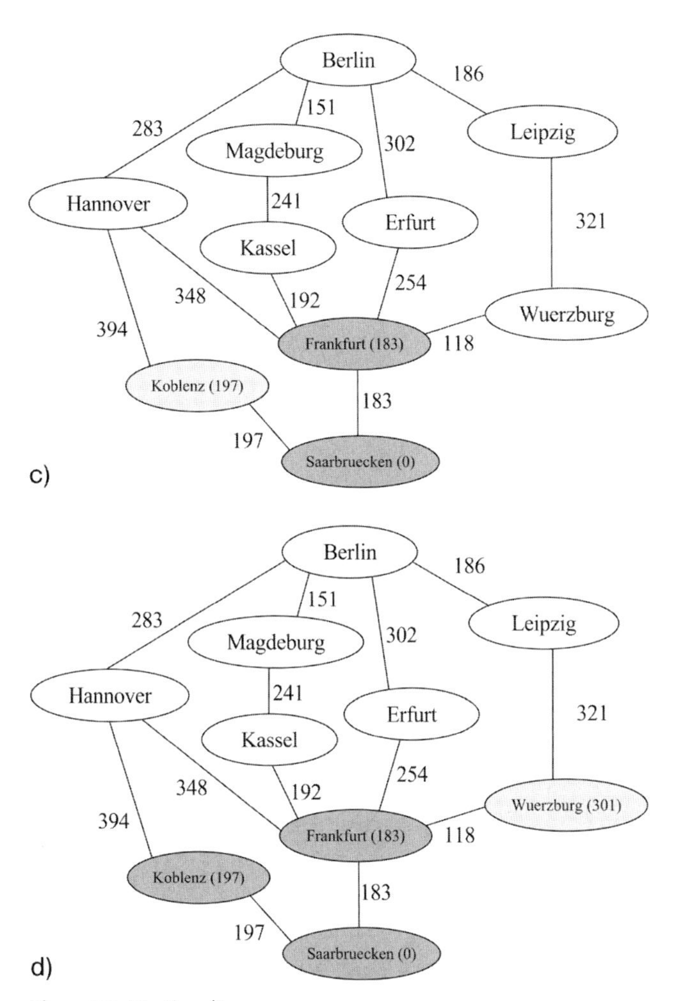

Figure 2.6 *(Continued)*

2.4.1 Description of the Algorithm

The algorithm works by keeping for each vertex v the cost $d[v]$ of the shortest path found so far between s and v. Initially, this value is 0 for the source vertex s ($d[s] = 0$), and infinity for all other vertices according to the fact that, so far, we do not know any path leading to those vertices ($d[v] = \infty$ for every v in V, except s). When the algorithm finishes, $d[v]$ will be the cost of the shortest path from s to v – or infinity, if no such path exists. The initialization involves the following steps:

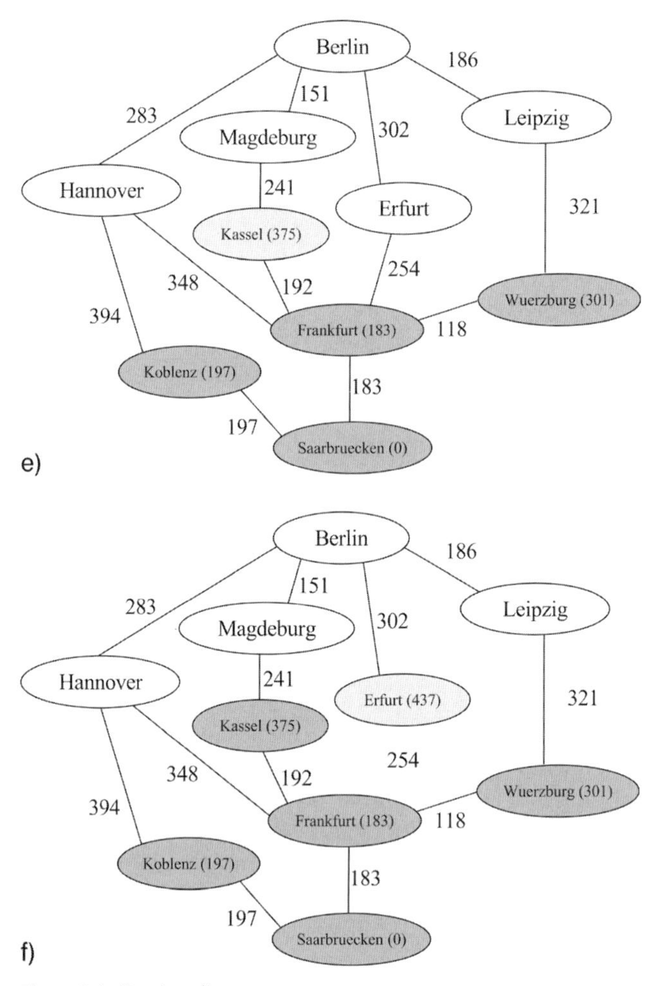

Figure 2.6 *(Continued)*

```
1 function Dijkstra(G,w,s)
2  for each vertex v in V[G]    //Initialization
3          do d[v] := infinity
4                  previous[v] := undefined
5 d[s] := 0
```

The basic operation of Dijkstra's algorithm is **edge relaxation**: if there is an edge from u to v, then the shortest known path from s to u ($d[u]$) can be extended to a path from s to v by adding edge (u,v) at the end. This path will have length $d[u] + w(u,v)$. If this is less than the current $d[v]$, we can replace the current value of $d[v]$ with the new value. Edge relaxation is applied until all values $d[v]$ represent the cost of the shortest path from s to v. The algorithm is organized so that each edge (u,v) is relaxed only once, when $d[u]$ has reached its final value.

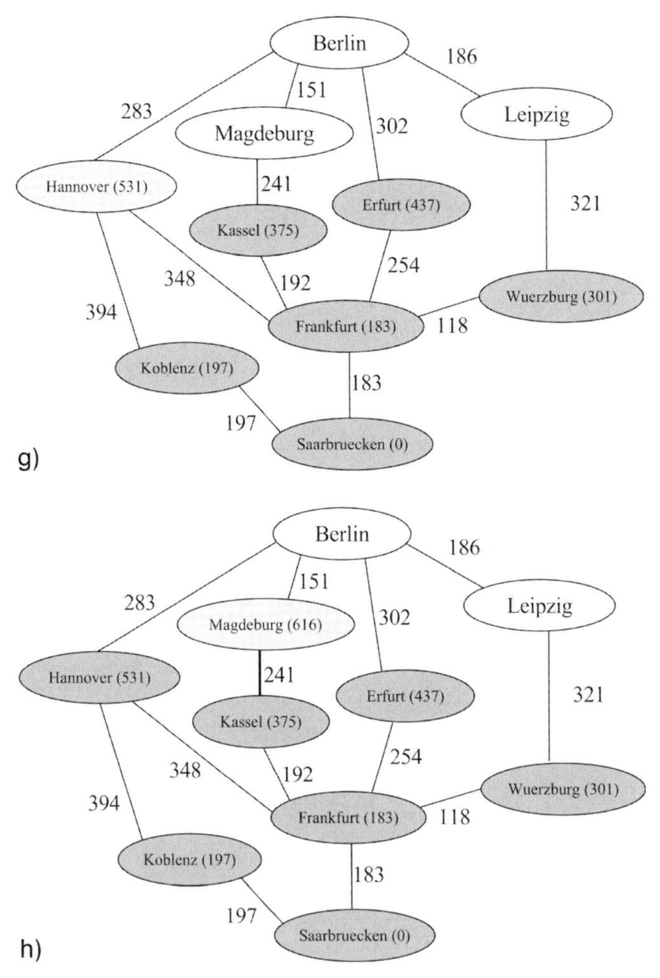

Figure 2.6 *(Continued)*

The algorithm maintains two sets of vertices S and Q. Set S contains all vertices for which we know that the value $d[v]$ is already the cost of the shortest path and set Q contains all other vertices. Set S starts empty and in each step one vertex is moved from Q to S. This vertex is chosen as the vertex with lowest value of $d[u]$. When a vertex u is moved to S, the algorithm relaxes every outgoing edge (u,v).

Figure 2.6 shows an example where the shortest road is found for traveling from the city of Saarbrücken to that of Berlin.

2.4.2 Pseudocode

In the following algorithm, `u := Extract-Min(Q)` searches for the vertex u in the vertex set Q that has the smallest $d[u]$ value. That vertex is removed from the set Q and

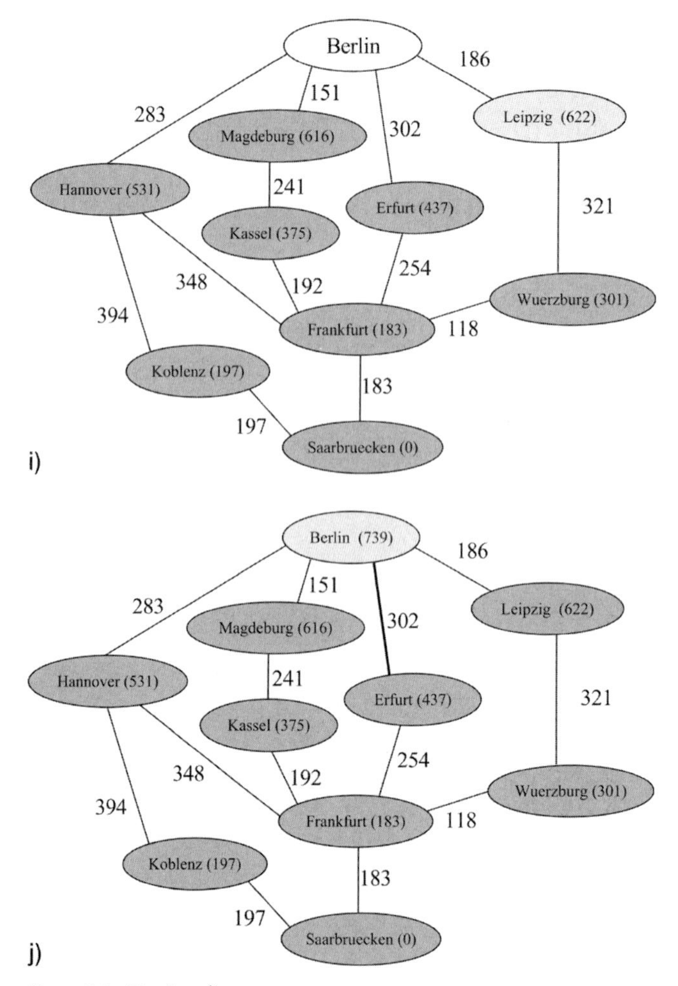

Figure 2.6 (*Continued*)

then returned. `Q:=update(Q)` updates the weight field of the current vertex in the vertex set Q.

```
1 function Dijkstra(G,w,s)
2   for each vertex v in V[G]    //Initialization
3          do d[v]  := infinity
4               previous := undefined
5   d[s] : = 0
6   S := empty set
7   Q := set of all vertices
8   while  Q is not an empty set
9          do u  := Extract-in(Q)
10           S := S union { u}
```

```
11          for each edge (u,v) outgoing from u
12               do if d[v] > d[u] + w(u,v)    //Relax (u,v)
13                  then d[v]  := d[u] +w(u,v)
14                       previous[v] : = u
15                       Q := Update(Q)
```

To keep an overview over the execution of the algorithm, it is quite helpful to represent the intermediate results as in Table 2.1.

If we are only interested in a shortest path between vertices s and t, we can terminate the search at line 9 by checking whether $u = t$. Now we can read the shortest path from s to t by iteration:

```
16 S := empty sequence
17 U := t
18 while defined u
19 do insert u to the beginning of S
20       u := previous[u]
```

Sequence S is the list of vertices on the shortest path from s to t.

2.4.3 Running Time

The simplest implementation of the Dijkstra's algorithm stores vertices of set Q in an ordinary linked list or array, and operation `Extract-Min(Q)` is simply a linear search through all vertices in Q. In this case, the running time is $O(n^2)$. For sparse graphs, i.e. graphs with much fewer than n^2 edges, Dijkstra's algorithm can be implemented more efficiently.

2.5 Minimum Spanning Tree

Given a connected, undirected graph G, a **spanning tree** of that graph is a subgraph which is a tree (see Figure 2.3) composed of all the vertices and some (or perhaps all) of the edges of G (Figure 2.7). A single graph can have many different spanning trees. We can also assign a *weight* to each edge and use this to assign a weight to a spanning tree by computing the sum of the weights of the edges in that spanning tree. A **minimum spanning tree** or **minimum weight spanning tree** is then a spanning tree with weight less than or equal to the weight of every other spanning tree.

As an example where such a concept would be quite useful, let us consider a cable TV company laying cable to a new neighborhood. If the cable can only be buried along certain paths, then there would be a graph representing which points are connected by those paths. Some of those paths might be more expensive, because they are longer or require the cable to be buried deeper. These paths would be represented by edges with larger weights. A spanning tree for that graph would be a

Table 2.1 Table that keeps track of events during execution of Dijkstra's algorithm on the road network between Saarbrücken and Berlin (in the zeroth iteration, all entries of the distance array *d* and of the array containing previous vertices *p* are set to infinity).

Iteration	set S	d[F], p[F]	d[Ko], p[Ko]	d[W], p[W]	d[Ka], p[Ka]	d[E], p[E]	d[H], p[H]	d[M], p[M]	d[L], p[L]	d[B], p[B]
0		∞	∞	∞	∞	∞	∞	∞	∞	∞
1	SB	183, SB	197, SB	∞	∞	∞	∞	∞	∞	∞
2	SB, F			301, F	375, F	437, F	531, W	∞	∞	∞
3	SB, F, Ko							616, Ka	622, F	739, E
4	SB, F, Ko, W									
5	SB, F, Ko, W, Ka									
6	SB, F, Ko, W, Ka, E									
7	SB, F, Ko, W, Ka, E, H									
8	SB, F, Ko, W, Ka, E, H, M									
9	SB, F, Ko, W, Ka, E, H, M, L									
10	SB, F, Ko, W, Ka, E, H, M, L,	183, SB	197, SB	301, F	375, F	437, F	531, F	616, Ka	622, W	739, E

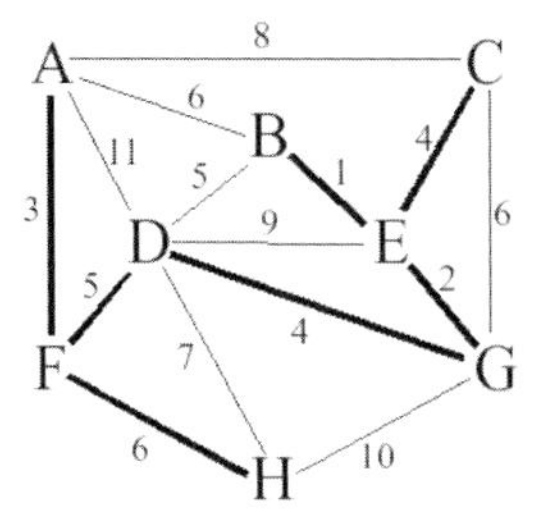

Figure 2.7 Example of a minimum spanning tree so that each pair of vertices is connected. Each edge is labeled with its weight.

subset of those paths that has no cycles but still connects to every house. There might be several spanning trees possible. A minimum spanning tree would be one with the lowest total cost.

In case of a tie, there could be several minimum spanning trees. In particular, if all weights are the same, every spanning tree is minimal. However, a mathematical theorem states that if each edge has a distinct weight, the minimum spanning tree is unique. This is true in many realistic situations, such as the one above, where it is unlikely that any two paths have *exactly* the same cost.

2.5.1 Kruskal's Algorithm

There are two algorithms commonly used to solve this problem, **Prim's algorithm** and **Kruskal's algorithm**. Both are greedy algorithms and both run in polynomial time. As an example, Figure 2.8 shows how Kruskal's algorithm works.

What is the fastest possible algorithm for this problem? That is one of the oldest open questions in computer science. There is clearly a linear lower bound, since we must at least examine all the weights. If the edge weights are integers with a bounded bit length, then deterministic algorithms are known with linear running time, $O(m)$. For general weights, randomized algorithms are known that run in linear expected time. Whether there exists a deterministic algorithm with linear running time for general weights is still an open question.

2.6 Graph Drawing

We will see in later chapters that biological networks often involve hundreds to thousands of vertices so that their interpretation becomes complicated. A powerful visualization concept is of crucial importance. While it is very hard to define what is the best way of representing a particular graph network in mathematical terms, it is generally agreed that aesthetic drawings have minimal edge crossing, emphasize symmetry if present and use an even spacing between vertices.

Many approaches have been proposed in the literature to create graphical images of networks, but few of them scale well to large networks and, at the same, provide a

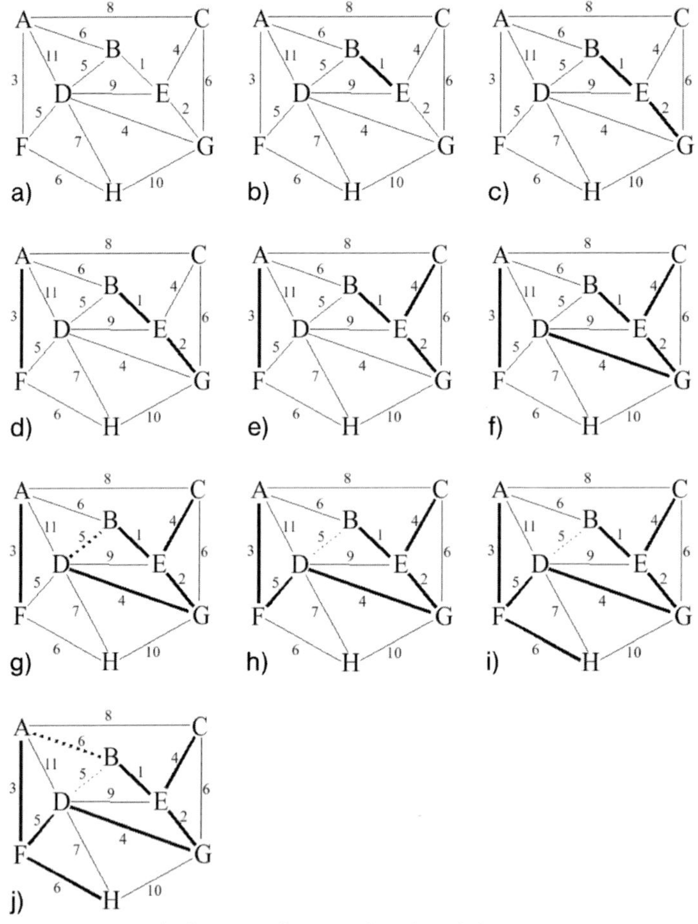

Figure 2.8 Example illustrating the principles of Kruskal's algorithm. At each iteration, the edge not used so far with the smallest weight is selected unless that would lead to a closed cycle. Note the dotted edge in step 7 (B,D) would lead to a closed cycle B–E–G–D–B. Therefore, this edge is not selected. The same applies to edge (A,B) in step 10 that would lead to a closed circle A–B–E–G–D–F–A and all remaining edges.

satisfactory representation. Here, we will introduce a physically motivated method – the **force-directed layout** that is based on a very simple but powerful concept and is therefore widely used. For this the vertices of a network are modeled as charged mass points that repel each other. Each link, on the other hand, is modeled by a spring, pulling the respective vertices closer together. When such a system is left alone it tries to organize into a state of minimal energy, where the vertices are as far apart from each other as possible – with the constraint that each pair of connected vertices has to stay close together. Thus the distances on the network (number of edges between two vertices) are transformed into spatial distances (Figure 2.9).

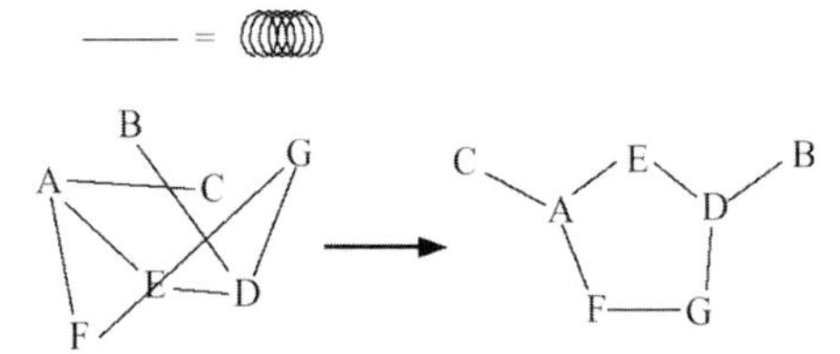

Figure 2.9 Example illustrating how the force-directed layout algorithm will disentangle the graph network shown on the left. By transforming edges into springs and assuming overall repulsion of all nonconnected vertices, the vertices will rearrange into the clean layout shown on the right that resembles a doubly protonated imidazole molecule. One can verify easily that the right graph contains exactly the same seven edges as the left graph. However, it gives a much clearer representation of the connectivity.

The uniform repulsion between nonconnected vertices can be conveniently modeled by analogy to the electrostatic repulsion of two like-charged particles. The Coulombic interaction energy between two charges q_1 and q_2 is given as:

$$E_c(r) = \frac{q_1 q_2}{r}.$$

The connections between vertices are modeled as harmonic springs with the Hooke potential:

$$E_h(r) = \frac{k}{2} r^2.$$

In order to rearrange particles, the algorithm computes the resulting force on each particle due to all its connections and repulsions with other particles. For this we use the fact that the physical force equals the negative gradient of the energy, i.e., the force F is a measure for how much the energy changes with an infinitesimal displacement:

$$\vec{F}(\vec{r}) = -\nabla E(\vec{r}),$$

with the gradient operator:

$$\nabla := \begin{pmatrix} \partial/\partial x \\ \partial/\partial y \\ \partial/\partial z \end{pmatrix}.$$

In one dimension this reduces to $\nabla = d/dr$, i.e. the simple derivative with respect to the distance r. Problems (3) and (4) below will make you more familiar with this concept. This is one of the intuitive methods that one has to see at work to appreciate how well they work in practice.

Summary

This chapter is meant as an introduction to mathematical graphs, algorithms and data structures for those readers who do not have a background in computer science.

Computer science has developed a remarkable repertoire of highly efficient strategies (algorithms and data structures) to solve certain problem classes. New problems are quickly being analyzed for whether any of the well-known recipes can be applied. If this is the case, the way to solve the problem and often the necessary software is already at hand. For all those interested, I can sincerely recommend picking up a bit more from one of the well-established introductory computer science textbooks on data structures and algorithms.

Problems

(1) Dijkstra's algorithm

Construct an example similar to Figure 2.6, e.g. for connecting your home city to the capital of your country. Use Dijkstra's algorithm iteratively to find the shortest path. After doing so, write a simple computer program using the programming language of your choice and try to reproduce the result you obtained manually.

(2) Minimal spanning tree

Construct a minimal spanning tree manually for the graph shown in Figure 2.10.

Force–directed layout of graphs

In the third exercise you will derive the equations of motion for the connected mass points, while the fourth exercise gives you the chance to do some nice network layouts.

(3) Energy, forces and equations of motion

(a) Configuration of minimal energy

Determine the equilibrium distance between two equally charged mass points, which are connected by a spring. At the equilibrium distance the total force vanishes. Verify that instead of calculating the forces explicitly, it is equivalent to determine the configuration of minimal energy.

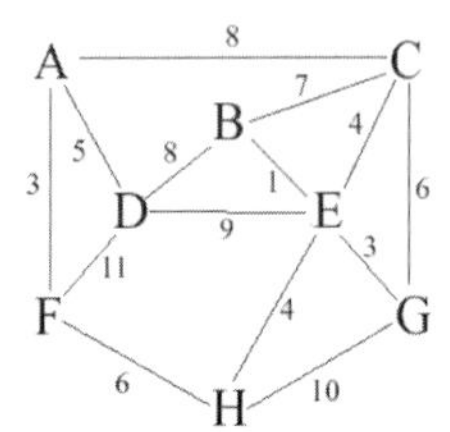

Figure 2.10 Weighted undirected graph (see Problem (2)).

Hint: To show the equivalence of vanishing force and minimal energy remember how the minimum of a function is defined. Also note that the distance between two particles is a one-dimensional measure.

(b) Force field from a spherically symmetric potential
Calculate the force fields:

$$\vec{F}(\vec{r}) = -\nabla E(\vec{r}),$$

for both the Coulomb interaction E_c and the harmonic potential E_h from (Section 2.6).

Hint: Write ∇ and the resulting force field:

$$\vec{F}(\vec{r}) = \begin{pmatrix} F_x(x) \\ F_y(x) \\ F_z(x) \end{pmatrix}.$$

in component form. Then you get one equation for x, y and z, each. This is the form that we need for the second part. Note that:

$$r = \sqrt{x^2 + y^2 + z^2}.$$

(4) Force-directed layout of graphs

Now we want to layout some graphs, first a few simple test cases, then two real interaction networks from the "Biomolecular Interaction Network Database" (BIND) database. Use a repulsive Coulomb-type potential $E_c(r_{ij}) = 1/r_{ij}$ between all vertices plus a harmonic attractive potential $E_h(r_{ij}) = r_{ij}^2/2$ between interacting vertices; r_{ij} is the distance between vertices i and j. Perform this layout in two dimensions, therefore:

$$r_{ij}^2 = (x_i - x_j)^2 + (y_i - y_j)^2.$$

(a) Test files
Start with the test files "star.dat", "square.dat", "star2.dat" and "dog.dat" (available online on our web server gepard.bioinformatik.uni-saarland.de). These should give you final configurations with final energies of about 9.4, 7.3, 59.4 and 101 units. The simple ones converge after about 300 iterations, while the more complex may take up to a 1000 iterations.

For each of the networks plot the final configuration and give the final energies. Also plot the energy versus iteration number and determine at which point the layout process can be considered converged.

Hint: To create the layouts follow these steps – and have a look at the supplied example code:

(a) Read in the interaction files and create a network from them. The files contain in the first line the number of vertices. The subsequent lines each contain the two endpoints of a link. From the network we only need the list of interactions $W[i][j]$.

(b) Choose initial positions for all the vertices in the x–y plane. A reasonable width is within ± 10 units from the origin.

(c) For the iterated layout perform at least 500 steps for the test files and at least 5000 steps for the "real" networks:
  (i) Calculate the distances r_{ij} between the vertices and from them the resulting forces as:

$$\vec{F}_{ij} = \vec{F}_c(\vec{r}_{ij}) + W[i][j]\,\vec{F}_h(\vec{r}_{ij}),$$

and sum them up. The total force on vertex i is:

$$\vec{F}_i = \sum_j \vec{F}_{ij}.$$

Note that $F_{ij} = -F_{ji}$, meaning that the forces are symmetric.
  (ii) Add a random force in the range $-0.3, \ldots, 0.3$ to the total force on each vertex. This additional "thermal" contribution improves the convergence as it helps to escape from local minima.
  (iii) Update the position of each vertex from the forces as:

$$\Delta \vec{r}_i = \alpha\,\vec{F}_i,$$

with the inverse friction coefficient:

$$\alpha = \Delta t/\gamma.$$

A reasonable value is $\alpha = 0.03$.
  (iv) Calculate the total energy as the sum of the individual interaction energies:

$$E = \sum_{j>i} E_c(r_{ij}) + W[i][j]\,E_h(r_{ij}).$$

Print out this energy together with the number of the iteration. You will see that the energy decreases fast in the beginning and then slower and slower.
  (v) Repeat from (i) until the total energy is essentially constant.

(b) "Real" networks

Now perform the same layout on the following networks. Give the final energies and configurations for each of them and the number of iterations after which you stopped the layout process.

(i) "sfnet_100.dat" is a thinned out scale-free network of 100 vertices, where every second edge has been left out.

(ii) "11309.txt" and "2287.txt" contain networks extracted from the BIND database for the taxon identifiers 11309 and 2287, respectively (what species are these?).

Hint: You may have to run a few trials with different initial placements of the vertices and then choose the best result.

What happens when you skip step (ii), i.e. do the optimization without the random forces? Plot the total energy versus iteration with and without the random forces for one of the "real" networks of your choice. Be careful to scale the axis so that the important difference can clearly be seen. These networks contain more than a single cluster. What happens to the different clusters? Why?

Further Reading

Cormen TH, Leiserson CE, Rivest RL, Stein C (2001) *Introduction to Algorithms*, MIT Press, Cambridge, MA.

3
Protein–Protein Interaction Networks – Pairwise Connectivity

The formidable advances in protein sciences in recent years have highlighted the importance of protein–protein interactions in biology. Well before the proteomics revolution, we knew that many proteins were capable of interacting with each other in a highly specific manner and that the function of certain proteins was regulated by interacting partners. However, the extent and degree of the protein–protein interaction network was not realized. It is now believed that not only are a majority of proteins in a eukaryotic cell involved in complex formation at some point in the life of the cell, but also that each protein may have on average six to eight interacting partners (Section 8.2).

In this chapter, we will start with the collection of protein–protein interaction data from high-throughput experiments followed by its bioinformatics interpretations. This topic will be continued in Chapters 8 and 9 that go deeper into analyzing individual protein–protein interfaces and the three-dimensional structures of protein complexes. Already at this point, we note that the structural details of molecular complexes are particularly important when considering simultaneous binding of more than two proteins. Imagine that protein A binds another protein B via a particular interface on its surface. Then, it can obviously not bind a third protein C at the same interface at the same time.

3.1
Principles of Protein–Protein Interactions

In biological systems proteins rarely act in isolation, but bind to other biomolecules to initiate cellular processes. Often, these binding partners are other proteins so that the proteins form homo- and hetero-dimers and -oligomers. Dimerization or oligomerization may provide several different structural and functional advantages to proteins such as improved stability, control over the accessibility and specificity of active sites, and increased complexity. Figure 3.1 illustrates the main functional consequences of dimerization and oligomerization

Principles of Computational Cell Biology – From Protein Complexes to Cellular Networks. Volkhard Helms

ISBN: 978-3-527-31555-0

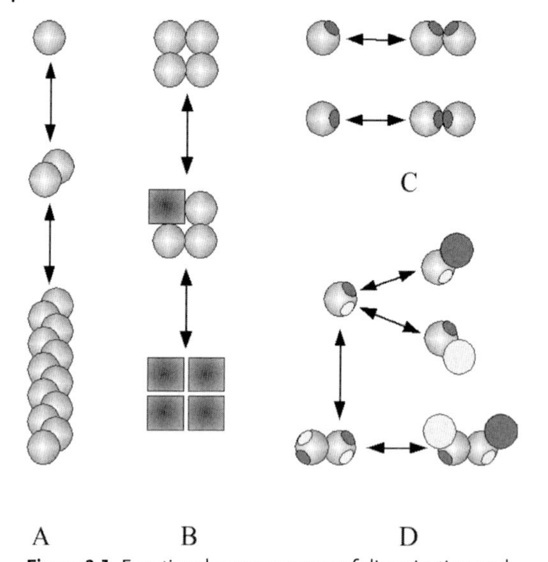

Figure 3.1 Functional consequences of dimerization and oligomerization. (A) Concentration, stability and assembly. (B) Cooperation and allostery. (C) Modification of the active site. (D) Dimerization yields increased diversity in the formation of regulatory complexes.

One important property of protein–protein interactions is the specificity of this type of interaction. The specificity of interactions is determined by structural and physicochemical properties of the two interacting proteins. Due to the rather crowded environment *in vivo*, protomers are not colocalized, and need to be highly specific in partner recognition and binding as it is the case, for example, for hormone–receptor and enzyme–inhibitor complexes. An early study by Jones and Thornton (1996) characterized interfaces in terms of residue propensities, conservation, interaction propensities, protrusions and planarity. These structural details of interfaces will be discussed in greater detail in Chapter 9.

3.2 Experimental High-Throughput Methods for Detecting Protein–Protein Interactions

We will begin our discussion of protein–protein interactions by introducing several experimental techniques that can be applied to “fish” in cell lysates for unknown interaction partners. An understanding of the basics of these techniques is essential to understand why different methods give different results. Also, we need to appreciate that different and sometimes even contradictory answers about the composition of a macromolecular complex can be correct depending on which properties of the complex are being probed.

3.2.1 Gel Electrophoresis

The term **gel electrophoresis** is used for a group of techniques that separate and partially purify molecules based on their physical characteristics such as size, shape or isoelectric point (p*I*). The first part, "gel", refers to the matrix used to separate the molecules. In most cases the gel is a crosslinked polymer of different porosity. The second part, "electrophoresis", refers to the electromotive force that is applied to push or pull the molecules through the gel matrix. By placing the molecules in wells in the gel and applying an electric voltage across the gel, the molecules will move through the matrix at different speeds, towards the anode if they are negatively charged or towards the cathode if positively charged.

As proteins can have different charges and complex shapes, they may not migrate equally well in the gel. Proteins are therefore usually denatured through the addition of a detergent such as sodium dodecyl sulfate (SDS) that coats the proteins with a negative charge. This method is therefore termed SDS-polyacrylamide gel electrophoresis (SDS–PAGE). Generally, the amount of SDS bound is proportional to the size of the protein, so that the resulting denatured proteins have an overall negative charge and all the proteins have a similar charge to mass ratio. Since denatured proteins act like they were long rods instead of having a complex tertiary shape, the speed at which the resulting SDS-coated proteins migrate in the gel is related only to their size and not their charge or shape. The distance a protein travels is approximately inversely proportional to the logarithm of the mass of the molecule.

If several mixtures are initially injected next to each other on the gel, they will run parallel to each other in individual lanes. Bands in different lanes that end up at the same distance from the starting point contain molecules that moved through the gel with the same speed, which usually means that they have approximately the same mass. There are special markers available (ladders) that contain a mixture of molecules of known masses. If such a marker was run on one lane in the gel parallel to the unknown samples, the bands observed can be compared to those of the unknown in order to determine their mass.

3.2.2 Two-Dimensional Gel Electrophoresis

As for one-dimensional electrophoresis, two-dimensional electrophoresis separates molecules by molecular weight in one direction using SDS–PAGE and, additionally, in a first step, also by their p*I* in the perpendicular direction. As it is unlikely that two molecules will be similar in both properties, molecules are more effectively separated in two-dimensional electrophoresis than in one-dimensional electrophoresis. The procedure starts with placing the sample in a gel manufactured with a stationary pH gradient in one direction and applying an electrostatic potential difference across it. At all pHs other than their isoelectric point, the proteins will be charged. If they are positively charged, they will be pulled towards the more negative end of the gel; if they are negatively charged, they will be pulled to the more positive end of the gel. The protein therefore migrates along the pH gradient until it carries no overall charge. This location of the protein in the gel corresponds to the apparent p*I* of the protein. In the second step, the proteins separated according to their p*I* are placed at the start of an SDS gel (see above) and are now separated by their mass as well.

3.2.3
Affinity Chromatography

Affinity chromatography is a biochemical separation method that combines the size fractionation capability of gel permeation chromatography with the ability to design a stationary phase that reversibly binds to a known subset of molecules. Usually the starting point is an undefined heterogeneous group of molecules in solution, such as a cell lysate, growth medium or blood serum. The molecule of interest will have a well known and defined property which can be exploited during the affinity purification process. The process itself can be thought of as an entrapment, with the target molecule (i.e. the molecule targeted for purification) together with its potentially bound interaction partners becoming trapped on a solid or stationary phase or medium. The nontarget heterogeneous mixture will not become trapped. After this purification step, the eluted remainder – the target protein and its interaction partners – are denatured and thus separated, and identified by a subsequent gel electrophoresis or mass spectroscopy step. The purification process is sketched in Figure 3.2.

In the area of protein complexes, a variant of the method is used that is termed **tandem affinity purification (TAP)**.

3.2.4
Yeast Two-Hybrid Screening

Yeast two-hybrid screening (Y2H) is a molecular biology technique used to discover protein–protein interactions by testing for physical interactions (such as binding) between two proteins. One protein is termed the **bait** and the other is a library protein (or **prey**).

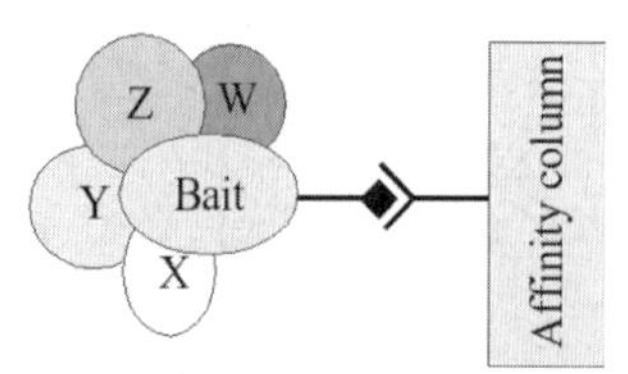

Figure 3.2 In affinity purification, a protein of interest (bait) is tagged with a molecular label (dark route in the middle of the figure) to allow easy purification. The tagged protein is then copurified together with its interacting partners (W–Z). This strategy can also be applied on a genome scale. Drawn after Aloy and Russell (2006).

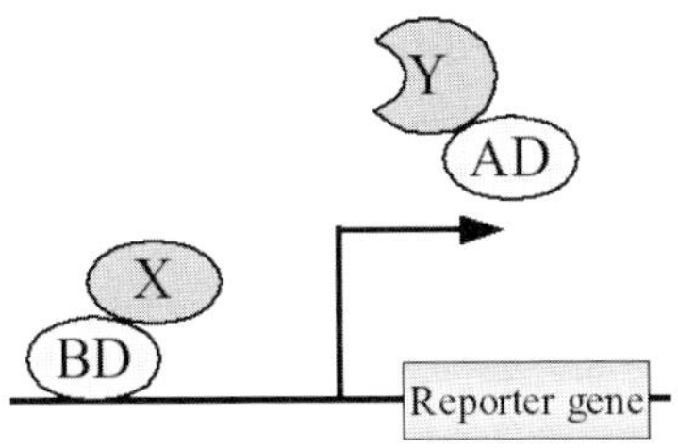

Figure 3.3 The Y2H system is one of the most widely used high-throughput systems to detect protein–protein interactions. In the most common variant, a bifunctional transcription factor (usually GAL4) is split into its BD and its AD. Each segment is then fused to a protein of interest (X and Y) and if these two proteins interact, the activity of the transcription factor is reconstituted. The system has been scaled up and applied in genome-scale screens.

The idea behind the test is the activation of downstream reporter gene(s) by the binding of a transcription factor to an upstream activating sequence (UAS). For the purposes of two-hybrid screening, the transcription factor is split into two separate fragments, called the binding domain (BD) and activating domain (AD). The BD is the domain responsible for binding to the UAS and the AD is responsible for activation of transcription. The key to the two-hybrid screen is that in most eukaryotic transcription factors, the AD and BD are modular, and can function in close proximity to each other without direct binding. This means that even though the transcription factor is split into two fragments, it can still activate transcription even if the two fragments are only indirectly connected.

In the Y2H screen, the BD fragment is fused onto the bait protein *X* and the AD fragment onto a library protein *Y*. If *X* and *Y* bind to each other (Figure 3.3) then the AD and BD of the transcription factor would be indirectly connected, and transcription of the reporter gene(s) could occur. If the two proteins do not interact, there would be no transcription of the reporter gene. Thus, a huge "library" of proteins can be tested for interaction with the bait. A common transcription factor used for Y2H screening is GAL4 that binds specifically to the UAS sequence and initiates activation of a downstream target gene. For example, this can be the gene coding for green fluorescent protein so that the expression of the target gene can be efficiently and quickly read out.

An advantage of this method is that the interactions are probed *in vivo*. As yeast is cheap and robust, the method can be applied on a large scale. Disadvantages are that the interactions need to be probed in the nucleus and some proteins, such as membrane proteins, may not be translocated easily into the nucleus. Also, it is possible that two proteins interact in Y2H experiments, although they are not simultaneously expressed during the cell cycle or in the particular compartment. In addition, a reported interaction may also be mediated through a third (or even more) protein that binds X and Y simultaneously. In this case, X and Y could be reported to interact directly (although they actually do not) and the mediating partners would remain undetected.

3.2.5
Synthetic Lethality

The synthetic lethality method produces mutations or deletions of two separate genes which are viable alone, but cause lethality for the cell when combined together under certain conditions. Since these mutations are lethal, they cannot be isolated directly and should be synthetically constructed. Synthetic interaction can point to the possible physical interaction between two gene products, their participation in a single pathway, or a similar function.

3.2.6
Gene Coexpression

Since the function of a protein complex depends on the functionality of all subunits, these should be present in stoichiometric amounts and gene expression levels of subunits in a complex should be related. Gene expression profiles can be provided, for example, from cell cycle experiments and expression levels of a gene under different conditions. Several studies demonstrated that interacting proteins in yeast are more likely to have their genes coexpressed compared with noninteracting proteins. Moreover, the expression levels of physically interacting proteins coevolve, which can be used as a predictor of protein interactions.

3.2.7
Mass Spectroscopy

The principle of the mass spectroscopy (MS) method is to produce ions either by electrospray ionization or matrix-assisted laser desorption ionization (MALDI). Detection of samples is based on their mass-to-charge ratios, thereby allowing the identification of polypeptide sequences. Different algorithms exist to analyze mass spectra, and to identify peptides and proteins from their fragments. MS has been applied on a large scale to detect protein–protein interactions.

There exist many other experimental methods such as surface plasmon resonance (SPR), nuclear magnetic resonance (NMR) or chemical cross-linking. On the one hand, these methods give more precise information whether two proteins really interact. They can provide binding constants and association rates in titration experiments. On the other hand, they require more work and are typically not applied on a large genome-wide scale.

When applying MS to identify components of protein complexes, the abbreviation HMS-PCI is used for high-throughput mass spectrometric protein complex identification.

3.2.8
Databases for Interaction Networks

The traditional major source for data on protein–protein interactions is the Protein Data Bank (PDB), which provides crystallographic data on protein–protein

Table 3.1 Some public databases compiling data related to protein interactions: (P) and (D) stand for proteins and domains (the number of interactions reflects the status of June 2007).

	URL	Number of interactions	Type	Proteins /domains
MIPS	mips.gsf.de/genre/proj/mpact	4300	curated	
BIND	bond.unleashedinformatics.com	200000	curated	P
MINT	160.80.34.4/mint/	103800	curated	P
DIP	dip.doe-mbi.ucla.edu	56000	curated	P
PDB	www.rcsb.org/pdb	800 complexes	curated	
HPRD	www.hprd.org	37500	curated	P, D
Scoppi	www.scoppi.org	102000	automatic	D
UniHI	theoderich.fb3.mdc-berlin. de:8080/unihi/home	209000	integrated data	P
STRING	string.embl.de	interactions of 1500000 proteins	integrated data from genomic context, high-throughput experiments, coexpression, previous knowledge	P
iPfam	www.sanger.ac.uk/Software/ Pfam/iPfam	3019	data extracted from PDB	D
YEAST protein complex database	yeast.cellzome.com	232 complexes	experimental	P
ABC	service.bioinformatik. uni-saarland.de/abc	13000 complexes	semiautomatic	P

complexes. Although atomic resolution structural data remains the gold standard for predicting and modeling protein–protein interactions, the recent development of other experimental techniques for determining interacting pairs of proteins has resulted in the development of a number of other protein–protein interaction databases. Table 3.1 lists some databases that contain information on protein interactions. Note that the number of important databases in this field and the number of interactions deposited in each of them are expanding at increasing speed.

3.2.9
Overlap of Interactions

Unfortunately, interaction data sets from high-throughput experiments discussed in the previous sections were quickly found to be incomplete in many regards and even contradictory. This means that two proteins may in reality interact with each other although their interaction was not detected in the high-throughput experiment ("false negatives") or that reported interactions may be artificial ("false positives"). Clearly, in the context of genome-wide analyses, these inaccuracies do not only result from intrinsic errors of the particular methods, but are also affected by the difficult statistical nature of these decisions. We must consider that these inaccuracies are likely greatly magnified in this case because the protein pairs that do not interact **(negatives)** by far outnumber those that do interact **(positives)**. In yeast, for example, the around 6000 proteins allow for $N(N-1)/2 \sim 18$ million potential interactions, whereas the estimated number of actual interactions is at most 100 000. This gives a ratio of true positives of roughly 1:200 or 0.5%. To obtain an equal number of false positives as true positives would require an experiment with an error rate of only 0.5%. This is obviously far below the accuracy of any of the existing techniques mentioned in the previous section. For example, it is estimated that the error rate of Y2H experiments is of the order of 50%.

Also, we need to be aware that various high-throughput methods even yield differing results on the same complex. For example, there are currently more than 80 000 interactions available for yeast, but only 2400 are supported by more than one method. Possible explanations for this discrepancy are (i) the experimental methods may not have reached saturation, (ii) many of the methods produce a significant fraction of false positives or (iii) some methods have difficulties for certain types of interactions. For example, it has turned out that each experimental technique produces a unique distribution of interactions with respect to functional categories (Figure 3.4). Due to probing interactions in solution, TAP and HMS–PCI naturally predict few interactions for proteins involved in transport and sensing because these categories are enriched with membrane proteins. On the other hand, Y2H detects few proteins involved in translation because these proteins operate in the cytosol, not in the nucleus.

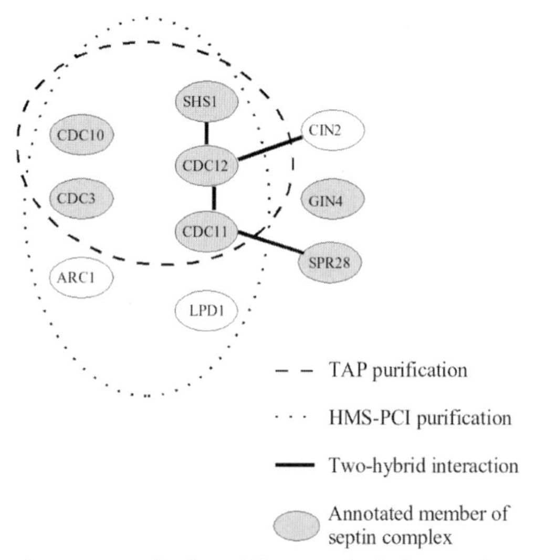

Figure 3.4 Results from different methods for complexes involving the cell cycle protein CDC11 located in the center of the figure. Drawn after von Mering (2002).

3.2.10
Criteria to Judge the Reliability of Interaction Data

Given this partly conflicting data, it would be very useful to have some confidence measures at hand to separate putative interactions into likely and unlikely ones. The following principles may provide some guidance in judging interaction data.

(1) **mRNA abundance** is a rough measure of protein abundance. By dividing the mRNA data for the yeast genome into mRNA abundance classes (bins) of equal size, from zero to maximum expression, it turned out that most data sets are heavily biased towards proteins of high abundance except for genetic techniques (Y2H and synthetic lethality). This means that, for example, a missing MS detection of the interaction between a pair of low-abundance proteins should be treated with less confidence than a missing interaction of two highly abundant proteins.

(2) Analyzing the interaction coverage for protein–protein pairs expressed in different cellular compartments would indicate whether a particular method shows a bias towards protein pairs from certain compartments. Indeed, comparing the protein localization from the MIPS (Munich Information Center for Protein Sequences) and TRIPLES (TRansposon-Insertion Phenotypes, Localization and Expression in *Saccharomyces cerevisiae*) databases against the interaction coverage showed that *in silico* predictions such as conserved gene neighborhood, co-occurrence of genes and gene fusion events (see Figures 3.5, 3.6 and 3.7) overestimate mitochondrial interactions.

Taken the other way around, an independent quality measure is to check whether interacting proteins belong to the same compartment (Figure 3.8). The Y2H method gave relatively poor results here. A possible complication of this approach is that proteins may actually be translocated between several compartments in reality and not all such multiple occurrences in different compartments may be annotated yet in databases or may even have been discovered.

(3) A certain degree of **cofunctionality** is to be expected for interacting protein–protein pairs. Whereas proteins from different groups of biological functions may certainly interact with each other, the degree to which interacting proteins are annotated to the same functional category is a measure of quality for predicted interactions. The predicted interactions should cluster fairly well along the diagonal, meaning that like pairs with like.

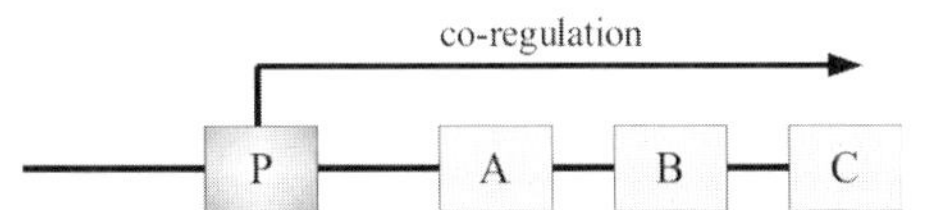

Figure 3.5 The **gene cluster method**. Genes A, B and C are arranged linearly as one operand. When transcription is activated at promoter P, all three genes are simultaneously transcribed.

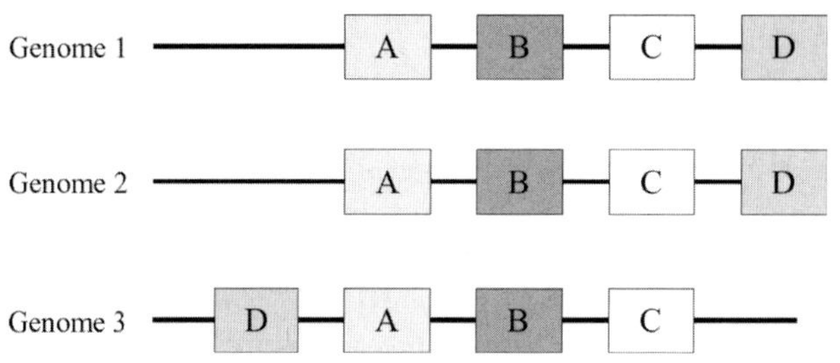

Figure 3.6 The **gene neighborhood method** analyzes the gene order in different evolutionarily related organisms. Genes that always occur in the same order (A, B, C) are likely to form an operon, meaning that they would be jointly regulated and are quite likely to interact. Gene D may occur at different locations, making it less likely to be part of the same operon.

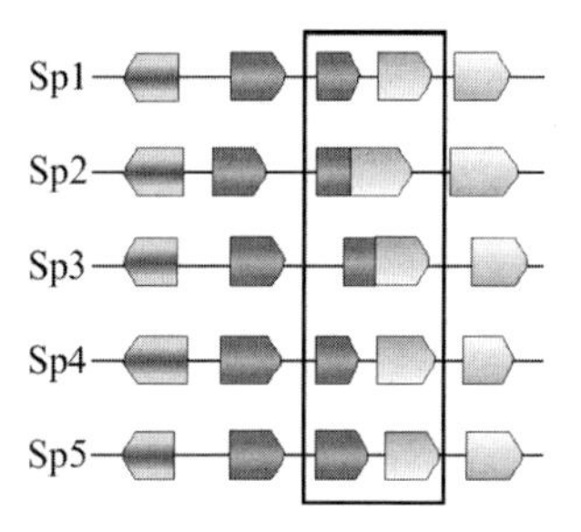

Figure 3.7 If two protein-coding genes are found to be separate in one species (Sp1, Sp4, Sp5) and fused to form a single gene in another (Sp2, Sp3), a physical interaction is probable. Analyzing the pairwise connectivity of genes to detect putative interactions is termed the **Rosetta Stone method**. The Ouzounis group actually found almost 40 000 predicted pairwise functional associations in this way from a search in 23 complete genomes (Enright and Ouzounis, 2001).

3.2.11
How Many Protein–Protein Interactions can be Expected in Yeast?

Estimating the rough total number of interactions in yeast based on the existing data for a subset of all yeast proteins is a typical task required for a bioinformatician.

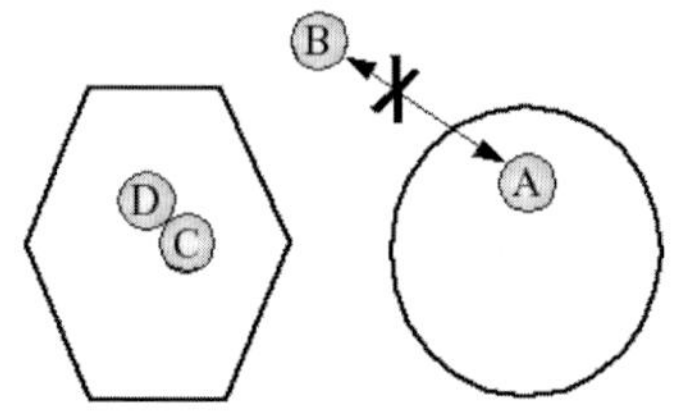

Figure 3.8 (Left) Proteins C and D are localized in the same compartment and may interact. (Right) On the other hand, proteins A and B belong to different compartments, making it very unlikely that they may physically interact *in vivo* unless one of them moves to the other compartment.

One approach is to compare the observed overlap of different protein–protein interaction data sets to the overlap that can be expected by chance. In fact, the overlap of the high-throughput data listed in Section 3.2.9 is 20 times larger than expected by chance. This means a good signal-to-noise ratio. For interactions discovered at least twice, usually both partners were annotated to the same functional category and cellular localization. Therefore, the overlap mainly consists of "true positives". On the other hand, less than one-third of the new interactions in the overlap set were previously known. Therefore, given 10 000 currently known interactions, one can give a lower boundary of a least 30 000 expected protein interactions in yeast.

3.3 Bioinformatic Prediction of Protein–Protein Interactions

Complementing the experimental methods introduced in Section 3.2, various purely computational methods have been developed for sequence-based predictions of protein–protein interactions that take into account the genomic content of pairs of genes or the occurrence of genes in related organisms. Table 3.2 provides an overview over some of the methods that will be discussed in this section.

3.3.1 Analysis of Gene Order

Genes with closely related functions encoding potentially interacting proteins are often transcribed as a single unit, an operon, in bacteria (Figure 3.5) and are coregulated in eukaryotes. Despite the effect of neutral evolution which tends to shuffle gene order between distantly related organisms, gene clusters or operons encoding for coregulated genes are usually conserved. Analysis of gene order conservation in various genomes found that about 70% of coregulated genes interact physically.

Table 3.2 Bioinformatics methods to predict protein–protein interactions (P and D stand for proteins and domains; F and PI stand for functional relationship and direct physical interaction, respectively).

Method name	Protein/domain interaction	Physical interaction/functional relationship
Gene cluster and gene neighborhood	P	F
Phylogenetic profile	P, D	F
Rosetta Stone	P	F
Sequence coevolution	P, D	F
Classification	P, D	PI
Integrative methods	P, D	PI

3.3.2
Phylogenetic Profiling/Coevolutionary Profiling

The method of **phylogenetic profiling** is based on the idea that the genes coding for two interacting proteins should either both be present in the genome of an organism or both be absent. While this may not sound as a strong indicator at first, with the availability of hundreds of sequenced genomes, this rather weak pattern may indeed become very powerful.

In contrast to methods predicting direct physical interactions, this computational method detects proteins that participate in a common structural complex or metabolic pathway. Proteins within these groups are termed *functionally linked*. The underlying hypothesis is that functionally linked proteins evolve in a correlated fashion and they, therefore, have homologs in the same subset of organisms. This property may be systematically exploited to map edges between all the proteins coded by a genome. For instance, one expects to find flagellar proteins in bacteria that possess flagella, but not in other organisms.

To represent the subset of organisms that contain a homolog, a phylogenetic profile is constructed for each protein (Figure 3.9). This profile is a string with n entries, each one bit long, where n corresponds to the number of genomes. The presence of a homolog to a given protein in the nth genome is indicated by an entry of unity at the nth position. If no homolog is found, the entry is zero. Proteins are then clustered according to the similarity of their phylogenetic profiles. Similar profiles show a correlated pattern of inheritance and, by implication, functional linkage. The method also predicts that the functions of uncharacterized proteins are likely to be similar to those of proteins with functional annotations within a cluster. This presence is most conveniently tabulated in a table (Table 3.3).

Each row is named a "profile" that characterizes the presence or absence of protein P_i in different organisms. From Table 3.3 we may compute the Hamming distances between all profiles as the number of columns with different entries (Table 3.4). For example, P1 and P3 differ in columns SC and BS. This yields a Hamming distance of 2. The entries of this table may again be presented in a graph as shown in Figure 3.10.

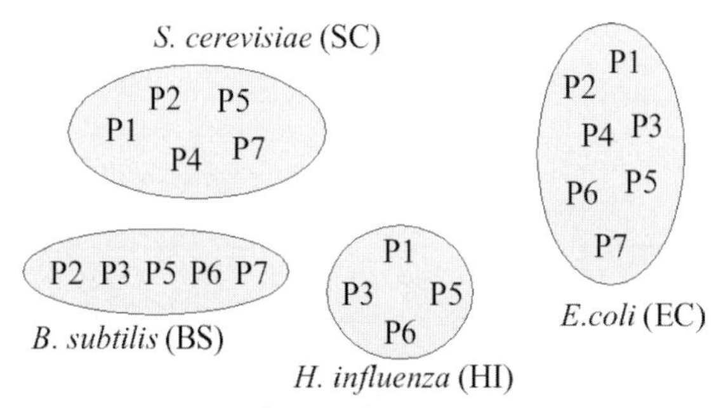

Figure 3.9 Presence of protein families in various organisms detected, for example, by conventional sequence alignment methods in the genome sequences of these organisms.

Table 3.3 Information on presence or absence of proteins P1–P7 in the four organisms of Figure 3.9.

Proteins	EC	SC	BS	HI
P1	1	1	0	1
P2	1	1	1	0
P3	1	0	1	1
P4	1	1	0	0
P5	1	1	1	1
P6	1	0	1	1
P7	1	1	1	0

Table 3.4 Hamming distances between profiles of proteins P1–P7.

	P1	P2	P3	P4	P5	P6	P7
P1	0	2	2	1	1	2	2
P2		0	2	1	1	2	0
P3			0	3	1	0	2
P4				0	2	3	1
P5					0	1	1
P6						0	2
P7							0

3.3.3 Coevolution

Apart from simply detecting the presence and absence of pairs of proteins among different genomes, one may also take into account the phylogenetic relationships among these organisms. Protein pairs that bind have a higher correlation between their phylogenetic distance matrices than other homologs drawn from the ligand and

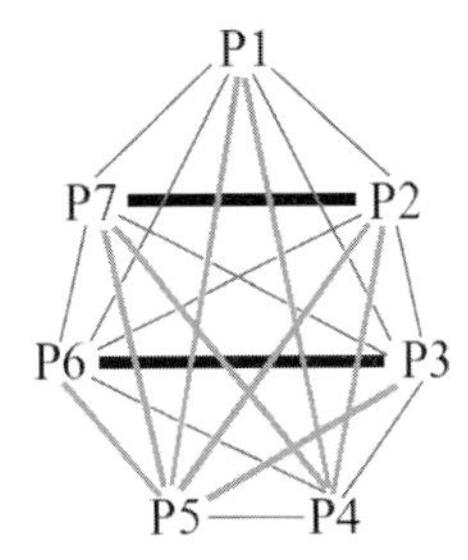

Figure 3.10 Graphical representation of the Hamming distances between the phylogenetic profiles of proteins P1–P7. The thick black lines connect proteins with a Hamming distance of 0, the thinner grey lines those with a distance of 1. The thinnest lines connect proteins with a distance of 2. Obviously, the two pairs P2 and P7, as well as P3 and P6, have identical profiles (simultaneous presence or absence in the four different organisms), compare Figure 3.9.

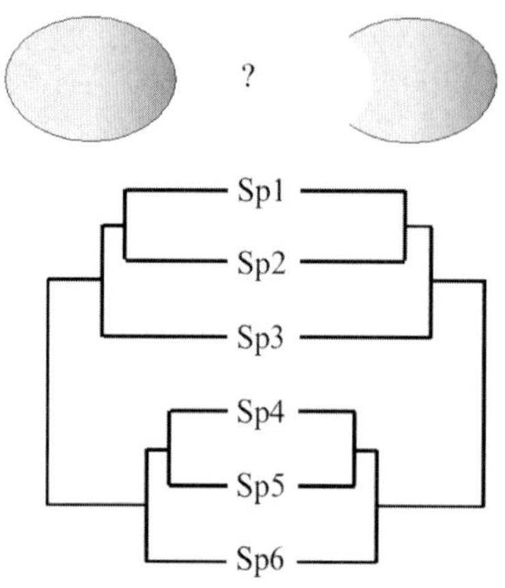

Figure 3.11 Potentially interacting proteins or functionally related proteins should show a similar pattern of evolution across several species.

receptor families that do not bind (Goh and Cohen, 2002) (Figure 3.11). One may also use this concept to identify by phylogenetic analysis proteins that can functionally substitute for another in various organisms. Such proteins are expected to have an anti-correlated distribution pattern across organisms. This allows discovery of non-obvious components of pathways, function prediction of uncharacterized proteins and prediction of novel interactions.

There now exists a small selection of promising experimental and theoretical methods to analyze the cellular interactome (by which we understand the set of all biomolecular interactions in a cell). We have encountered a first problem in that each method detects too few interactions (as seen by the fact that the overlap between predictions of various methods is very small), and a second problem in that each method has an intrinsic error rate producing "false positives" and "false negatives". Although this may not sound too encouraging, there is hope that everything will converge to a big picture eventually. The first problem can be partially solved by simply producing more experimental data. Also, solving the first problem will help solve the second problem by combining predictions. We will now introduce such a statistical approach for combining results from various experimental and prediction methods that can be used to estimate the quality of the interactions by statistical methods.

3.4
Bayesian Networks for Judging the Accuracy of Interactions

Given the considerable uncertainties of the available experimental data and of the *in silico* predictions on protein–protein interactions, it would be highly desirable to develop an indicator of confidence for every suggested interaction or even for every possible interaction. Can one integrate evidence from many different sources to increase the predictivity of true and false protein–protein predictions? One class of techniques to do this is made up of Bayesian approaches that allow for the probabilistic combination of multiple data sets. For illustration, this strategy will be applied to real yeast interaction data. These approaches can be used for combining noisy genomic interaction data sets given as input. For **normalization**, each source of evidence for interactions is compared against samples of known positives and negatives ("gold standard"). The Bayesian network then outputs for every possible protein pair its likelihood of interaction.

3.4.1 Bayes' Theorem

Bayes' theorem (named after the English mathematician Thomas Bayes) is a result in probability theory. It yields the **conditional probability distribution** of a random variable A, assuming we know the information about another variable B in terms of the conditional probability distribution of B given A, and the marginal probability distribution of A alone. Bayes' theorem relates conditional and marginal probabilities. To derive the theorem, we will start from the definition of conditional probability:

$$P(A|B)P(B) = P(A, B) = P(B|A)P(A),$$

where $P(A,B)$ is the **joint probability** of A and B. In words, *the probability of A given B* times *the probability of B* equals *the probability of both events A and B occurring together* and also equals *the probability of B given A* times *the probability of A*.

Dividing the left- and right-hand sides by $P(B)$ providing that it is nonzero gives:

$$P(A|B) = \frac{P(B|A)P(A)}{P(B)},$$

which is known as Bayes' theorem. It reads: "*The probability of A given B* equals *the probability of B given A* times *the probability of A*, divided by *the probability of B*".

$P(A)$ is also called the **prior probability** of A where "prior" means that it precedes any information about B. $P(A|B)$ is the **posterior probability** of A, given B. "Posterior" means that the probability is derived from or entailed by the specified value of B. For example, $P(A)$ could be the likelihood of an arbitrary person you bump into to hold a PhD degree in bioinformatics. This likelihood is probably quite low these days. $P(A|B)$ could then be the likelihood for some other person of holding a PhD in bioinformatics if you are given the evidence (B) that they regularly read the journal *Nature*, they work long but irregular hours and show activities in running a start-up company. This likelihood $P(A|B)$ is probably higher than $P(A)$ alone. So "*posterior*" means determining the likelihood after you are provided with evidence B. $P(B|A)$, for a specific value of B, is the **likelihood function** for A given B. It may also be written as $L(A|B)$. $P(B)$ is the prior probability of B. As it appears in the denominator, it acts as a **normalizing constant**.

With this terminology, the theorem may be paraphrased as:

$$\text{posterior} = \frac{\text{likelihood} \times \text{prior}}{\text{normalizing constant}}.$$

In addition, the ratio $P(B|A)/P(B)$ is known as the standardized likelihood and the theorem may be written:

$$\text{posterior} = \text{standardized likelihood} \times \text{prior}.$$

3.4.2 Bayesian Network

A **Bayesian network** is a directed acyclic graph of vertices representing variables and arcs representing dependence relations among the variables (Figure 3.12). If there is an arc from vertex A to another vertex B, then we say that A is a *parent* of B. If a vertex has a known value, it is said to be an *evidence* vertex. A vertex can represent any kind of variable, be it an observed measurement, a parameter, a latent variable or a hypothesis. Vertices are not restricted to representing random variables – this is what is "Bayesian" about a Bayesian network. A Bayesian network can be considered a mechanism for automatically constructing extensions of Bayes' theorem to more complex problems. Learning the structure (topology) of a Bayesian network is a very important part of machine learning.

Bayesian networks include a quantitative measure of dependency. For each variable and its parents this measure is defined using a conditional probability function or a table. Together, the graphical structure and the conditional probability functions/tables completely specify a Bayesian network probabilistic model. In Figure 3.12, one such measure is the probability $Pr(E_1|Y)$. The probability for the combined occurrence of E_1, E_2 and E_3 is described by $Pr(Y,E_1,E_2,E_3) = Pr(E_1|Y)$ $Pr(E_2|Y)$ $Pr(E_3|Y)$ $Pr(Y)$. In this case, the individual probabilities simply multiply as E_1, E_2 and E_3 do not depend on each other. Obviously, for E_1, E_2 and E_3 to occur, Y has to be "on" as well.

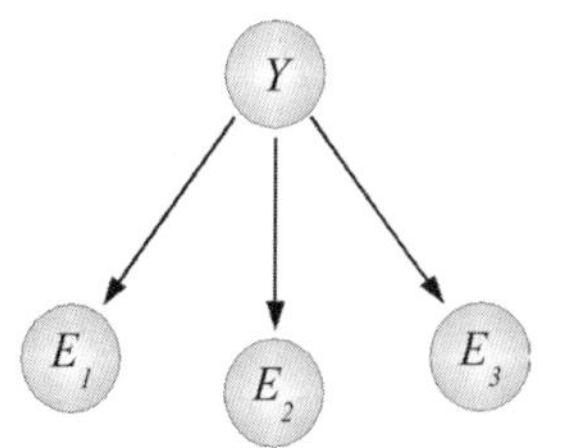

Figure 3.12 Example of a Bayesian network. A directed arc, e.g. between variables Y and E_1, denotes conditional dependency of E_1 on Y, as determined by the direction of the arc.

3.4.3 Application of Bayesian Networks to Protein–Protein Interaction Data

Gerstein and co-workers (Jansen *et al.*, 2003) applied the concept of a Bayesian network to combine data on protein–protein interactions from *Escherichia coli* obtained by different sources using three different types of data:

(1) Interaction data from high-throughput experiments comprising large-scale two-hybrid screens (Y2H) and *in vivo* pull-down experiments.
(2) Genomic features involving expression data, the biological function of proteins [from the Gene Ontology (GO) biological process and the MIPS functional catalog] and data about whether proteins are essential.
(3) "Gold standards" of known interactions and of non-interacting protein pairs.

As **positives** they used the MIPS catalog of complexes, which is a hand-curated list of complexes (8250 protein pairs that are within the same complex) from biomedical literature. The **negatives** are harder to define as it is hard to say for sure which protein–protein pairs definitely do not interact. However, a properly defined negatives set is as essential for the successful training of the Bayesian network as the set of positives. *In lieu* of an experimentally verified set, Gerstein *et al.* assumed that proteins localized to different compartments do not interact (Figure 3.9). As noted before, this choice may not be perfectly true because certain proteins may be exchanged between different compartments. However, the power of this approach derives from its statistical ansatz and it is probably true that, on average, proteins localized to different compartments are much less likely to interact than those that belong to the same compartment.

3.4.3.1 Measurement of reliability "likelihood ratio"

Considering a genomic feature f expressed in binary terms (i.e. "absent" or "present"), the standardized likelihood ratio $L(f)$ introduced above may also be written as:

$$L(f) = \frac{\text{fraction of gold} - \text{standard positives having feature} f}{\text{fraction of gold} - \text{standard negatives having feature} f},$$

$L(f)=1$ means that the feature has no predictability: the same fractions of positives and negatives have feature f. The larger $L(f)$, the better its predictability. For two features f_1 and f_2 with uncorrelated evidence, one may use a "naïve" Bayesian network where the likelihood ratio of the combined evidence is simply the product:

$$L(f_1, f_2) = L(f_1) \times L(f_2).$$

For correlated evidence $L(f_1, f_2)$ cannot be factorized in this way and a non-naïve Bayesian network is an appropriate formal representation of such relationships between features. The combined likelihood ratio is proportional to the estimated odds that two proteins are in the same complex, given multiple sources of information, or

$$L(f) = \frac{\text{fraction of gold} - \text{standard positives having features } f_1 \text{ and } f_2}{\text{fraction of gold} - \text{standard negatives having features } f_1 \text{ and } f_2}.$$

3.4.3.2 Prior and posterior odds

A "**positive**" result should refer to a pair of proteins that are in the same complex. Given the number of positives among the total number of protein pairs, the "prior" odds of finding a positive are:

$$O_{\text{prior}} = \frac{P(\text{pos})}{P(\text{neg})} = \frac{P(\text{pos})}{1 - P(\text{pos})}.$$

The "posterior" odds are the odds of finding a positive after considering N data sets with values $f_1, \ldots, f_N$:

$$O_{\text{post}} = \frac{P(\text{pos}|\, f_1 \cdots f_N)}{P(\text{neg}|\, f_1 \cdots f_N)}.$$

Again, the terms "prior" and "posterior" refer to the situation before and after knowing the information in the N data sets. In the case of protein–protein interaction data, the **posterior odds** describe the odds of having a protein–protein interaction given that we have the information from the N experiments, whereas the **prior odds** are related to the chance of randomly finding a protein–protein interaction when no experimental data is known.

If $O_{\text{post}} > 1$, the chances of having an interaction are higher than those of having no interactions. The likelihood ratio L defined as:

$$L(f_1 \cdots f_N) = \frac{P(f_1 \cdots f_N|\text{pos})}{P(f_1 \cdots f_N|\text{neg})},$$

relates prior and posterior odds according to Bayes' rule:

$$O_{\text{post}} = L(f_1 \cdots f_N)\, O_{\text{prior}}.$$

In the special case that the N features are conditionally independent (i.e. they provide uncorrelated evidence), L can be simplified as before to:

$$L(f_1 \cdots f_N) = \prod_{i=1}^{N} L(f_i) = \prod_{i=1}^{N} \frac{P(f_i|\text{pos})}{P(f_i|\text{neg})}.$$

L can be computed from contingency tables relating positive and negative examples with the N features (by binning the feature values $f_1, \ldots, f_N$ into discrete intervals), see below. Determining the prior odds O_{prior} is somewhat arbitrary in that it requires an assumption about the number of positives. Assuming that 30 000 is a conservative lower bound for the number of positives (i.e. pairs of proteins that are in the same complex) and 18 million possible protein pairs for yeast (Section 3.2.2):

$$O_{\text{prior}} = \frac{P(\text{pos})}{P(\text{neg})} = \frac{\dfrac{3 \times 10^4}{18 \times 10^6}}{\dfrac{17.97 \times 10^6}{18 \times 10^6}} = \frac{1}{600}.$$

This means that $O_{\text{post}} > 1$ can be achieved with $L > 600$.

3.4.3.3 A worked example: parameters of the naïve Bayesian network for essentiality

We will present only one data set from Jansen *et al.* (2003) that is the simplest to present and discuss. Proteins are termed essential or nonessential by considering whether a deletion mutant where this protein is knocked out from the genome has the same phenotype. We expect that it should be more likely that both of two proteins in a complex are essential or nonessential, but not a mixture of these two attributes. Deletion mutants of either one protein should impair the function of the same complex. Remarkably, data about essentiality was available for approximately 4000 out of the 6000 yeast proteins from individual gene knockout experiments. These 4000 proteins can form about 8 million interactions. In Table 3.5, the essentiality data is compiled for all possible protein–protein pairs and for those from the gold standard of interacting proteins. Column 1 describes the genomic feature. In the "essentiality data" protein pairs can take on three discrete values (EE: both essential; NN: both nonessential; NE: one essential and one not). Column 2 gives the number of protein pairs with a particular feature (i.e. "EE") drawn from the whole yeast interactome (around 18 million pairs). Columns "pos" and "neg" contain the overlap of these pairs with the 8250 gold standard positives and the 2 708 746 gold standard negatives.

Jansen *et al.* performed a similar analysis for mRNA expression data by computing a correlation index for the expression of each protein pair and for the functional similarity. How can these different sorts of information be combined? The way of linking the different types of information depends on whether the information is independent or whether it is dependent. Here, a simple "naïve" Bayesian network may be used to connect such independent sorts of data because these information sets hardly overlap.

3.4.3.4 Fully connected experimental network

The binary experimental interaction data sets from high-throughput experiments contain correlated evidence and should be combined in a fully connected Bayesian

Table 3.5 *P*(feature value|pos) and *P*(feature value|neg) give the conditional probabilities of the feature values.

		Gold standard overlap				
Essentiality	**No. of protein pairs**	**Pos**	**Neg**	***P(E|pos)***	***P(E|neg)***	***L***
EE	384126	1114	81924	5.18E − 01	1.43E − 01	3.6
NE	2767812	624	285487	2.90E − 01	4.98E − 01	0.6
NN	4978590	412	206313	1.92E − 01	3.60E − 01	0.5
Sum	8130528	2150	573724	1.00E + 00	1.00E + 00	1.0

L is the ratio of these two conditional probabilities. For example, $P(E|\text{pos}) = 5.18 \times 10^{-1}$ in the first row, column 6, is obtained as the ratio of 1114 to 2150. $P(E|\text{neg}) = 1.43 \times 10^{-1}$ is obtained as the ratio of 81 924 to 573 724. The value of 3.6 is then obtained as the ratio of 0.518 over 0.143. Although *P*(*E*|pos) seems only 50% predictive, this feature becomes more powerful by the small value of *P*(*E*|neg). The number of protein pairs for each category listed in the second column is not needed for the computation of likelihood factors. It is given here to emphasize the importance of sufficient coverage.

Table 3.6 The first four columns contain results from high-throughput experiments on protein–protein interactions ("1" means that an interaction was detected in this experiment, "0" that is was not detected; the meaning of the other columns is equivalent to those in Table 3.3).

Exp Gavin (g)	Exp Ho (h)	Exp Uetz (u)	Exp Ito (i)	No. of protein pairs	Gold standard overlap		P(g,h,u,i|pos)	P(g,h,u,i|neg)	L
					Pos	Neg			
1	1	1	0	16	6	0	7.27E − 04	0.00E + 00	–
1	0	0	1	53	26	2	3.15E − 03	7.38E − 07	4268.3
1	1	1	1	11	9	1	1.09E − 03	3.69E − 07	2955.0
1	0	1	1	22	6	1	7.27E − 04	3.69E − 07	1970.0
1	1	0	1	27	16	3	1.94E − 03	1.11E − 06	1751.1
1	0	1	0	34	12	5	1.45E − 03	1.85E − 06	788.0
1	1	0	0	1920	337	209	4.08E − 02	7.72E − 05	529.4
0	1	1	0	29	5	5	6.06E − 04	1.85E − 06	328.3
0	1	1	1	16	1	1	1.21E − 04	3.69E − 07	328.3
0	1	0	1	39	3	4	3.64E − 04	1.48E − 06	246.2
0	0	1	1	123	6	23	7.27E − 04	8.49E − 06	85.7
1	0	0	0	29221	1331	6224	1.61E − 01	2.30E − 03	70.2
0	0	1	0	730	5	112	6.06E − 04	4.13E − 05	14.7
0	0	0	1	4102	11	644	1.33E − 03	2.38E − 04	5.6
0	1	0	0	23275	87	5563	1.05E − 02	2.05E − 03	5.1
0	0	0	0	2702284	6389	2695949	7.74E − 01	9.95E − 01	0.8
Sum					8250	2708746			

network (Table 3.6). Here, the four data sets can be combined in at most $2^4 = 16$ different ways (subsets). For each of these 16 subsets, a likelihood ratio is computed from the overlap with the gold standard positives ("pos") and negatives ("neg") in the same way as for the essentiality data. This representation generates a transformation of the individual binary-valued interaction sets into a data set where every protein pair is weighted according to the likelihood that it exists in a complex.

First of all, Table 3.6 reveals some interesting effects obtained when applying plain statistics to real data. Row 2 has a higher L score than row 3. This seems odd. Is it a stronger indication for true interaction if an interaction is only found by two experimental groups (row 2) than if it was found by all four experimental groups (row 3)? Our intuition would say that the opposite should be the case. However, this is what results from the performance of the gold standard in these experiments. In row 2, 26 members of the gold standard were positive and two negative. In row 3, there were only nine members positive and one negative. Thus, the positive ones outnumber the negative ones by a smaller ratio. In this case, the differences arise from the small number of interactions. Having one or two negative members makes quite a difference here. After all, this tells us that the difference between $L = 4268.3$ and $L = 2955.0$ is not meaningful.

Another important fact about the values in Table 3.6 is that one needs to account for **statistical fluctuations of** these values. The Neg column contains a zero entry in the first row reflecting that none of the gold standard negatives had positive g, h and u experiments, and a negative i experiment (which is comforting!). Consequently, the likelihood L is not defined in this column as division by zero is not allowed. However, all experimental values are subject to statistical fluctuations on the order of:

$$\pm\sqrt{N+1},$$

L for this column can therefore adopt values of:

$$\frac{(6\pm\sqrt{7})\cdot 2,708746}{(0\pm\sqrt{1})\cdot 8250},$$

so that L can be between 1101 and infinity. In this case, one would use the smallest possible value of 1101 as a most conservative estimate.

The combined experimental data in Table 3.6 yields a much higher likelihood factor for interactions that were confirmed multiple times than the essentiality data in Table 3.5. However, this is not the end of the story. To achieve a stronger predictivity, the essentiality data may be combined with other factors such as mRNA coexpression. As argued before, the likelihood values of features with uncorrelated evidence are multiplicative. Consequently, we may also achieve predictions with $L_{\text{combined}} > 600$ from such combined data sets.

In summary, the Bayesian approach allows making relatively reliable predictions of protein–protein interactions by combining weakly predictive genomic features. In a similar manner, the approach could have been extended to include a number of other features related to interactions (e.g. phylogenetic co-occurrence, gene fusions, gene neighborhood).

3.5 Protein Domain Networks

So far in this chapter we have always considered entire proteins. Yet, investigation of three-dimensional protein structures suggests that the fundamental structural unit of proteins is a "domain". In the following, we will focus on simple domains formed by a continuous stretch of a peptide chain, although we note that there exist more complicated domains, e.g. in the catalytic subunits of protein kinases, where the large helical domain is not formed from one, but from several stretches of the protein sequence. In the case of simple domains, this region of a polypeptide chain folds into a distinct structure that often carries out the biological functionality of the protein domain independent of neighboring sequences. With regard to protein complexity, eukaryotes increasingly tend to have multidomain proteins, while the proteomes of bacteria or archaea mostly contain single-domain proteins. Such domain architectures govern interactions among proteins (Figure 3.13), offering a framework for prediction models.

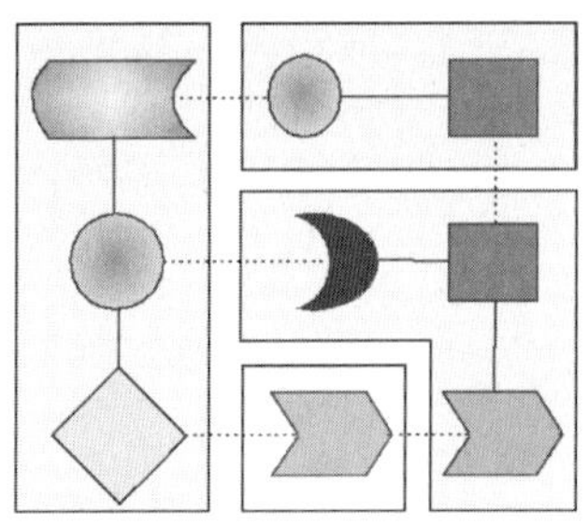

Figure 3.13 Nine domains (geometrical figures) are the fundamental units of these four proteins (shaded areas), mediating a distinct structure and biological functionality.

One method to predict domain–domain interactions is the **association method**. It assigns an interaction probability:

$$\Pr(d_m, d_n) = \frac{I_{mn}}{N_{mn}},$$

to each domain pair (d_m, d_n), where I_{mn} is the number of interacting protein pairs that contain (d_m, d_n) and N_{mn} is the total number of protein pairs that contain (d_m, d_n) (cf. Section 9.3.1).

As the experimental coverage of the human interactome is still low, one often resorts to domain–domain interactions for predicting protein interaction networks. The databases iPfam and 3did (the "database of 3D interacting domains") provide experimental sets of domain–domain interactions. Alternatively, the database InterDom derives potential domain interactions by combining data from multiple sources, ranging from domain fusions, protein interactions and complexes to scientific literature.

One may assume that interacting domains share a common biological process (BP) (annotation according to the GO terminology) and a common molecular function (MF) annotation. Statistical analysis (Schlicker *et al.*, 2007) showed that the BP similarity of interacting domains is generally higher than the corresponding MF similarity. This suggests that interacting domains or proteins may perform different functions even though they act in similar processes. Another reason may be that GO terms are more densely connected in the top levels of the BP ontology than of the MF ontology. Incorporating statistical checks for functional similarity may thus be useful for improving the prediction of protein–protein interactions.

Sequences of large proteins often seem to have evolved by joining preexisting domains in new combinations, "domain shuffling", that involves domain duplication or domain insertion. Remarkably, it was found that the number of domain–domain connections follows a power law (Figure 3.14). The majority of highly connected InterPro domains, the highly connected *hub* domains, appear in signaling pathways. Table 3.7 shows a list of the 10 best linked domains in various species. From left to right, the number of edges increases with increasing complexity of the organisms. At

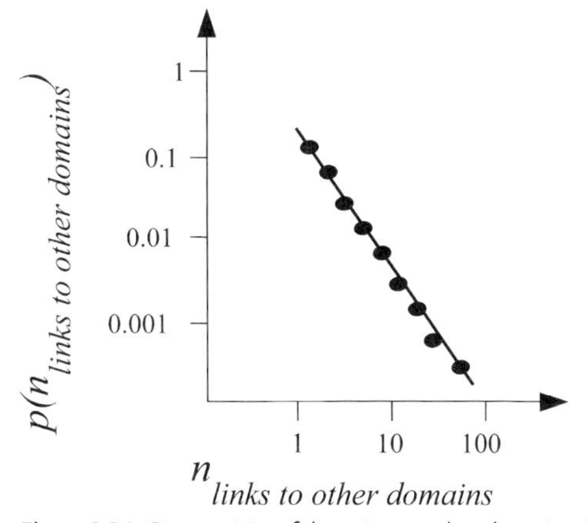

Figure 3.14 Connectivity of domains to other domains. Idealized picture drawn after Wuchty (2001).

the same time, the number of signaling domains (pleckstrin homology, SH3), their ligands (proline-rich extensions) and their receptors (G-protein-coupled receptor/rhodopsin) increases. This shows that the evolutionary trend toward compartmentalization of the cell and multicellularity lead to a higher degree of organization.

According to the Barabási–Albert model (BA) (Section 4.4), a so-called scale-free network is constructed by preferential attachment of newly added vertices to already well connected ones. Along these lines, it was argued that vertices with many connections in metabolic network are metabolites originating very early in the course of evolution where they shaped a core metabolism. If we transfer this concept to the connectivity of domains, highly connected domains should also have originated very early. Is this true? In the simple organisms *Methanococcus* and *E. coli*, the majority of highly connected domains are concerned with maintenance of metabolism. None of the highly connected domains of higher organisms are found here. On the other hand, helicase C has roughly similar degrees of connection in all organisms. Therefore, the BA model for network growth appears not applicable to this evolutionary scenario although the domain–domain connectivity clearly follows a power-law dependence.

The expansion of protein families in multicellular vertebrates coincides with a higher connectivity of the respective domains. Extensive shuffling of domains to increase combinatorial diversity might give rise to a large repertoire of multidomain proteins. These could then carry out the complex variety of cellular functions without dramatically expanding the absolute size of the protein complement. Therefore, the greater proteome complexity of higher eukaryotes is not simply a consequence of the genome size, but must also be a consequence of innovations in domain arrangements. Highly linked domains represent functional centers in various different cellular aspects. They could be treated as “evolutionary hubs” that help to organize the domain space.

Table 3.7 The 10 most highly connected InterPro domains of *Methanococcus*, *E. coli*, yeast, *Caenorhabditis elegans*, *Drosophila* and humans (drawn after Wuchty, 2001).

Methanococcus		***E. coli***		**Yeast**		***C. elegans***		***Drosophila***		**Humans**	
Domain	k_ν	**Domain**	k_ν	**Domain**	k_ν	**Domain**	k_ν	**Domain**	k_ν	**Domain**	k_ν
SAM	13	NAD-binding	20	Pkinase	18	Pkinase	57	Prich-extension	101	ATP-GTP-A	169
Fer4	11	Esterase	16	P-kinase-ST	18	EGF	57	Pkinase	70	GPCR-rhodopsin	162
FMN-enzymes	10	SAM	15	PH	16	PH	46	Zf-C2H2	53	Prich-extension	110
NAD-binding	9	Fer4	13	Zf-C3HC4	14	Efhand	45	Ank	52	EGF	98
AA-TRNA-ligase-1	8	AA-TRNA-ligase-II	12	AA-TRNA-ligase-II	14	Ank	37	EGF	50	Pkinase	89
Intein	7	FMN	12	Efhand	14	P-kinase-st	35	SH3	48	Ig	79
Pyr-redox	7	HIS-KIN	11	C2	13	EGF-CA	34	Antifreeze1	46	PH	72
ATP-GTP-A	6	AA-TRNA-ligase-1	11	CPSase-L-chain	13	Zf-C3HC4	33	Efhand	45	Efhand	64
CBS	6	HIS-REC	10	GATase	13	Ig	30	PH	45	SH3	61
N6-MTASE	6	PAS	9	WD40	13	SH3	30	P-kinase-ST	44	Zf-C2H2	58

Summary

A number of different experimental high-throughput methods have been developed over the past 10 years to probe the "interactome" of a cell, i.e. all biomolecular interactions occurring under certain conditions. Unfortunately, the results of some methods have an error of the order of about 50%. Complementing these experimental techniques are computational methods that can be employed genome-wide. Also, these methods are not very reliable *per se*. A statistical approach, Bayesian networks, was introduced to calibrate the accuracy of the individual approaches and obtain much more reliable results from combinations of various features. The protein–protein interaction network is nowadays considered as a highly connected assembly of individuals. Biological functions emerge from the activity of individual proteins as well as from their coordinated interactions.

Problems

Bayesian analysis of (fake) protein complexes

One way to estimate whether a given combination of proteins is a potential complex or not is to use a Bayesian analysis. It allows determining probabilities (likelihood ratios) from the properties of known protein complexes. These likelihood ratios can then be used to estimate the probability whether a candidate is a potential complex.

For problems (1) – (3), we use fake binary complexes where each of the two proteins has two properties: it belongs to one of three compartments "A", "B" or "C" and it has a mass. These two properties are encoded in the protein names as "Compartment" "_" + "Mass", e.g. a protein labeled "A_86" is, consequently, found in compartment "A" and has a mass of 86 units (you may think of kDa).

To determine the likelihood ratios you should download two "gold standard" data sets, GoldPos.dat and GoldNeg.dat, from the textbook website (gepard.bioinformatik.uni-saar-land.de/ccB_book) which contain complexes that either definitely occur or do not occur, respectively, plus two "experimental" sets Exp1.dat and Exp2.dat. These sets, which have a certain overlap with the gold standard data sets, contain both true and false complexes at a variable ratio, i.e. the experiments were performed at different levels of accuracy. Consequently, these sets are of different size, too. As an estimate of the initial probability O_{prior} you may assume that the relative sizes of the gold standard sets resemble the natural distribution of complexes and noncomplexes.

(1) Likelihood ratios from the theoretical properties

Use the gold standard data sets to determine likelihood ratios for the following properties.

(a) Compartment

For each complex in the gold standard data sets, determine the respective compartments of the two partners. Sort them into six categories AA, BB, CC, AB, AC and BC, where both proteins belong to compartment A, B or C, or one is in A and the other in B, etc. Note that it does not matter which of the two partners is found in A and which in B. From the relative occurrences of the gold standard data sets in these six classes determine the likelihood ratio L_{comp}.

(b) Mass difference

For the gold standard sets determine the absolute mass difference $\Delta m = |m1 - m2|$ between the two proteins, and sort Δm into four categories with $0 \leq \Delta m < 11$, $11 \leq \Delta m < 23$, $23 \leq \Delta m < 35$ and $35 \leq \Delta m$. From their numbers determine the respective values for L_{mass}.

(2) Likelihood ratios from the "experiments"

Use a fully connected Bayesian scheme to determine the likelihood ratios L_{exp} from the "experimental" data sets Exp1.dat and Exp2.dat. This gives you four categories for various combinations of positive and negative outcome for the complexes from the gold standard.

Try to judge the quality of the experiments by determining the likelihood ratios for both experiments independently, i.e. from the overlap between only one experiment and the gold standard sets.

(3) Identifying complexes

(a) Test set

For all the potential complexes in the test set of test_small.dat give the likelihood ratios for the properties L_{comp}, L_{mass} and L_{exp} and the final probability O_{post} that it is a true complex. Start from a reasonable O_{prior}.

Indicate for each potential complex whether you classify it as a true or as a false complex.

(b) Re-evaluation with the gold standard sets

Now run the Bayesian analysis with the theoretical ratios L_{comp} and L_{mass} alone on the gold standard positive and negative sets, and count how many false negatives and false positives you find in the two sets, respectively.

Give the percentage how many of the complexes in these two sets are classified correctly.

Is the classification biased; if yes, in which direction?

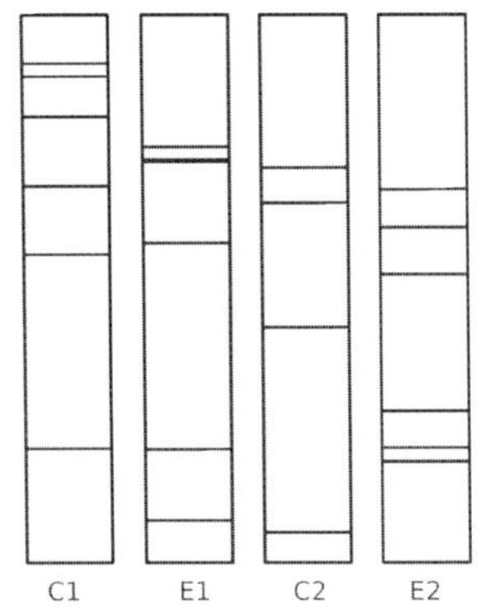

Figure 3.15 Resulting gel from a fictitous TAP experiment (see problem 5).

(4) Analyze a TAP experiment

Analyze the gels and determine the masses of the fragments from a protein with a total weight of 100 kDa. For better resolution, the gels C1 and E1 were run a second time, three times as long (Figure 3.15).

C1 and C2 are two controls with a calibration set that contained the masses 5, 12, 20, 42, 90 and 120 kDa, while for E1 and E2 the fragmented complex was used. For E1 and E2 two different preparation protocols were used, which lead to different sets of fragments.

First, assign the masses in the control gels C1 and C2, and quantify the relation between distance traveled and mass. With this relation, determine, assign and tabulate the masses of the fragments in E1 and E2. Name the fragments with capital letters starting from A for the smallest fragment. Note that some of the bands for larger masses may be due to parts of the protein that still consist of multiple subunits. Describe your observations and try to figure out which subunits of the complex stick together better.

Further Reading

Protein Interactions

Jones S, Thornton JM (1996) Principles of protein–protein interactions, *Proceedings of the National Academy of Sciences USA*, **93**, 13–20.

Experimental Methods

Shoemaker BA, Panchenko AR (2007) Deciphering protein–protein interactions. Part I. Experimental techniques and data-bases, *PLoS Computational Biology*, **3**, e42.

Protein–Protein Interaction Databases

Mathivanan S, Periaswamy B, Gandhi TKB, Kandasamy K, Suresh S, Mohmood R, Ramachandra YL amd Pandey A (2006) An evaluation of human protein–protein interaction data in the public domain. *BMC Bioinformatics*, **7**, S19.

von Mering C, Krause R, Snel B, Cornell M, Oliver SG, Fields S, Bork P (2002) Comparative assessment of large-scale data sets of protein–protein interactions, *Nature*, **417**, 399–403.

Bioinformatic Prediction Methods

Shoemaker BA and Panchenko AR (2007) Deciphering protein–protein interactions. Part II. Computational methods to predict protein and domain interaction partners. *PLoS Computational Biology*, **3**, e43.

Aloy P. and Russell R.B. (2006) Structural systems biology: modelling protein interactions. *Nature Reviews Molecular Cell Biology*, **7**, 188–197.

Enright AJ, Ouzounis CA (2001) Functional associations of proteins in entire genomes by means of exhaustive detection of gene fusions, *Genome Biology 2*, 341–347.

Goh CS, Cohen FE (2002) Co-evolutionary Analysis Reveals Insights into Protein–Protein Interactions, *Journal of Molecular Biology*, **324**, 177–192.

Bayesian Network

Jansen R, Yu H, Greenbaum D, Kluger Y, Krogan NJ, Chung S, Emili A, Snyder M, Greenblatt JF and Gerstein M (2003) A Bayesian networks approach for predicting protein–protein interactions from genomic data, *Science*, **302**, 449–453.

Protein Domain Networks

Wuchty S (2001) Scale-free behavior in protein domain networks, *Molecular Biology and Evolution*, **18**, 1694–1702.

Schlicker A, Huthmacher C, Ramirez F, Lengauer T. and Albrecht M (2007) Functional evaluation of domain–domain interactions and human protein interaction networks, *Bioinformatics*, **23**, 859–865.

4
Protein–Protein Interaction Networks – Structural Hierarchies

This chapter follows up on the previous chapter that provided an introduction to protein interaction networks. Now is the time to organize all the various data, detect and categorize topological properties of such networks, and possibly relate them to biological function.

4.1
Protein Interaction Graph Networks

As discussed in Chapter 3, a large amount of information has become available on protein–protein interactions in biological cells over the past few years. As organisms contain between 400 and 30 000 protein-coding genes, a large number of pairwise interactions are possible in principle. How can all this information be conveniently stored for subsequent analysis?

As was already introduced in Section 2.3, one way of doing this is by representing the known interactions as a two-dimensional $n \times n$ matrix with entries "1" for the known interactions and entries "0" everywhere else (Figure 4.1, middle). Considering that the 6000 yeast proteins form about 30 000 suspected interactions, less than 0.1% of all fields would be filled. Representing the data as a matrix would be quite wasteful in terms of computer memory and would make finding interaction partners quite inefficient. Instead, it is preferable to use a mathematical data structure that also allows for visualization of the data connectivity. For example, a biologist may suspect that a certain protein is related to cancer and wonders which the interaction partners of this protein are. Mathematical graphs introduced in Chapter 2 are an ideal data structure to address such questions. Biomolecular interactions are typically bidirectional. When molecule A binds molecule B, B also binds A. Therefore, they should be represented by undirected graphs where vertices (nodes) represent proteins and edges represent known interactions (Figure 4.1, left).

Principles of Computational Cell Biology – From Protein Complexes to Cellular Networks. Volkhard Helms

ISBN: 978-3-527-31555-0

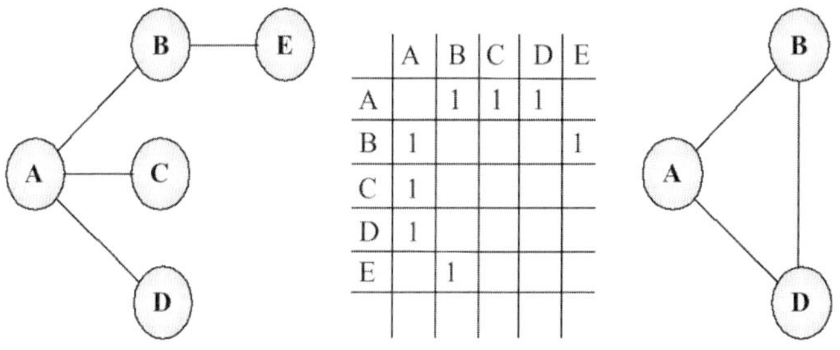

	A	B	C	D	E
A		1	1	1	
B	1				1
C	1				
D	1				
E		1			

Figure 4.1 (Left) Protein A interacts with proteins B, C and D, and B also interacts with protein E. In this example, it is not clear whether any of these interactions occur simultaneously or independently. (Middle) Interaction matrix representing the connectivity of the network on the left. (Right) Proteins A, B and C all interact with each other, suggesting that they may form a permanent complex ABC.

4.1.1 Degree Distribution

The **degree** of a vertex specifies the number of edges by which it is connected to other vertices. The **degree distribution** of a graph is a function measuring the **total number** of **vertices** in a graph with a given **degree**. Formally, the degree distribution is:

$$p(k) = \frac{1}{N} \sum_{v_i \in V | \deg(v_i) = k} 1,$$

where v_i is a vertex in the set V of the graph's N vertices and $\deg(v_i)$ is the degree of vertex v_i. This expression simply counts how many vertices have degree k. Figure 4.2 shows a simple example.

In the two-dimensional lattice of Figure 4.2, each vertex has exactly four neighbors. $p(k)$ is zero except for $k = 4$. What is the virtue of this representation? It is an abstraction of the network topology. If you were given $p(k)$ of a square lattice – with its single peak at 4, respectively 6, in three dimensions – you would quickly be able to draw the corresponding network. While this is not very exciting for simple examples like this one, it becomes extremely valuable for complicated networks involving thousands of vertices and edges.

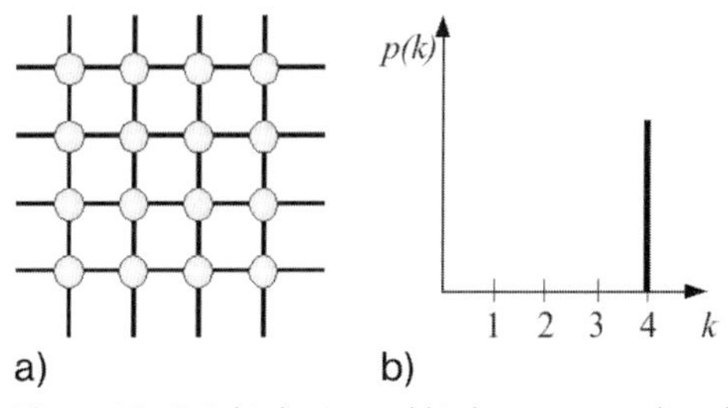

Figure 4.2 a) Cubic lattice and b) the corresponding distribution $p(k)$.

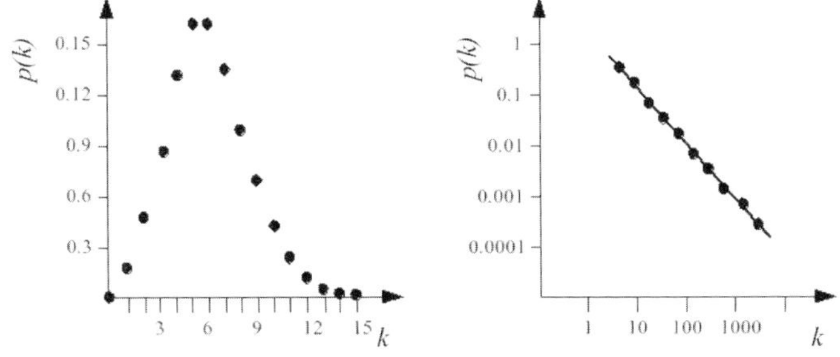

Figure 4.3 Degree distribution in random network (left) showing a Poisson distribution and for a scale-free network (right).

The degree distribution $p(k)$ is a common way of classifying graphs into categories. $p(k)$ of a random graph has the shape of a Poisson distribution – most vertices have an average number of connections; sparsely connected as well as densely connected vertices are equally unlikely (Figure 4.3). In scale-free networks, $p(k)$ follows a power law:

$$p(k) = k^{-\gamma}, \quad \gamma \in \Re^{+}.$$

Therefore, $p(k)$ decays much slower for large k than the exponential decay of a Poisson distribution. Highly connected "hubs" occur at a much larger frequency than expected in random graphs. One simple way of classifying protein networks is to analyze the distribution of connections made by specific proteins.

Technical comment

Why does a scale-free network appear as a straight line with negative slope on a log-log plot? Let us start from the given power-law dependence:

$$p(k) = k^{-\gamma}.$$

Taking the logarithm on both sides gives:

$$\begin{aligned} \log p(k) &= \log k^{-\gamma} \\ &= -\gamma \cdot \log k. \end{aligned}$$

Therefore, plotting log $p(k)$ on the y-axis against log(k) on the x-axis will give a straight line with the slope $-\gamma$ as seen in Figure 4.3 (right).

4.1.2 Clustering Coefficient

In 1998, Duncan J. Watts and Steven Strogatz introduced the **clustering coefficient** graph measure to determine whether or not a graph is a small-world network. As an

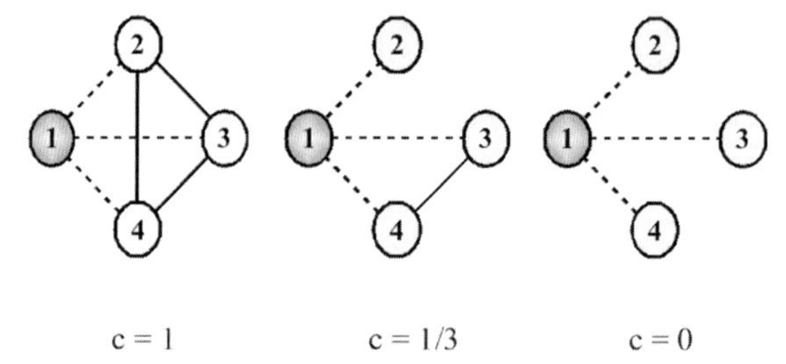

c = 1 c = 1/3 c = 0

Figure 4.4 Example illustrating the clustering coefficient on an undirected graph for the shaded vertex *i*. Dotted lines connect the shaded vertex with three white vertices, its neighbors. The black edges connect neighbors of *i* among each other. In the left example of a fully connected subgraph, three out of three possible connections are formed between the white neighbor vertices. Thus, the clustering coefficient of the black vertex equals 1. Vertices 1–2–3, 1–2–4 and 1–3–4 form triangles in the left picture. The middle picture contains one triangle (1–3–4) and the same three triples as the left picture.

example, consider the group formed by all your N friends. The clustering coefficient describes whether your friends are also friends among themselves. When your group of friends is a "clique" (see below), all of its members will also be friends of each other. The clustering coefficient is defined for vertex i (which is you in the example) and considers the connectivity of its neighbors (here, your friends) among each other (Figure 4.4).

To define the clustering coefficient of a vertex, let us first define a graph in terms of a set of n vertices $V = v_1, v_2, \ldots, v_n$ and a set of edges E, where e_{ij} denotes an edge between vertices v_i and v_j. Below we assume that v_i, v_j and v_k are members of V. We define the **neighborhood** N_i of a vertex v_i as its immediately connected neighbors:

$$N_i = \{v_j\} : e_{ij} \in E.$$

As introduced before, the degree k_i of vertex v_i is the number $|N_i|$ of vertices in its neighborhood N_i. The clustering coefficient C_i for a vertex v_i is then the number of edges between the vertices within its neighborhood divided by the maximum possible number of edges that could exist between them, $k_i\ (k_i - 1)/2$ (Figure 4.4):

$$C_i = \frac{2|\{e_{jk}\}|}{k_i(k_i - 1)} : v_j, v_k \in N_i, e_{jk} \in E_{\text{undirected}}.$$

By convention, $|\cdot|$ counts the numbers of edges in the set $\{e_{jk}\}$. Consequently, $0 \le C_i \le 1$. For a directed graph, e_{jk} is distinct from e_{kj} and therefore for each neighborhood N_i there are $k_i(k_i - 1)$ arcs that could exist among the vertices within the neighborhood. Thus, the clustering coefficient is given as:

$$C_i = \frac{|\{e_{jk}\}|}{k_i(k_i - 1)} : v_j, v_k \in N_i, e_{jk} \in E_{\text{directed}}.$$

We can also describe the clustering coefficient in an alternative way. If vertices v_j and v_k are neighbors of vertex v_i, and if they are connected by an edge (arc) (j,k), then the three edges (i,j), (j,k) and (i,k) form a triangle (Figure 4.4, left). Let $\lambda_G(v_i)$ be the number of triangles involving $v_i \in V(G)$ for undirected graph G. That is, $\lambda_G(v_i)$ is the number of subgraphs of G with three edges and three vertices, one of which is v_i. Let $\tau_G(v_i)$ be the number of triples on $v_i \in V$. That is, $\tau_G(v_i)$ is the number of subgraphs (not necessarily induced) with two edges and three vertices, one of which is v_i and such that v_i is end point of both edges (Figure 4.4, middle). Obviously, each triangle is also a triple. Then we can also define the clustering coefficient as:

$$C_i = \frac{\lambda_G(v_i)}{\tau_G(v_i)}.$$

It is simple to show that the two preceding definitions are the same, since:

$$\tau_G(v_i) = \frac{1}{2}k_i(k_i - 1),$$

and:

$$\lambda_G(v_i) = |\{e_{jk}\}|,$$

as defined above.

These measures are 1 if every neighbor connected to v_i is also connected to every other vertex within the neighborhood and 0 if no vertex that is connected to v_i connects to any other vertex that is connected to v_i.

The clustering coefficient for the whole system is given by Watts and Strogatz as the average of the clustering coefficients for each of the n vertices:

$$\bar{C} = \frac{1}{n}\sum_{i=1}^{n} C_i.$$

This average clustering coefficient can be used as another measure of network topology.

4.2 Finding Cliques

In graph theory, a **clique** in an undirected graph G is a set of vertices V such that for every two vertices in V there exists an edge connecting the two (Figure 4.5). This is equivalent to saying that the subgraph induced by V is a **complete graph**. The size of a clique equals the number of vertices it contains.

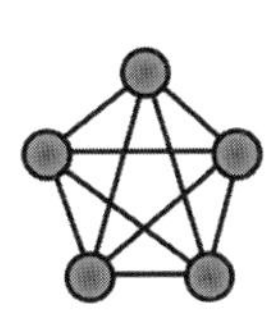

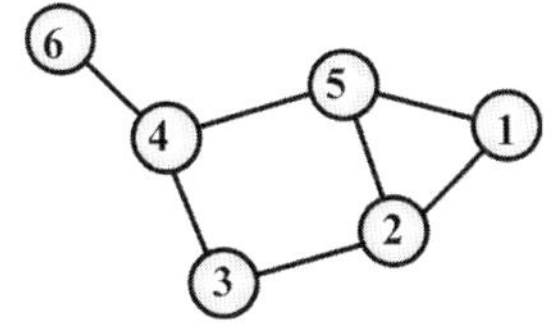

Figure 4.5 A clique in a graph is a set of pairwise adjacent vertices or a fully connected subgraph. The left example shows a clique of size 5. In the graph at the right, vertices 1, 2 and 5 form a clique, because each vertex is connected to the two other vertices.

The "clique problem" denotes the problem of finding the largest clique of a graph G. This problem is NP-complete (Section 2.3) and, as such, many consider it to be unlikely that an efficient algorithm for finding the largest clique of a graph exists. In a protein interaction network, cliques usually correspond to permanent multiprotein complexes as will become clearer in later chapters.

A ***k*-clique** is a clique of size k. The k-clique problem, therefore, is the task of finding a clique of size k, i.e. a complete subgraph $G'(V',E')$ of G with $|V'| = k$. Obviously, a k-clique can be found using a brute-force algorithm in $O(n^k)$ time. One simply starts with all vertices $[O(n)]$ and tests whether they have one connection $[O(n^2)]$ up to $k-1$ connections $[O(n^k)]$ which yields a clique of size k when including the vertex itself.

A **brute-force algorithm** to find a clique in a graph is to examine each subgraph with at least k vertices and check to see if it forms a clique. Another algorithm by Spirin and Mirny starts by considering each vertex to be a clique of size one and to merge cliques into larger cliques until there are no more possible merges. Two cliques A and B may be merged if each vertex in clique A is joined to each vertex in clique B. This requires only linear time, but may fail to find a large clique because two or more parts of the large clique have already been merged with vertices that are not in the clique. It does, however, find at least one *maximal clique*, which is a clique not contained in any larger clique.

4.3 Random Graphs

As mentioned in Section 1.1, a **random graph** describes a graph $G_{n,p}$ of n vertices joined by edges that all have been chosen and placed between pairs of vertices at random. In $G_{n,p}$ each possible edge is present with probability p and absent with probability $1-p$. As each possible edge was established with probability p, the average number of edges in $G_{n,p}$ is:

$$\frac{n(n-1)}{2}p.$$

As each edge connects two vertices, the average degree of a vertex is:

$$\frac{n(n-1)p}{n} = (n-1)p \underset{n\to\infty}{\rightarrow} np.$$

In the 1950s, Erdös and Renyi studied how the expected topology of a random graph with n vertices changes as a function of the number of edges m. When m is small, the graph is likely fragmented into many small connected components (Section 2.1) having vertex sets of size at most $O(\log n)$. As m increases, the components grow at first by linking to isolated vertices and later by fusing with other components. A transition happens at $m = n/2$, when many clusters cross-link spontaneously to form a unique largest component called the **giant component**. Its vertex set size is much larger than the vertex set sizes of any other components. It contains $O(n)$ vertices, whereas the second largest component contains $O(\log n)$ vertices. In statistical physics, this phenomenon is called **percolation** and describes, for example, at what density of edges between lattice points in a planar lattice, a continuous path exists between two opposite walls. The shortest path length between any pairs of vertices in the giant component grows like log n. In a network with $n = 1\,000\,000$ vertices, $\log 10^6 = 6$. Therefore, any two vertices have only a few edges in between them and this is why these graphs are called "**small worlds**".

The properties of random graphs have been studied very extensively (Bollobas, 2004). However, random graphs are not adequate models for real-world networks because real networks appear to have a power-law degree distribution (while random graphs have Poisson distribution) and real networks show strong clustering while the clustering coefficient of a random graph is $C = p$.

Can one cure this deficiency by manipulating the connectivities in a random graph to generate a power-law degree distribution while leaving all other aspects as in the random graph model? Yes, given a degree sequence (e.g. a power-law distribution) one can generate such a tweaked random graph by assigning to a vertex i a degree k_i from the given degree sequence. Then pairs of vertices are chosen uniformly at random to make edges so that the assigned degrees remain preserved. When all degrees have been used up to make edges, the resulting graph is a random member of the set of graphs with the desired degree distribution. However, this method does not allow specifying the clustering coefficient. On the other hand, this property makes it possible to exactly determine many properties of these graphs in the limit of large n. For example, it can be shown that almost all random graphs with a fixed degree distribution and no vertices of degree smaller than 2 have a unique giant component.

4.4 Scale-Free Graphs

In 1999, Albert and Barabási published a paper in the journal *Science* that completely changed our understanding of complex networks. In this paper, a new network growth algorithm was introduced to generate networks with a scale-free topology. The **Barabási–Albert (BA) algorithm** starts with n_0 isolated vertices. At every step t, a new vertex v is added and m edges ($m < n_0$) will be linked from v to the existing vertices with a **preferential attachment** probability:

$$\Pi_i(k_i) = \frac{k_i}{\sum\limits_{i=1}^{N-1} k_i},$$

where k_i is the degree of the ith vertex and the summation runs over all vertices present so far. Eventually, the graph will have $(n_0 + t)$ vertices and (mt) edges. It was found that if either growth or preferential attachment is eliminated from the BA algorithm, the resulting network does not exhibit scale-free properties anymore.

Considering the properties of networks generated by the BA model, the **average path length** in the BA model is proportional to $\ln n / \ln \ln n$ which is shorter than in random graphs. Therefore, scale-free networks are termed **ultrasmall worlds**. As the degree distribution decays polynomial, many more **hubs** exist having a very high degree than is expected for random graphs.

In the BA model $p(k) \propto k^{-\gamma}$ with $\gamma = 3$, whereas real networks often show a softer decay with $\gamma \approx 2.1$–2.4. The numerical result for the clustering coefficient of the BA model is $C \approx n^{-0.75}$.

Scale-free networks are resistant to random failures of arbitrary hubs ("**robustness**") because a few high-degree hubs dominate their topology. Intuitively, we understand that an arbitrary failing vertex probably has a small degree and thus will not severely affect the rest of the network. On the other hand, scale-free networks are quite **vulnerable** to dedicated **attacks** on the hubs. These properties have also been confirmed numerically and analytically by studying the average path length and the size of the giant component.

A classical result from the work of Barabási is shown in (Figure 4.6). The authors investigated the protein interaction network of the yeast *Saccharomyces cerevisiae* from yeast two-hybrid experiments. Figure 4.6 shows the giant component of this interaction network. The authors color-coded all proteins by information about their essentiality – green for nonessential, yellow for unknown and red for essential proteins. In agreement with the vulnerability hypothesis, hub proteins were much more likely to be essential than nonhub proteins.

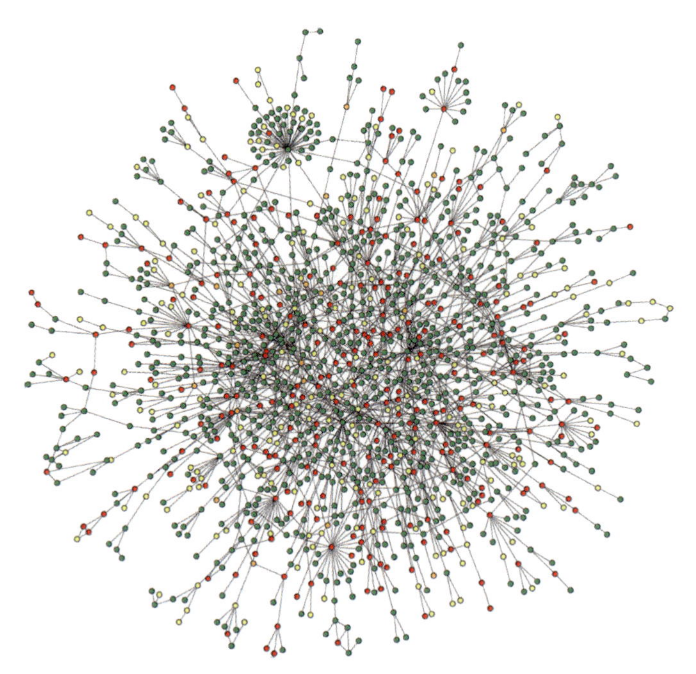

Figure 4.6 Result of one of the first analyses of a large-scale biological network. Each vertex corresponds to a protein from *S. cerevisiae*. The plot shows the largest connected cluster that contains about 78% of all proteins. The vertex color reflects the phenotypic effect of a gene-deletion mutant where the gene coding for the respective protein was knocked out. Green circles reflect nonlethal gene deletions, red ones lethal mutations, orange ones lead to slow growth and the yellow vertices reflect gene deletions where the effect is unknown. Reprinted from Jeong *et al.* (2001) by permission from Macmillan Publishers Ltd.

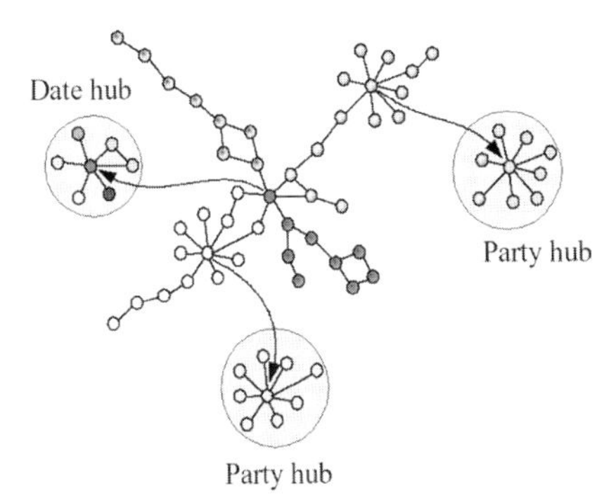

Figure 4.7 In this schematic of protein interaction networks, proteins are colored according to mutual similarity in their mRNA expression patterns. "Party" hubs are highly correlated in expression with their partners, and presumably interact with them at similar times. The partners of "date" hubs exhibit reduced coexpression, and presumably the corresponding physical interactions occur at different times and/or different locations. Drawn after Han *et al.* (2004).

Recent investigations addressed how hubs might contribute to robustness and other cellular properties for protein–protein interactions dynamically regulated both in time and in space. Two types of hubs were found: "**party hubs**", which interact with most of their partners simultaneously, and "**date hubs**", which bind their partners at different times or locations (Figure 4.7). This picture supports a model of organized modularity in which date hubs organize the proteome, connecting biological processes (or modules) to each other, whereas party hubs function inside modules. We will come back to this issue in Chapter 10 when we will deal with integrated networks.

We will end this subsection by noting that the BA model is a minimal model that captures the mechanisms responsible for the power-law degree distribution observed in real networks. However, there may be other models that explain the same phenomenon equally well. A discrepancy is the fixed exponent of the predicted power-law distribution ($\gamma = 3$). Recent efforts have involved studying variants of the BA construction algorithm with cleaner mathematical properties. These may include effects of adding or rewiring edges as this allows vertices to age so that they can no longer accept new edges or vary the forms of preferential attachment. These models also predict exponential and truncated power-law degree distributions in some parameter regimes.

4.5 Detecting Communities in Networks

Many seemingly unrelated types of networks, including social networks (such as acquaintance networks and collaboration networks), technological networks (such

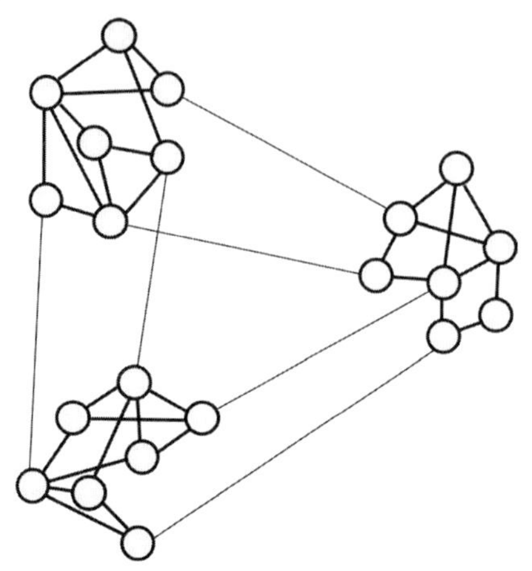

Figure 4.8 Three communities of densely connected vertices (circles with solid lines) with a much lower density of connections (gray lines) between them.

as the Internet, the World Wide Web and power grids) and biological networks (such as neural networks, food webs and metabolic networks), have several distinct statistical properties in common. One such property is the "small-world effect" (Section 1.1). Another is the right-skewed degree distribution which often follows a power-law distribution (Figure 4.3). A third property shared by many networks is clustering.

In this chapter we will focus on another property common to many networks – the property of community structure. "Community" may stand here for module, class, group, cluster, etc. We define **community** as a subset of vertices within the graph such that connections between the vertices are denser than connections with the rest of the network (Figure 4.8). The detection of community structure is generally intended as a procedure for mapping the network into a tree ("dendogram" in social sciences). The leaves of the tree represent vertices; branches join vertices or (at higher level) groups of vertices. A traditional method to perform this mapping is **hierarchical clustering**. For every pair i,j of vertices in the network, a weight W_{ij} is computed that measures how closely connected the vertices are. Starting from the set of all vertices and no edges, edges are iteratively added between pairs of vertices in order of decreasing weight. In this way vertices are grouped into larger and larger communities, and the tree is built up to the root which represents the whole network.

What measures could be used together with hierarchical clustering to detect communities in a given network?

1. The number of **vertex-independent paths** between vertices. Two paths that connect the same pair of vertices are said to be vertex-independent if they share none of the same vertices other than their initial and final vertices.

2. The number of **edge-independent paths** defined in the same spirit. A classical result from graph theory is that the number of vertex-independent (edge-independent) paths between two vertices i and j in a graph is equal to the minimum number of vertices (edges) that must be removed from the graph to disconnect i and j from one another (theorem of Menger, 1927). These numbers are, therefore, a measure of the robustness of the network to deletion of vertices (edges).

3. One may also count the **total number of paths** that run between vertices i and j (not just those that are vertex- or edge-independent). As the number of paths between any two vertices is either 0 or infinite, one typically weights paths of length l by a factor α^l with small α so that the weighted count of their number of paths converges. In doing so, long paths contribute exponentially less weight than short paths.

Although these vertex- or edge-dependent path definitions for weights work okay for certain community structures, there generally exist some problems with all these measures. In particular, counting of both vertex- and edge-independent paths has a tendency to separate single peripheral vertices from the communities to which they should rightly belong. If a vertex is, for example, connected to the rest of a network by only a single edge then, to the extent that it belongs to any community, it should clearly be considered to belong to the community at the other end of that edge. Unfortunately, both the numbers of independent paths and the weighted path counts for such vertices are small and hence single vertices often remain isolated from the network when the communities are constructed. This and other pathologies make the hierarchical clustering method, although useful, far from perfect.

Therefore, in 1977, Freeman introduced the measure of "**betweenness**" as a new concept to measure the centrality of a vertex within a graph. Vertices that occur on many shortest paths between other vertices have a higher betweenness than those that do not (Figure 4.9).

For a graph $G := (V, E)$ with n vertices, the **betweenness** $C_B(v)$ for vertex v is:

$$C_B(v) = \frac{\sum_{s \neq v \neq t \in V} \sigma_{st}(v)}{(n-1)(n-2)},$$

where $\sigma_{st}(v) = 1$ if the shortest path from s to t passes through v and 0 otherwise. Alternatively, we may focus on those edges that are least central, that are "between" communities. **Edge betweenness** of an edge is defined as the number of shortest paths between pairs of vertices that run along this particular edge. If there is more than one shortest path between a pair of vertices, each path is given equal weight such that the total weight of all of the paths is 1. If a network contains communities or groups that are only loosely connected by a few intergroup edges, then all shortest paths between different communities must go along one of these few edges. The edges connecting communities will have high edge betweenness. By removing these edges first, we will separate groups from one another and so reveal the underlying community structure of the graph.

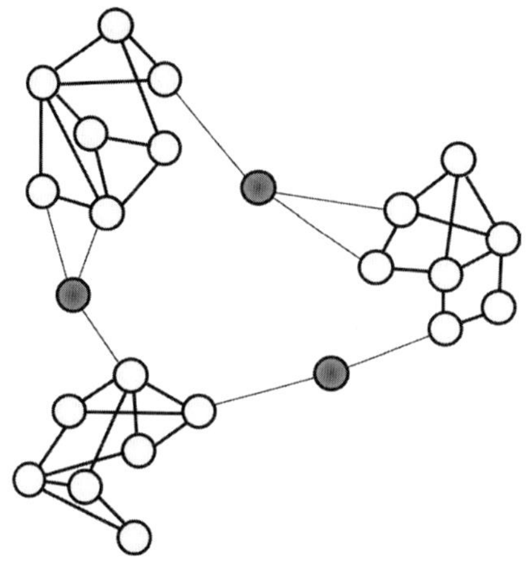

Figure 4.9 Modified version of Figure 4.8 where three dark vertices mediate the contacts between the three communities. Obviously, all shortest pathways between members of different communities would run through one or two of these vertices making them very "between" in the sense of the definition given in the text.

This is the basis of the **Girven–Newman algorithm** introduced in 2002:

1. Calculate betweenness for all m edges in a graph of n vertices (can be done in $O(mn)$ time).
2. Remove the edge with the highest betweenness.
3. Recalculate betweenness for all edges affected by the removal.
4. Repeat from step 2 until no edges remain.

Because step 3 has to be done for all edges, the algorithm requires in the worst case $O(m^2n)$ operations.

4.5.1 Divisive Algorithms for Mapping onto Tree

To construct a tree from the computed edge betweenness, the order of tree construction compared to agglomerative algorithm is reversed. By starting with the whole graph and iteratively cutting the edges, the network is progressively divided into smaller and smaller disconnected subnetworks identified as the communities. The crucial point in this recipe is how to select those edges to be cut. The Girvan–Newman algorithm is one way of doing so.

Figure 4.10 shows an example given in Girvan and Newman (2002) for the clustering of a social network drawn from the well-known karate club study of Zachary who observed the personal relationships between 34 members of a karate club over a period of 2 years. During the course of the study, a disagreement developed between the administrator of the club and the club's instructor, which

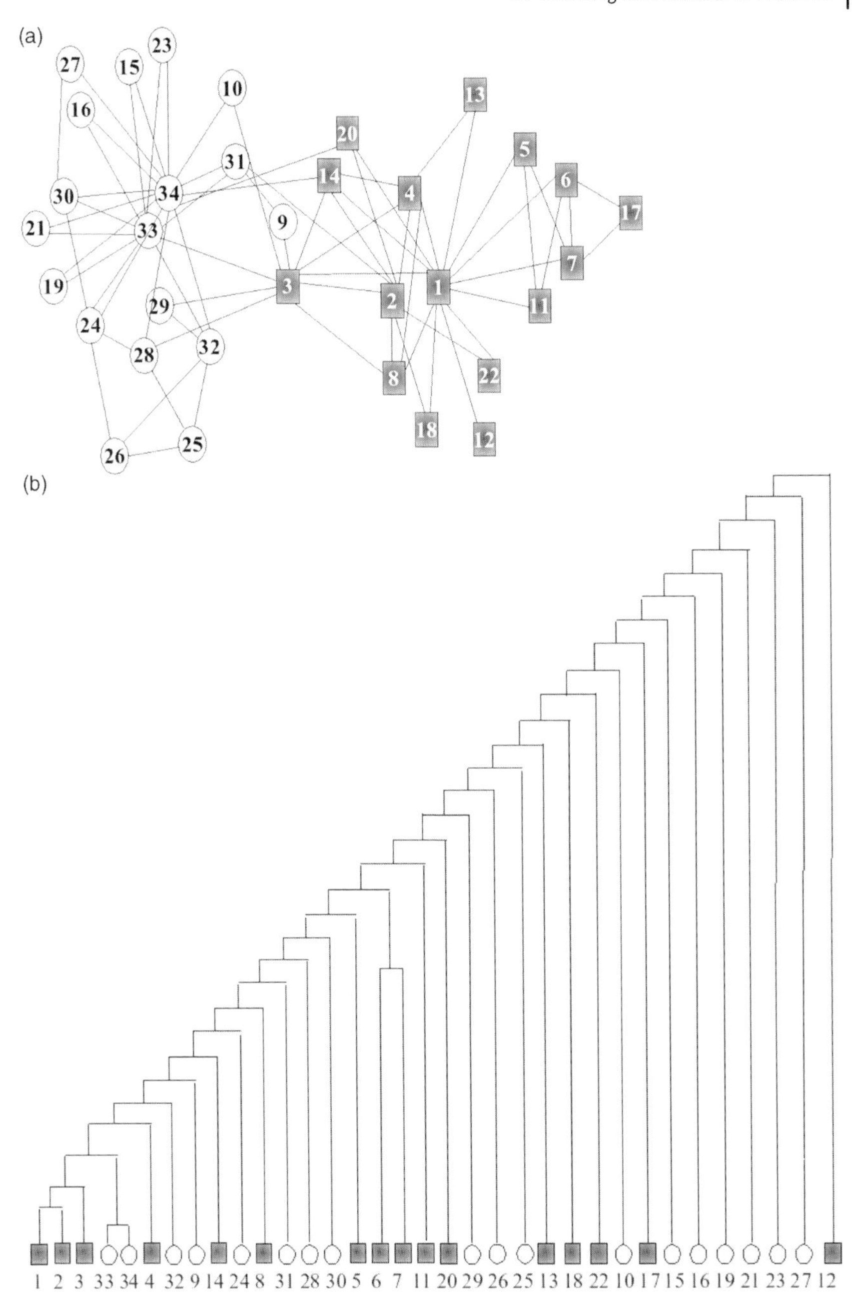

Figure 4.10 (a) The friendship network from Zachary's karate club study. The instructor and the administrator are represented by vertices 1 and 34. Vertices associated with the club administrator's fraction are drawn as white circles, those associated with the instructor's fraction are drawn as grey squares. (b) Hierarchical tree calculated by using edge-independent path counts, which fails to extract the known community structure of the network. (c) Hierarchical tree showing the complete community structure for the network calculated by using the Girven–Newman algorithm. The initial split of the network into two groups is in agreement with the actual factions observed by Zachary, except for the misclassified vertex 3. Redrawn after Girvan *et al.* (2002).

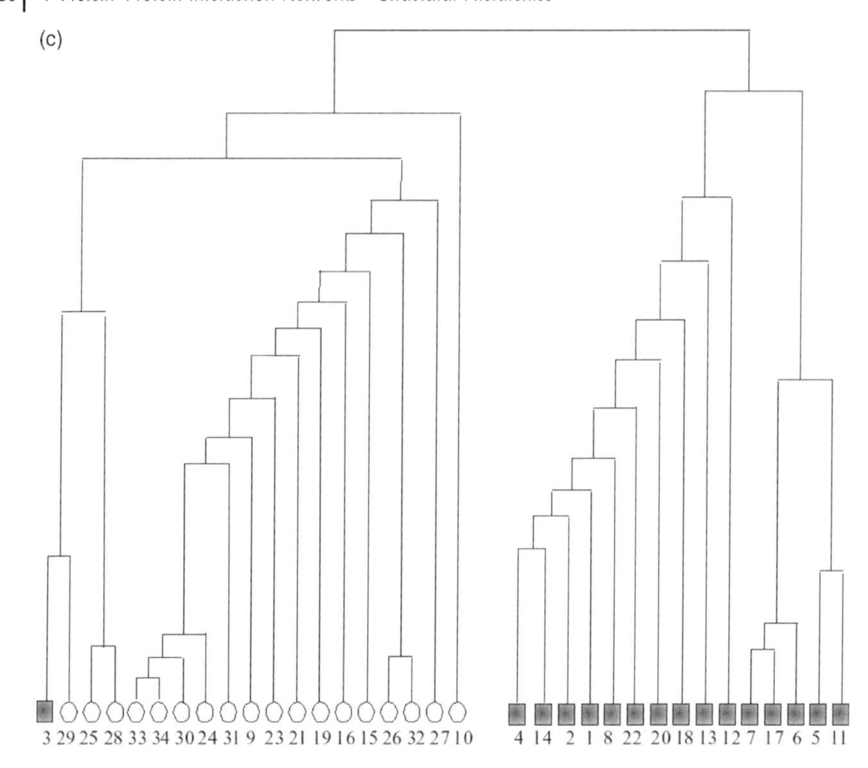

Figure 4.10 *(Continued)*

ultimately resulted in the instructor's leaving and starting a new club, taking about half of the original club's members with him. Zachary constructed a network of friendships between members of the club. As is shown in Figure 4.10b, hierarchical clustering performs quite poorly and cannot identify the two "camps" based on the knowledge of pairwise friendships. The Girvan–Newman algorithm, however, solves this task almost perfectly. Only the grey vertex 3 which is sort of in between the camps is clustered closest with the white vertex 29.

The Girvan–Newman algorithm suffers from a speed problem that results from requiring the repeated evaluation of a global property, the betweenness, for each edge whose value depends on the properties of the whole system. This becomes computationally very expensive for networks with, for example, more than 10 000 vertices. A **faster algorithm** was devised in Radicchi *et al.* (2004) who introduced a divisive algorithm that only requires the consideration of local quantities. To single out edges connecting vertices belonging to different communities, the **edge-clustering coefficient** is used. This measure is defined as the number of triangles to which a given edge belongs to divided by the number of triangles that might potentially include it, given the degrees of the adjacent vertices. For the edge connecting vertex i to vertex j, the modified edge-clustering coefficient is:

$$C_{i,j}^{(3)} = \frac{z_{i,j}^{(3)} + 1}{\min[(k_i - 1), (k_j - 1)]},$$

where $z_{i,j}^{(3)}$ is the number of triangles built on that edge and $\min[(k_i - 1), (k_j - 1)]$ is the maximal possible number of them; 1 is added to $z_{i,j}^{(3)}$ to remove the degeneracy for $z_{i,j}^{(3)} = 0$. With $z_{i,j} = 0$, the actual values of k_i and k_j would otherwise be irrelevant. Edges connecting vertices in different communities are included in few or no triangles and tend to have small values of $C_{i,j}^{(3)}$. On the other hand, many triangles exist within clusters (Figure 4.11). By considering higher order cycles one can define coefficients of order g:

$$C_{i,j}^{(g)} = \frac{z_{i,j}^{(g)} + 1}{s_{i,j}^{(g)}},$$

where $z_{i,j}^{(g)}$ is the number of cyclic structures of order g the edge (i,j) belongs to and $s_{i,j}^{(g)}$ is the number of possible cyclic structures of order g that can be built given the degrees of the vertices. For every g, a detection algorithm is used that works exactly as the Girvan–Newman method with the difference that, at every step, the removed edges are those with the smallest value of $C_{i,j}^{(g)}$. By considering increasing values of g, one can smoothly interpolate between a local and a nonlocal algorithm.

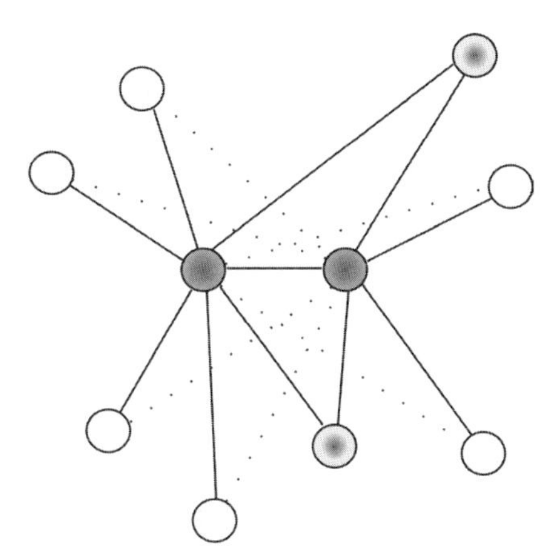

Figure 4.11 The left vertex has a degree of 7 and the right vertex has a degree of 5. The continuous lines denote the edges formed and dashed lines are those that could be formed additionally between the neighbors. Two triangles are currently formed including both vertices out of the $\min[(k_i - 1), (k_j - 1)] = 4$ that would be maximally possible for the given degrees of the two vertices. In both cases, one is subtracted as the central edge is already required to connect the two vertices. In this example, $C_{i,j}^{(3)} = (2+1)/4 = 0.75$.

4.6
Modular Decomposition

A very important aspect of biology is categorization. Grouping the millions of various plants and animals into categories, subcategories, etc., made up most part of zoology and botany. On the one hand, collectors like to be able to quickly get access to a particular species and a categorization scheme facilitates finding this particular species. Another, scientifically deeper, aspect is that a well-planned grouping provides insight into the organization of the individual elements. By uncovering which plants are related, we may understand which of them have a common ancestor and may then have similar metabolism, etc. Likewise in biological cells, grouping/categorization of the individual elements provides a much deeper level of understanding. The previous Chapter 3 was mostly concerned with discovering and enumerating all protein–protein interactions. In this chapter, we have started to develop a hierarchical way of looking at the interactome. Here, we will now address the aspect of **modularity**.

When we first hear somebody claim that biological cells are organized into separate functional modules, we may react surprised or even negatively. We have just started appreciating the fact that almost every protein seems to interact with every other protein, and now we are being told that they are rather organized in a strictly hierarchical way and interactions are mostly confined to within distinct modules.

The notion of a **biological module** is still heavily debated and its definition may undergo further refinement in the coming years. Hartwell *et al.* (1999) defined a functional module as a *discreted entity whose function is separable from those of other modules.* This separation depends on chemical isolation, which can originate from spatial localization or from chemical specificity.

In this textbook, we like to deal with well-defined concepts. For example, after formulating protein interaction networks as graph networks, it comes as some relief that there is a well-defined concept of **modular decomposition** for graph networks. We will first have a look into this principle and then discuss its applicability to biological networks.

4.6.1
Modular Decomposition of Graphs

Most cellular processes result from cascades of events mediated by proteins acting in a cooperative manner. Tandem affinity purification (TAP) showed that protein complexes can share components. Proteins can be reused and participate in several complexes.

Methods for analyzing high-throughput protein interaction data are mainly based on clustering techniques (Section 4.5). They are being applied to assign protein function by inference from the biological context as given by their interactors and to identify complexes as dense regions of the network. The logical organization into shared and specific components, and its representation, remains elusive.

Shared components, which are proteins or groups of proteins occurring in different complexes, are fairly common. A shared component may be a small part

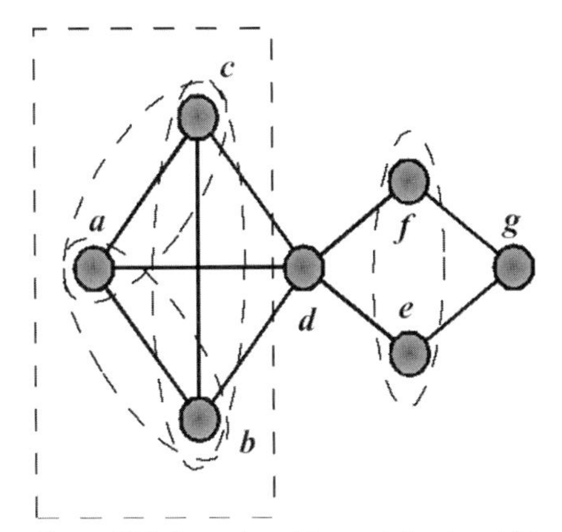

Figure 4.12 A graph and its modules. In addition to the **trivial modules** $\{a\}$, $\{b\}$, ..., $\{g\}$ and $\{a,b,c, \ldots, g\}$, this graph contains the modules $\{a,b,c\}$, $\{a,b\}$, $\{a,c\}$, $\{b,c\}$ and $\{e,f\}$. Drawn after Gagneur *et al.* (2004).

of many complexes, acting as a unit that is constantly reused for its function. Also, it may be the main part of the complex, e.g. in a family of variant complexes that differ from each other by distinct proteins that provide functional specificity. Here, we wish to identify and properly represent the modularity of protein–protein interaction networks by identifying the shared components and the way they are arranged to generate complexes. For this, we will follow Gagneur *et al.* (2004).

In graph theory, vertices connected by an edge are called **neighbors**. A **module** is a set of vertices that have the same neighbors outside the module (Figure 4.12).

To reveal the hierarchical organization of the graph, all elements of a module that have exactly the same neighbors outside the module can be substituted for a **representative vertex**. In such a **quotient**, all elements of the module are replaced by the representative vertex and the edges with the neighbors are replaced by edges to the representative (Figure 4.13). Quotients can be iterated until the entire graph is merged into a final representative vertex. The sequence of iterating quotients can be captured in a tree, where each vertex represents a module, which is a subset of its parent and the set of its descendant leaves.

The modular decomposition of the example graph in Figure 4.13 gives a labeled tree that represents iterations of particular quotients, here the successive quotients on the modules $\{a,b,c\}$ and $\{e,f\}$. The modular decomposition is a unique, canonical tree of iterated quotients. The vertices of the modular decomposition are categorized in three ways. In a **series module**, the direct descendants are all neighbors of each other (labeled by an asterisk within a circle). In a **parallel module**, the direct descendants are all nonneighbors of each other (labeled by two parallel lines within a circle). In a **prime module**, the structure of the module is neither series nor parallel (labeled by a P within a circle)

The graph can be retrieved from the tree on the right by recursively expanding the modules using the information in the labels. Therefore, the labeled tree can be seen as an exact alternative representation of the graph. This modular decomposition has

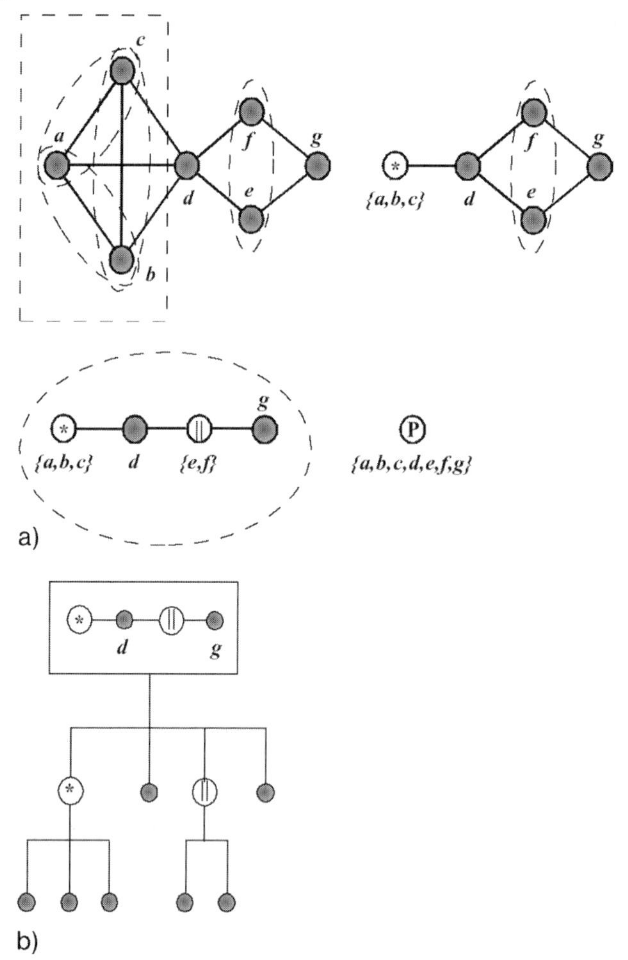

Figure 4.13 a) Modular decomposition and b) resulting tree. Vertices *a*, *b* and *c* are collapsed into one representing vector {*a*,*b*,*c*}. The edges with their common neighbors (*a*,*d*), (*b*,*d*) and (*c*,*d*) are replaced by one edge between *d* and {*a*,*b*,*c*}. Then, *f* and *e* are replaced by {*e*,*f*} and, finally, the whole graph into one vertex. Drawn after Gagneur *et al.* (2004).

been successfully applied to real protein interaction networks. The aim is that complexes are identified as modules. We have discussed in Section 3.2.7 that different experimental techniques probe different aspects of protein–protein interactions. For example, the yeast two-hybrid screen detects direct physical interactions between proteins, and protein complex purification (PCP) by TAP with mass spectrometric identification of the protein components identifies multiprotein complexes. Therefore, the molecular decomposition will have a different meaning due to different semantics of such graphs. One problem faced by this approach is the

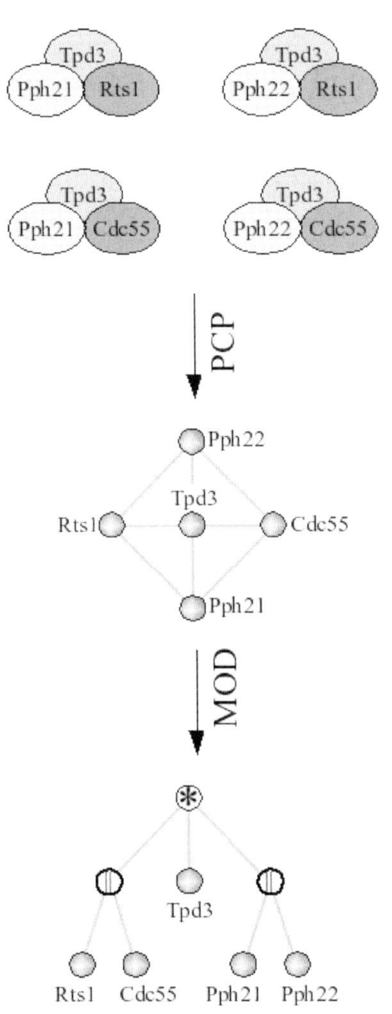

Figure 4.14 Modular decomposition of phosphatase 2A complex. Parallel modules group proteins that do interact, but are functionally equivalent. Here these are the catalytic Pph21 and Pph22 (module 2) and the regulatory Cdc55 and Rts1 (module 3). Drawn after Gagneur *et al.* (2004).

currently incomplete nature of interaction networks. This hierarchical scheme is most powerful when applied to almost complete interaction sets or subsets.

What is the value of this approach? Parallel modules typically occur when related complexes exist as combinatorial variants. Such a case is represented by protein phosphatase 2A, which is a family of distinct, yet related, serine/threonine phosphatase complexes. Each complex is composed of a trimer that consists of the structural scaffold Tpd3, either one of the two regulatory subunits Rts1 or Cdc55, and one of the two catalytic subunits Pph21 or Pph22 (Figure 4.14). Modular decomposition of a simulated PCP experiment revealed the logical organization of the complex family. To derive the individual complexes as maximal cliques, the tree shows that Tpd3 combines (module 1, series) with either (module 2, parallel) Rts 1 or Cdc55 and either (module 3, parallel) Pph21 or Pph22. Modular decomposition groups together proteins with similar function: the catalytic subunits Pph21 and Pph22 as alternatives in a parallel module and the regulatory subunits Cdc55 and Rts 1 in another parallel module. Such a functional relationship is not obvious from the initial network of interactions.

4.7
Network Growth Mechanisms

What is the "true" growth mechanism of real biological networks? Is it at all important to know this? Yes, we believe so, but it will be quite difficult to find this out as with any phylogenetic analysis.

Traditionally, the growth of protein networks has been modeled by the duplication–divergence mechanism. Here, a duplication of a vertex reflects the duplication of the corresponding gene and a divergence or loss of redundant edges or functions is a consequence of gene mutations. Although this view emerges from our understanding about the evolution of genomes, it is unclear whether this growth mechanism also applies to the evolution of protein–protein interaction networks. One way of finding this out may be to analyze topological properties such as the degree distribution, clustering coefficient, mean path length of a large number of artificial networks resulting from this growth principle. However, it turned out that networks generated by different growth algorithms (see below) all have the same overall topological properties.

A next strategy in exploring the properties of growth algorithms is to investigate their modular structure and distribution of various subgraphs or motifs. As an example, we will use here the results by a study of Middendorf *et al.* (2005) who analyzed the high-throughput data on the protein–protein interaction map for *Drosophila* by Giot *et al.* (2003). One generally occurring problem with data from high-throughput experiments, which we also mentioned in Chapter 3, is that the data set is subject to numerous false positives. Fortunately, Giot *et al.* (2003) were able to assign a confidence score $p \in [0,1]$ to each interaction measuring how likely the interaction occurs *in vivo*. However, what threshold p^* should be used to distinguish true from questionable interactions? With our background in topological analysis of networks gained so far, we expect that the network should contain one giant component. Therefore, Middendorf *et al.* (2005) measured the size of the components for all possible values of p^*. For $p^* = 0.65$, the two largest components were connected. Therefore, they used this value as threshold. Edges in the graph correspond to interactions for which $p > p^*$. After removing self-interactions and isolated vertices, this resulted in a network with 3359 (4625) vertices and 2795 (4683) edges for $p^* = 0.65$ (0.5).

The **duplication–mutation–complementation (DMC)** algorithm is based on a model proposing that most of the duplicate genes observed today have been preserved by functional complementation. If either the gene or its copy loses one of its functions (edges), the other becomes essential in assuring the organism's survival. In the algorithm, a duplication step is followed by mutations that preserve functional complementarity. At every time step a vertex v is chosen at random. A twin vertex v_{twin} is introduced copying all of v's edges. For each edge of v, either the original edge is deleted with probability q_{del} or its corresponding edge of v_{twin}. Twins cojoin themselves with independent probability q_{con} representing an interaction of a protein with its own copy. No edges are created by mutations. The DMC algorithm therefore assumes that the probability of creating new advantageous functions by random mutations is negligible.

A variant of DMC is the **duplication–random mutations (DMR)** algorithm where possible interactions between twins are neglected. Instead, edges between v_{twin} and the neighbors of v can be removed with probability q_{del}, and new edges can be created at random between v_{twin} and any other vertices with probability q_{new}/N, where N is the current total number of vertices. DMR emphasizes the creation of new advantageous functions by mutation.

Other models tested involved the linear preferential attachment (**LPA**) (Barabási), the random static networks (Erdös–Renyi) (**RDS**), random growing networks (**RDG** – growing graphs where new edges are created randomly between existing vertices), aging vertex networks (**AGV** – growing graphs modeling citation networks, where the probability for new edges decreases with the age of the vertex) and small-world network (**SMV** – interpolation between regular ring lattices and randomly connected graphs).

We recall that we started off this analysis to find out which network graph algorithm generates networks which best resemble the "true" *Drosophila* network. (We will disregard, for the moment, that the experimental network is error-prone too.) As was mentioned before, networks generated by these different algorithms could all be parameterized to produce networks having the same overall topological properties as the real network. The idea was therefore to quantify the fine structure of the generated networks. Therefore, 1000 graphs were created as training data for each of the seven different models. Every graph is generated with the same number of edges and vertices as measured in *Drosophila*. The topology of a network was quantified by counting all possible subgraphs up to a given cut-off, which could be the number of vertices, number of edges, or the length of a given walk. Middendorf *et al.* (2005) counted all subgraphs that can be constructed by a walk of length 8 (148 nonisomorphic subgraphs) (Figure 4.15).

These counts were then used as input features for a classifier. Note that the average shortest path between two vertices of the *Drosophila* network's giant component is 11.6 (9.4) for $p^* = 0.65$ (0.5). This means that walks of length 8 can traverse large parts of the network. It turned out that for 60% of the subgraphs (S1–S30), the counts for *Drosophila* are closest to the DMC model. All of these subgraphs contain one or more cycles, including highly connected subgraphs (S1) and long linear chains ending in cycles (S16, S18, S22, S23, S25). The DMC algorithm was the only mechanism that produces such cycles with a high occurrence. The protein interaction network of *Drosophila* was thus confidently classified as DMC network.

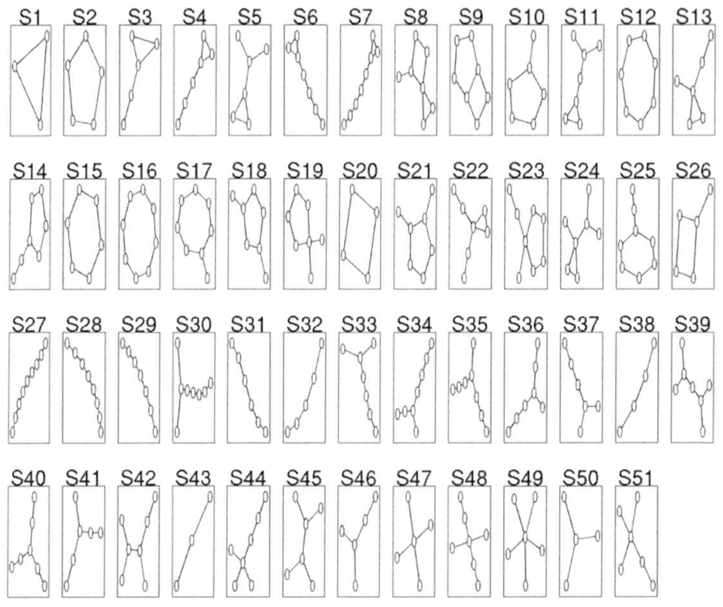

Figure 4.15 Fifty-one representative subgraphs of length 8 (out of 148 total). Reprinted from Middendorf *et al.* (2005) by permission from the National Academy of Sciences, USA.

This method of quantifying the fine structure of graph networks allows inferring growth mechanisms for real networks with confidence. The method is robust against noise and data subsampling, and requires no prior assumption about network features/topology. Also, the learning algorithm does not assume any relationships between features such as orthogonality. Therefore, the input space could be augmented in future with various features in addition to subgraph counts.

Summary

This chapter introduced several topological descriptors of graph networks. The biological protein interaction network is characterized by its small-world property, by the occurrence of network hubs, by the effect of clustering and by the property of community structure. The concept of communities (or formal modules) aims at connecting groups of proteins with a common biological function. As mentioned, the notion of a "biological module" has been used in different ways. It is unclear at the moment "where to draw the line" when defining a module. However, we have already noticed that there must be something to this concept. Most researchers in this field believe that this hierarchical concept of modules responsible for jointly carrying out biological functions will become more and more established. On the other hand, there exist direct physical interactions as well as functional interactions. In some analyses, molecules belonging to the same module need not even be present in the cell at the same time. Therefore, this area will certainly continue to develop over the coming years.

Problems

(1) Existing networks

Characterize with a short explanation the following examples of networks into the categories introduced in Section 4.1.

Hint: Some of the examples might fit into more than one category. If so, explain your choice.

– telephone system
– network of highways
– the physical backbone of the internet (cables, routers, computers, etc.)
– the world wide web
– European airports connected by direct flights
– your own social network . . .
. . . when you were in school (living at home)
. . . now as a student

(2) The random network

(a) Implement the algorithm given in Section 4.3 to generate random graphs. Start from a given number of vertices and add one edge after the other. Take care not to add the same edge twice. Store the constructed network in a file.

Hint: A simple yet efficient representation of the network is a list with a cell for each vertex, which itself holds a list of the vertices connected to this vertex (and vice versa – why both?). In standard Python you can use the following code to define and initialize a two-dimensional list:

```
net = [0]
net = node* net
for i in range(node):
net[i] = [ ]
```

Now you can assign values to the individual sublists with `net[i].append(k)` for a edge between vertices i and k. Do not forget the entry for $k \geq i$.

Hint: It is a good idea to read the variable parameters (number of vertices and edges) from the command line.

Hint: to store the network topology save the edges into a file. Each line then lists the (index of the) two vertices. Take care not to list the same edge twice (only print an edge when $i < k$).

(b) Determine and plot the degree distribution of the random network created above.

Hint: Implement a tool that reads in the network from part (a). You can avoid to explicitly save the network, when the network creation tool writes to `stdout` *(standard output via, e.g. print) and this tool reads from* `stdin` *(standard input). Then you just need to connect the two tools via a pipe. To plot the degree distribution pipe the output of this tool into a file. On a Mac or Linux box the easiest way to plot this file is using gnuplot. ("*`createNetwork[ parameters] | degreeDist[ parms] > outputfile`*")*

Hint: First determine the largest number of edges that occurs and initialize an array of that size. This array then holds the degree distribution. Now loop over the network list with the edge count (degree) of the vertices and increment the corresponding cells of the second array. Finally print out the degree distribution from this array with proper normalization.

(c) Verify that the degree distribution obeys the Poisson distribution $P(k)$ with a mean value of λ:

$$P(k) = \frac{\lambda^k}{k!} e^{-\lambda}.$$

Calculate λ from the numbers of vertices and edges and determine the Poisson distribution for a sufficiently large range of k.

Plot $P(k)$ together with the degree distributions of the random network for the following numbers of vertices and edges (each given in the form vertices/edges):

Plot 1: 50/100, 500/1000, 5000/10000, 50 000/100 000
Plot 2: 20 000/5000, 20 000/17 000, 20 000/40 000, 20 000/70 000
Plot 3: 5000/40 000, 13 000/40 000, 25 000/40 000, 50 000/40 000

Describe each plot and *explain* the difference (or the trend) between the different parameter sets. Do not forget to label the axes.

(3) The scale-free network

(a) Implement the BA algorithm given in Section 4.4 to generate a scale-free network. Start with one vertex and add vertices sequentially by preferentially connecting the new edges to those vertices that already have more edges.

Hint: To implement the preferential attachment use a list that contains the vertices that are already connected (each vertex occurs in that list as often as its degree). When you add a new edge attach( ) the indices of the two vertices to the end of this (growing) list. For the next edge randomly choose a vertex from that list. Why does this recipe give the desired preferential attachment?

(b) Determine the degree distribution of the scale-free network created above. Plot the degree distribution for a network of 100 000 vertices with double logarithmic axes and determine the exponent γ. Which region of degrees can you use to fit the exponent γ? Which problems do you encounter?

To extend the range for fitting the exponent, sum up $P(k)$ and compare:

$$S(k) = \sum_{k' \leq k} P(k'),$$

to the corresponding integral of the power law:

$$F(k) = \int dk c k^{-\gamma}.$$

First verify that

$$F(k) = \frac{c}{1-\gamma} k^{1-\gamma} + F_0$$

where F_0 is an integration constant.

Plot $(1 - S(k))$ and $F(k)$ into the same plot. How many vertices in the network are now necessary for a comparable fitting range (to fit the exponent, also adjust c and F_0)? Which main difference between the two ways to plot $P(k)$ (directly versus integrated) do you observe. Why does this difference help?

(c) Create a random network with the same number of vertices and edges as the scale-free network of 100 000 vertices and plot the degree distributions of both networks into the same plot. For which combination of logarithmic and/or linear axes is the difference between the degree distributions of the two networks seen best?

Identify and explain the two major differences between the degree distributions of the random and of the scale-free network.

(4) Biological interaction networks

The “Biomolecular Interaction Network Database” (BIND) contains many known interactions between proteins and small molecules for many different species. To set up a protein interaction network for a given species, perform the following steps.

(a) From the download area of the BIND database (ftp.bind.ca) get the flat text file with the interactions (/pub/BIND/data/bindflatfiles/bindindex/20060525.ints.txt). This directory contains a README file which explains the format of all files in this directory. You also need the file which contains the taxonomy index

(20060525.taxon.txt). To access the download site you have to register with your email address at http://bond.unleashedinformatics.com.

(b) Write a python script that creates a histogram of the taxon identifiers from those interactions, where the two partners are either "protein" or "small-molecule", i.e. how often each taxon identifier occurs in this subset of interactions registered in BIND. Which are the top five species that have the largest number of these interactions in BIND? Give their taxon identifiers, their scientific names and the respective number of occurrences.

Hint: Check the taxon identifier file for the highest occurring number and initialize an array of that size. Then read the interaction file line by line and split each line at the tabs to get the two taxon IDs. For each of them increment the corresponding entry of the array due to whether the types are protein or small-molecule. Alternatively, you can use a python dictionary. Note that a handful of taxon identifiers in the interaction file are not listed in the taxon overview.

Hint: There is the class of small molecules, which have an ID of 0. They are not a species on their own!

(c) For each of the top five species from (b), parse the interaction file to pick all interactions that belong to this species. Pick those interactions where both partners belong to the chosen species or where one belongs to the chosen species and the other is a "small molecule".

Count the total number of interaction partners that are known for this species, regardless of their type.

Hint: Look for one species at a time, i.e. run the script once for each species that you want to extract from the interaction file.

Hint: Read footnote 5 of the README file – there may be either missing accession codes or missing molecule IDs for the proteins. A workaround is to create a new label for the proteins by concatenating the accession code and the molecule ID. This ensures that each protein is labeled uniquely. Convert the labels into integers, print out the interactions one per line, and pipe the output into a file for each of the top five species.

Hint: Use a python dictionary (or a hash in Perl) to translate the protein identifiers into integers. For each new interaction check whether any of the two proteins is already stored in that dictionary to avoid duplicate entries. For any protein not yet in the dictionary, increase a counter and store the counter value with the protein identifier as the key.

(d) Use the tool implemented in problem (2) to determine the degree distribution of the protein interaction networks of the top five species. Do you see any difference between the degree distributions of their interaction networks? Are they more like the scale-free network or more like the random graph? Explain your observations.

Hint: Plot the integrated $P(k)$.

(5) Clustering coefficient: scale-free versus random network

(a) The average clustering coefficient in a random or a scale–free network is claimed to be independent of the degree. Do an *in silico experiment* to verify this statement and proceed as follows.
Implement a tool that:
(i) determines the cluster coefficient $C(k)$ of each vertex of a given network,
(ii) averages the cluster coefficients of vertices of the same degree, i.e. determines the average cluster coefficient $\langle C(k) \rangle$, and
(iii) averages over all $C(k)$ to get the average degree independent cluster coefficient $\langle C \rangle$ of the network.
Create two plots, one for the scale-free network á la Barabási and one for the random graph with the same number of edges, and plot $C(k)$ (as a scatter plot), $\langle C(k) \rangle$ and $\langle C \rangle$ against k for networks of 400 000 vertices each.

(b) Use the probability $p = 2L/(N(N-1))$ for an edge between two arbitrary vertices and calculate an estimate for the probabilities that a vertex has k edges and for the cluster coefficient $C(k)$ of a vertex of degree k. Interpret the results. Does this estimate of $C(k)$ reproduce the numerical results of (a)?

Hint: consider how many possible realizations of n links can occur between the k neighbors of a node. What is the probability for any of these configurations?

(6) Clusters of the scale-free network

How many clusters are contained in a scale-free network constructed by the BA algorithm starting from a single node? What is the size of the largest cluster? Why?

(7) Cluster sizes and numbers

(a) Determine the number of clusters N_{cl} and the size of the largest cluster N_{max} for a random graph of $N = 100$, 1000, 10 000, 100 000 and 1 000 000 vertices and twice as many edges, and plot them against the size of the network, i.e. the number of vertices. To get more accurate results create ten networks of each size and average the results.
What trends do you observe in the plots? Explain your observations.

Hint: To identify the clusters start from the first vertex and assign it to the first cluster. Then follow all edges from there and assign the vertices connected to this first vertex to the same cluster. Repeat from these vertices until you find no more connected but unassigned vertices. Then repeat this procedure from the first unassigned vertex, which you assign to the second cluster. Repeat until all vertices are assigned to a cluster. Note that a vertex without any edges forms a cluster on its own.

Hint: If you implemented a recursive algorithm to identify the clusters you may run into a "maximum recursion depth exceeded" runtime error (or a similar error message). An iterative algorithm will work, though.

(b) Check for the existence of the "spanning cluster" of a random graph. To do so determine the size of the largest cluster N_{max} and the number of clusters N_{cl} of a random network for different values of $\lambda = 2L/N$, i.e. for different average degrees.
For a random graph of 100 000 vertices vary λ between 0 and 4 and create two plots, one with N_{max} versus λ and one with N_{cl} versus λ. Do you observe any transition – and if, at which value of λ? To interpret your findings, determine the values and behavior at $\lambda \to 0$ and at $\lambda \to \infty$.

Hint: The spacing between the values of λ need not be constant. Just choose enough (and sensible) values of λ so that the trend of N_{max} and N_{cl} in the plots is clear.

(c) Clusters in random networks. Create 10 networks, each, with $N = 200$ and $L = 400$, 1200 and 2400, and determine the average numbers of clusters $C(nc)$ for each $\lambda = 2L/N$.

(8) Clusters in biological networks

(a) Read in the interactions listed in the BIND database for the fruit fly, for the mouse and for *E. coli*, and determine the histogram of cluster sizes $P(C(k))$, the size of the largest cluster N_{max} and the average cluster coefficient $\langle C \rangle$ (see (4)).

Hint: You can start from the interactions extracted in (4), but do not limit them to proteins and small-molecules this time.

(b) To check the stability of these biological networks against directed attacks take the interaction network of the mouse and determine (the labels of) the 200 vertices with the highest degrees. Compare the size of the largest cluster N_{max} and the number of clusters N_{cl} of the original network to networks where you delete the 10, 20, 50, 100 or 200 vertices with the highest degrees and also to networks, where you randomly delete the same numbers of vertices. How does the network behave?

(9) Theoretical estimate of the number of cliques

Use the probability for a single edge in a random network of N vertices and L edges and derive an analytical estimate for the number of cliques $C(nc)$ of size nc assuming that the network is large enough so that the cliques do not overlap.

Tabulate the values of $C(nc)$ for $nc = 2, \ldots, 10$ with $N = 1326$ and $L = 2548$. How many of the vertices of the network belong to a clique?

(10) Cliques in model networks

Adapt and implement the algorithm by Spirin and Mirny (end of Section 4.2) to search for cliques. Test the algorithm by searching the cliques in the supplied `testNetz.dat` file. List all cliques found. (The network contains four cliques of size 2, two cliques of size 3 and one clique of size 5.)

> *Hint: When reading the network file, check for duplicates. In the input files, the vertices are not necessarily sorted.*

Use the following modifications of the algorithm:

- Start from cliques of size 2, not 4. This is both a runtime constraint (why?) and gives more data points to compare. You can save a lot of time by collecting the list of 2-cliques while you read in the network.
- Combine the removal of redundant cliques and the search for additional vertices into the same step: if you find a new vertex to extend an existing clique, then immediately mark the original (smaller) clique as obsolete.
- After each extension step, remove the nonredundant smaller cliques, sort the list of new cliques and get rid of duplicates.

(11) Cliques in BIND networks

Read in the supplied interaction networks that were extracted from the BIND database (interaction_*.dat), determine the numbers of cliques, and tabulate $C(nc)$ for the occurring values of nc. List the members of the largest clique(s) only.

> *Hint: In the interactions filtered from the BIND database for each species – encoded in the filename via the taxon ID – either both molecules belong to the species or one is from the species and the other is a small molecule. The labels of the molecules are generated from the accession code and the molecule ID as in problem (4).*

For each network, record the run-time of the clique search. Which dependency do you expect? Explain your answer. Create different plots where you plot the run-time versus N, L, $N^* L$ and $N^* C(2)$. Do these plots confirm your expectation?

(12) Network Communities

A community is a part of a network that has more internal connections than to the rest of the network. In this problem you will use the algorithm of Radicchi *et al.* (2004) to identify the communities of a given network (see Section 7.5.1).

The **edge-clustering coefficient** of an edge between vertices i and j was defined in Section 4.5.1.

(a) If one of the vertices has a degree of 1 then $\tilde{C}_{i,j}^{(3)}$ is infinite. What is the maximal *finite* value that the edge-clustering coefficient can take? For which configuration does this occur?
(b) Read the given scale-free network in the file sfnetz_1000.txt, determine the edge-clustering coefficient for all occurring edges, and create a frequency histogram of the occurring values. Use a reasonable binsize.

Hint: To denote an infinite $\tilde{C}_{i,j}^{(3)}$ *use a value of ten times the maximal occurring finite value from (a).*

Hint: The file includes the number of vertices on the first line and then one edge per line.

(13) Network communities

To determine the communities of the supplied network given in the file HighSociety.txt proceed in two steps (parts (a) and (b)). The overall process is easier to implement if you create a separate script for each of the two parts and save the intermediate results into a file.

(a) Decomposition of the network. Iteratively delete the edges with the smallest $\tilde{C}_{i,j}^{(3)}$:
   (i) Read in the network file.
   (ii) Calculate the edge–clustering coefficient $\tilde{C}_{i,j}^{(3)}$ for each edge.
   (iii) Find the edge with the smallest $\tilde{C}_{i,j}^{(3)}$ and delete it from the network. Print out this edge.

   Hint: When you encounter multiple edges with the same $\tilde{C}_{i,j}^{(3)}$*, choose any of them. Does this actual choice make any difference in the final result?*

   (iv) Repeat from (ii) until there is no edge left.
(b) Buildup of the communities and of the dendogram. There are two criteria for a community (see Radicchi *et al.*, 2004):
   (i) In a *community in a strong sense* every single member of the subgraph V has more edges to the inside of the community (k^{in}) than to the outside (k^{out}):

$$k_i^{\text{in}}(V) > k_i^{\text{out}}(V) \qquad \forall i \in V.$$

   (ii) In a *community in a weak sense* the total number of edges inside the subgraph V is larger than to the outside:

$$\sum_{i \in V} k_i^{\text{in}} > \sum_{i \in V} k_i^{\text{out}}.$$

Now use the edges deleted in (a) in reverse order, i.e. the edge that was deleted last is now used first, to construct the communities. To do so read in one edge after the other and check whether they have vertices in common with the already included edges. During this composition stage you do not need to keep track of the edges, but only of the vertices that belong to the same subgraph.

(i) If the latest edge is disjoint from the already processed edges then start a new subgraph (list of vertices of this subgraph) from this one.

(ii) If the latest edge has a single vertex in common with one of the existing subgraphs then add the other vertex of this edge to that (list of the vertices of the) subgraph, too.

(iii) If the two vertices of the latest edge belong to two different subgraphs then join the two subgraphs to form a single one from them. Print out the two lists of vertices that are joined in this step. After adding the last edge, you should end up with a single graph that contains all vertices of the network and a listing of the subgraphs just before they were joined to form larger ones. To draw the dendrogram of the network, look at the above choice (iii), the joining of two groups: start from the individual vertices and every time that this happens connect two subgraphs.
Hint: You may want to draw the dendrogram by hand.

To identify communities determine the two community criteria explained above each time after you added a new edge. If one of the two criteria (weak or strong) is met for one of the subgraphs, print out the list of vertices. For the weak criterion also give the sums of the internal and external edges. Highlight these communities in the dendrogram. Do the communities obtained by the two criteria differ? If so, what is the reason for this specific network?

(c) Visualization of the communities. The supplied file societyPositions.txt contains the final positions from a force directed layout of the "High Society" network. Use these coordinates to plot the communities that you identified in the previous exercise. Identify the vertices in at least one of the plots.

Hint: The hierarchy of the communities that is captured in the dendrogram is best explained by creating multiple plots, starting from the smallest communities.

Hint: Discern between the communities from the weak and from the strong criterion.

Further Reading

Network Topology

Barabási AL, Oltvai ZN (2004) Network biology: understanding the cell's functional organization, *Nature Reviews Genetics* 5, 101–113.

Han J-DJ, Bertin N, Hao T, Goldberg DS, Berriz GF, Zhang LV, Dupuy D, Walhout

AJM, Cusick ME, Roth FP, Vidal M (2004) Evidence for dynamically organized modularity in the yeast protein–protein interaction network, *Nature* **430**, 88–93.

Jeong H, Mason SP, Barabási, Oltvai N (2001) Lethality and centrality in protein networks, *Nature* **411**, 41–42.

Random Graphs

Bollobas B (2004) *Random Graphs.* Academic, London.

Finding Cliques and Clusters

Spirin V, Mirny LA (2003) Protein complexes and functional modules in molecular networks, *Proceedings of the National Academy of Sciences USA* **100**, 12123–12128.

Communities

Girvan M, Newman MEJ (2002) Community structure in social and biological networks, *Proceedings of the National Academy of Sciences USA* **99**, 7821–7826.

Radicchi F, Castellano C, Cecconi F, Loreto V, Parisi D (2004) Defining and identifying communities in networks, *Proceedings of the National Academy of Sciences USA* **101**, 2658–2663.

Modular Decomposition

Hartwell LH, Hopfield JJ, Leibler S, Murray AW, (1999) From molecular to modular cell biology, *Nature* **402**, C47–C52.

Gagneur J, Krause R, Bouwmeester T, Casari G (2004) Modular decomposition of protein–protein interaction networks, *Genome Biology* **5**, R57.

Network Evolution

Middendorf M, Ziv E, Wiggins CH (2005) Inferring network mechanisms: the *Drosophila melanogaster* protein interaction network, *Proceedings of the National Academy of Sciences USA* **102**, 3192–3197.

Giot L *et al.* (2003) A Protein Interaction Map of Drosophila melanogaster, *Science* **302**, 1727–1736.

5 Gene Regulatory Networks

Transcriptional regulatory networks describe the interactions between transcription factor proteins and genes that they regulate. Transcription factors respond to biological signals and accordingly change the transcription rates of genes, allowing cells to make the proteins they need at the appropriate times and amounts. [This transcriptional regulation occurs at a different, lower level than the posttranscriptional regulation that is now ascribed to RNA snippets termed small interfering RNAs (siRNAs) that bind to mRNA and also involves other mechanisms such as alternative splicing.] Here, we will start by briefly reviewing the process of transcription and gene regulatory networks followed by discussing three mathematical techniques, i.e. Boolean and Bayesian networks as well as differential equation models, that are popular methods to model gene regulatory networks.

Experimental data of mRNA levels obtained with the use of high-throughput technologies yield snapshots of the molecular state of cell populations at the transcript level and are rich in information about gene networks. Uncovering gene networks from this rich data source by "reverse engineering" techniques is presently the most widely adopted approach. A popular method, sometimes called "guilt by association", assumes that genes with similar expression patterns are functionally related to each other. These associations are usually explored with the use of clustering algorithms and principal component analysis. However, this method may only work when the underlying networks are modular containing a small number of connections. When applied to heavily connected networks, it may provide ambiguous results.

Interestingly, recent work by the group of Uri Alon indicated that transcription networks contain a small set of recurring regulation patterns called *network motifs* reminiscent of the standard elements of electronic circuits. Network motifs can be thought of as basic circuits of interactions from which the networks are built. They were first systematically defined in *Escherichia coli*, in which they were detected as patterns that occurred in the transcription network much more often than would be expected in random networks. They have since been found in organisms from bacteria and yeast to plants and animals.

Principles of Computational Cell Biology – From Protein Complexes to Cellular Networks. Volkhard Helms

ISBN: 978-3-527-31555-0

5.1 Regulation of Gene Transcription at Promoters

Transcription is the process through which a DNA sequence is enzymatically copied by an RNA polymerase to produce a complementary piece of RNA sequence. Transcription is divided into the three stages of initiation, elongation and termination. Here, we will be only concerned with the initiation process. As transcription exclusively proceeds in the $5' \rightarrow 3'$ direction, the DNA template strand that is used must be oriented in the $3' \rightarrow 5'$ direction. In prokaryotes, transcription begins with the binding of RNA polymerase to the promoter sequence in the DNA. An RNA core polymerase is a multisubunit complex composed of $\alpha_2\beta\beta'$ subunits that catalyze the elongation of RNA (see Figure 8.1). At the start of initiation, the core enzyme is associated with a sigma factor that aids in finding the appropriate -35 and -10 base pairs downstream of promoter sequences (Figure 5.1).

Eukaryotic transcription initiation is far more complex than in prokaryotes as eukaryotic polymerases do not recognize directly their core promoter sequences. Here, additional proteins termed transcription factors regulate the binding of RNA polymerase to DNA. Many eukaryotic promoters, but by no means all, contain a TATA box which in turns binds a TATA binding protein that assists in the formation of the RNA polymerase transcriptional complex. The TATA box typically lies very close to the transcriptional start site (often within 50 bases).

A eukaryotic **transcription factor** is a protein needed to activate or repress the transcription of a gene, but is not itself a part of the enzymes responsible for the chemical steps involved in transcription. Some transcription factors bind to *cis*-acting DNA sequences only; some bind to each other; others bind to DNA as well as to other transcription factors. Regulation of gene transcription in an organism involves a complex network, of which the DNA-binding transcription factors are a key component. Based on sequence and structural homologies, DNA-binding regions of the prokaryotic transcription factors have been assigned to a number of structural families of DNA binding protein domains, including the three most well characterized ones:

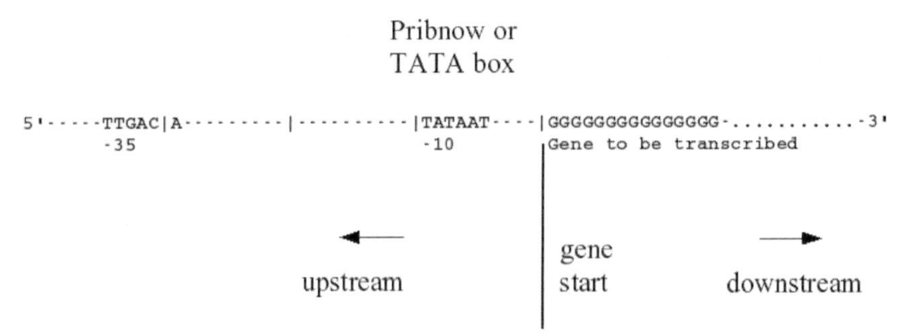

Figure 5.1 Typical promoter region of a prokaryotic gene. The TTGACA and TATAAT sequences at positions –35 and –10 nucleotides are not essential. The preference for the corresponding nucleotide at each position is between 50 and 80%.

the helix-turn-helix, the winged helix and the β ribbon. Determining homologies of DNA-binding domains and protein families of transcription factors and regulated genes and proteins of known three-dimensional structure showed that about three-quarters of all transcription factors are two-domain proteins. Analysis of domain architecture revealed that the same ratio of transcription factors have arisen by gene duplication.

5.2
Gene Regulatory Networks

The regulation of transcription is a highly complex process as it depends upon factors such as considering which transcription factors and other coregulatory proteins are present within a particular cell as well as the local three-dimensional structure of the DNA. A **gene regulatory network** represents relationships between genes that can be established from measuring how the expression level of each one affects the expression level of the others. Note that, in any global cellular network, genes do not interact directly with other genes (neither do the corresponding mRNAs). Instead, gene induction or repression occurs through the action of specific proteins, which are, in turn, products of certain genes as well. Gene expression can also be affected directly by metabolites. Figure 5.2 represents a model of a global biological network in which the three players (genes, proteins and metabolites) are arranged on different levels. Conceptually, it is often useful to abstract the action of proteins and metabolites and concentrate on the projection of all interactions to the "gene space" that is, here, located on the lowest level.

Consequently, gene networks are abstracted models that display causal relationships between gene activities, usually at the mRNA level, and are commonly represented by directed graphs (Figure 5.3).

5.2.1
Gene Regulatory Network of *E. coli*

The amount of experimentally validated knowledge for the *E. coli* K-12 regulatory network is the largest currently available for any organism. Over the past 10 years, the research group of Julio Collado-Vides has compiled all the available experimental data in the relational database RegulonDB (http://regulondb.ccg.unam.mx) (Figure 5.4). Release 6.0 as of January 2008 contained data on 3199 transcription units affecting 4578 genes. There are 1372 promoters, 1480 transcription factor binding sites, 2171 regulatory interactions and 159 transcription factors. It was found that seven regulatory proteins [cyclic AMP receptor protein (CRP), factor for inversion stimulation (FIS), leucine-responsive regulatory protein (Lrp), aerobic respiration regulatory protein (ArcA), fumarate and nitrate reductase regulatory protein (FNR), histone-like protein or nucleoid-associated protein (Hns) and integration host factor (IHF)] are sufficient to directly modulate the expression of more than half of all *E. coli* genes. Such **global transcription factors** are being defined on the basis of a collection of

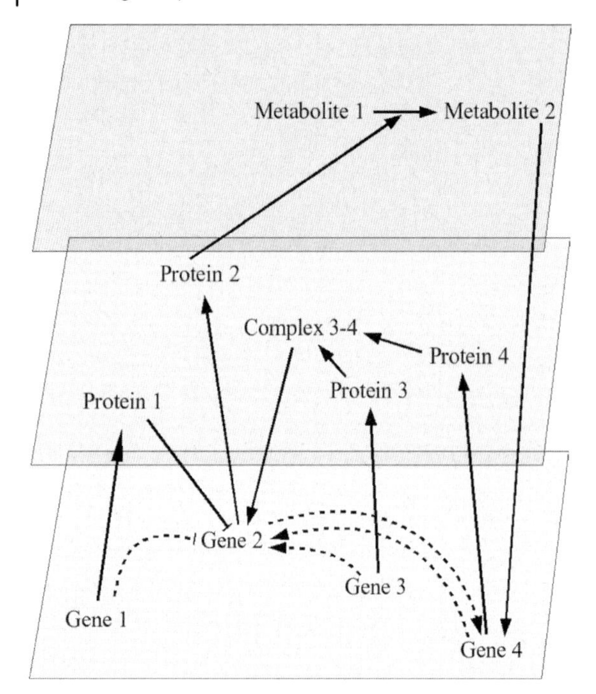

Figure 5.2 An example of a gene regulatory network. Solid arrows indicate direct associations between genes and proteins (via transcription and translation), between proteins and proteins (via direct physical interactions), between proteins and metabolites (via direct physical interactions or with proteins acting as enzymatic catalysts), and the effect of metabolite binding to genes (via direct interactions). Lines show direct effects, with arrows standing for activation and bars for inhibition. The dashed lines represent indirect associations between genes that result from the projection onto "gene space". For example, gene 1 deactivates gene 2 via protein 1 resulting in an indirect interaction between gene 1 and gene 2. Drawn after Brazhnik *et al.* (2002).

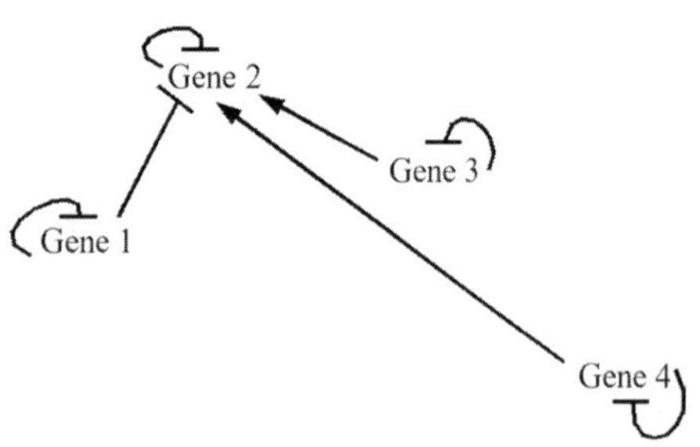

Figure 5.3 Graph representation of the gene network corresponding to the biochemical network in Figure 5.2. This figure corresponds to the lowest tier of Figure 5.2. Most genes in gene networks will have a negative effect on their own concentration because the degradation rate of their mRNA is proportional to their concentration. Drawn after Brazhnik *et al.* (2002).

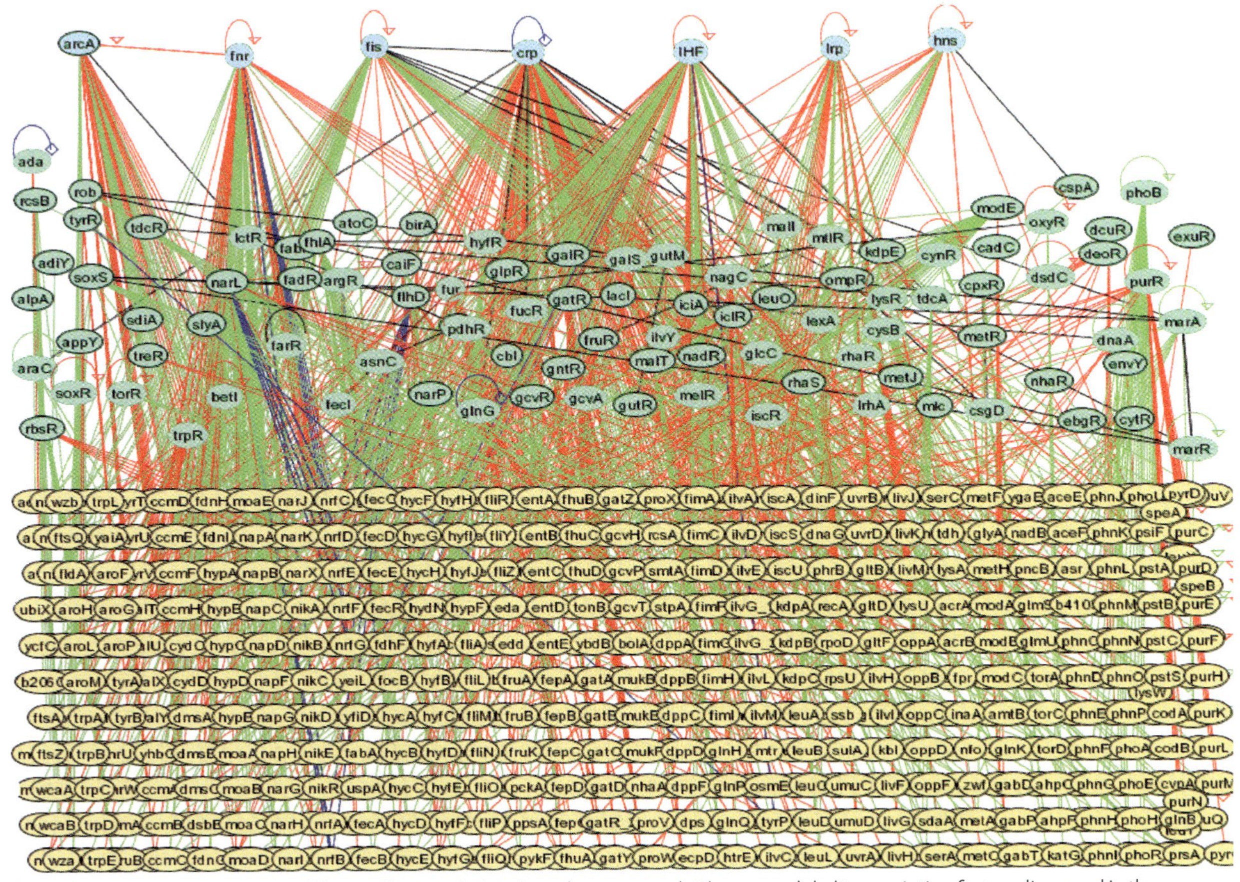

Figure 5.4 Graph representation of the *E. coli* transcriptional regulatory network. The seven global transcription factors discussed in the text are shown at the top of the figure as blue ovals. Other transcription factors are shown in green and regulated genes at the bottom in yellow. Reprinted from Collado-Vides *et al.* (2003) by permission from Elsevier.

diagnostic criteria: (1) they regulate many genes, (2) they regulate several genes encoding transcription factors, (3) they cooperate with numerous transcription factors and together regulate other genes, (4) they directly affect gene expression from a variety of promoters that use different sigma factors, and (5) their regulated genes belong to different classes. Considering the connectivity in this network (Section 4.1), the out-going connectivity was shown to follow a power-law distribution, whereas the in-coming connectivity follows an exponential distribution. This may be simply explained by considering the three-dimensional structures of protein and DNA. It may well happen that one particular transcription factor binds to hundreds of different promoter regions at different times turning it into a *hub* transcription factor. One the other hand, one can hardly imagine that one gene is regulated by the binding of hundreds of transcription factors. How should they all bind in a coordinated fashion to its promoter region?

In the gene regulatory network of *E. coli*, about half of the genes are regulated by binding of one transcription factor and the other half is regulated by multiple transcription factors. In most of these cases, a "global" regulator (with more than 10 interactions) works together with a more specific local regulator. However, in a process of decisions and flow of information, the number of controlled or affected elements is not the only important factor. In general, global regulators work together with other global regulators. The dynamics of decision making is a cooperative process of different subsets of the network put into action at certain moments. Another function of transcription factors is to sense changes in environmental conditions or other internal signals encoding changes. Thus, often a global transcription factor acts together with an environmentally sensitive transcription factor.

After having compiled all the regulatory information, Collado-Vides *et al.* used a hierarchical clustering approach to study the modularity of the gene regulatory network of *E. coli*. Between each pair of genes, they computed the shortest path length d_{ij} of gene i and gene j from the connectivity matrix. A d_{ij} value of 1 means that two genes have a direct connection and higher values denote indirect connections mediated by other genes/transcription factors. These distances are then converted into an association function ($1/d_{ij}^2$) and a value of 0.0 is assigned to pairs of genes that are not connected. Fifty-five out of the 320 estimated transcription factors regulated the expression of other transcription factors. Hierarchical clustering on the basis of their association functions grouped them into eight modules, where M1 contains genes involved in respiration, M2 contains genes in stress response, and M3 contains genes involved in chemotaxis, motility and biofilm formation. M4 contains 23 transcription factors that are involved in the regulation of the different preferential carbon sources. It can be divided into smaller submodules related to alternative carbon sources. Four smaller modules of interacting transcription factors are fully disconnected from the other modules and among themselves. These involve genes regulating relevant cellular responses such as sulfur assimilation, metabolism of nitrogen sources, fermentative metabolism and chromosome replication.

The seven global transcription factors mentioned before are found to be evenly distributed within the major modules. CRP is the transcription factor with the largest number of interactions. It belongs to the module of carbon metabolism, together with

FIS and Lrp. ArcA and FNR belong to the respiration-response module. Hns and IHF belong to the module involved in chemotaxis, motility and biofilm formation.

5.3 Graph Theoretical Models

Graph theoretical models may be used to describe the topology and architecture of a gene network. We will follow in part the presentation in (Filkov, 2005). These models feature causal relationships between genes, but yield no dynamic information how fast the system responds, for example, to external stimuli or how the gene expression levels change during a cell cycle. Such models are particularly useful for knowledge representation in databases. Together with the Boolean network models, graph theoretical models belong to the group of qualitative network models because they do not yield quantitative predictions of gene expression in the system. Graph theoretical models represent gene networks as graph structures, $G(V,E)$, where the vertices $V = \{1, 2, \ldots, n\}$ represent the gene regulatory elements, e.g. genes, proteins, etc., and the edges $E = \{(i,j) / i,j \in V\}$ represent the interactions between them, e.g. activation, inhibition, causality, binding specificity, etc. Most often, G is a simple graph and the edges represent relationships between pairs of vertices, although hyperedges, connecting three or more vertices at once, are sometimes appropriate. Many biologically pertinent questions about gene regulation and networks have direct counterparts in graph theory, and can be answered using well established methods and algorithms on graphs. For example, the task of identifying highly interacting genes corresponds to finding high degree vertices, that of resolving cascades of gene activity corresponds to determining the topological vertex ordering and the task of comparing gene networks for similarity corresponds to finding graph iso(homo)-morphisms in graphs. Inferring gene networks under graph theoretical models requires identifying the edges and their parameters (coexpression, regulation, causality) from given expression data.

5.3.1 Coexpression Networks

The idea behind analyzing the coexpression of genes is that coexpressed genes may be coregulated and may have a similar function. In other words, if genes show the same expression profiles, e.g. at various stages during the course of a cell cycle or when exposed to different environmental conditions, they are expected to follow the same regulatory regimes. Expression profiles are typically given as vectors $\mathbf{X} = (X_1, \ldots, X_p)$ where the elements of each entry X_i are the expression profiles of all genes in the experiment i. From this, correlation coefficients ρ_{ij} between the expression of genes i and j are computed. A correlation graph is then drawn showing the genes as vertices connected by edges (i,j) when their correlation coefficient is larger (or lower) than a given positive (or negative) threshold. These results are then easy to interpret. However, coexpression-based approaches also face certain problems. For example,

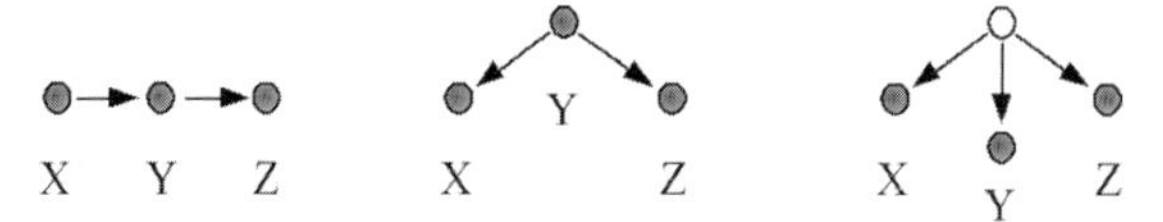

Figure 5.5 Three different gene connectivities may lead to similar observed coexpression patterns.

Figure 5.5 shows an example where three genes *X*, *Y* and *Z* are experimentally found to be coexpressed. However, this finding alone does not tell us whether *X* activates *Y*, which then activates *Z* (left picture), or whether *Y* activates *X* and *Z* (middle picture), or whether there even exists a fourth protein that simultaneously activates all three proteins (right picture). Therefore, one has to search for correlations that cannot be explained by other variables.

5.3.2
Bayesian Networks

Bayesian networks that we already encountered in Section 3.4 are a class of graphical probabilistic models that combine two established mathematical areas, i.e. probability and graph theory. A Bayesian network for a gene regulatory model consists of an annotated directed acyclic graph $G(X,E)$, where the vertices, $x_i \in \mathbf{X}$, are random variables for the expression of genes and the edges indicate the dependencies between the vertices. The random variables are drawn from conditional probability distributions $P(x_i|P_a(X_i))$, where $P_a(x_i)$ is the set of parents for each vertex which are those genes that affect the expression of gene x_i. A Bayesian network implicitly encodes the Markovian assumption that, given its parents, each variable is independent of its nondescendants.

The problem with learning Bayesian networks is their combinatorial complexity. If the graph model is not known then the space of all graph models has to be explored. However, this space is super-exponential and exploring it completely is impossible even with the fastest heuristics. In practice, many different Bayesian networks may fit the data just as well. To lower the number of high-scoring networks, simplifying assumptions regarding the topology of the graph or the nature of the interactions have to be used.

5.4
Dynamic Models

5.4.1
Boolean Networks

Boolean networks allow dynamic modeling of synchronous interactions between vertices in a network. They belong to the simplest models that possess some of the biological and systemic properties of real gene networks. In Boolean logic, a Boolean variable *x* is a variable that can assume only two values. The values are denoted usually as 0 and 1, and correspond to the logical values `true` and `false`. The logic operators `and`, `or` and `not` are defined to correspond to the intuitive notion of truthfulness and composition of those operators. A Boolean function is a function of Boolean variables connected by logic operators. A **Boolean network** is a directed graph $G(X,E)$, where the vertices, $x_i \in X$, are Boolean variables. To each vertex, x_i, is associated a Boolean function, $b(x_{i1}, x_{i2}, \ldots, x_{il})$, $l \leq n$, $x_{ij} \in X$, where the arguments x_j are limited to the parent vertices of x_i in G. Together, at any given time, the values of all vertices represent the state of the network, given by the vector:

$$S(t) = (x_{i1}(t),\ x_{i2}(t), \ldots, x_{il}t)).$$

For gene networks, the vertex variables correspond to levels of gene expression, discretized to either 0 or 1. The Boolean functions at the vertices model the aggregated regulation effect of all their parent vertices. The states of all nodes are updated at the same time according to their respective Boolean functions:

$$x_i(t+1) = b(x_{i1}(t),\ x_{i2}(t), \ldots, x_{il}t)).$$

The transitions of all states together correspond to a *state transition* of the network from $S(t)$ to the new network state, $S(t+1)$. A series of state transitions is called a trajectory. Since there is a finite number of network states, all trajectories are periodic. This simply follows from the fact that as soon as one state is visited a second time, the trajectory will take exactly the same path as for the first time. The repeating parts of the trajectories are called **attractors** and can be one or more states long. All the states leading to the same attractor are the *basin of attraction* for this attractor.

The dynamic properties of Boolean networks make them attractive for modeling of gene networks. Namely, they exhibit complex behavior, and are characterized with stable and reproducible attractor states, resembling many biological situations, like steady expression states. In addition, the range of behaviors of the system is completely determined and amenable to analysis, and is much smaller than that of other dynamic models. In terms of topology, it has been shown that high connectivity yields chaotic behavior, whereas low connectivity leads to stable attractors, which again corresponds well to real biological networks.

5.4.2
Reverse Engineering Boolean Networks

Clustering is a relatively easy way to extract useful information out of large-scale gene expression data sets. However, it typically only tells us which genes are coregulated, not which gene is regulating what other gene(s) (Figure 5.5). The goal in reverse engineering Boolean networks is to infer both the underlying topology (i.e. the edges in the graph) and the Boolean functions at the vertices from observed gene expression data. The actual observed data can come either from gene expression experiments conducted at different time intervals or when the expression of various genes is perturbed. For time-course data, measurements of the gene expressions at two consecutive time points simply correspond to two consecutive states of the network, $S(i)$ and $S(i+1)$. Perturbation data come in pairs, which can be thought of as the input/output states of the network, I_i/O_i, where the input state is the one before the perturbation and the output the one after it.

Given the observations of the states of a Boolean network, in general many networks may be constructed that are consistent with that data. Hence the solution network is ambiguous. There are several variants of the reverse engineering problem: (1) finding one network consistent with the data, (2) finding all networks consistent with the data and (3) finding the "best" network consistent with the data (according to some prespecified criteria). The first task is the simplest one and efficient algorithms exist.

The reverse engineering problems are intimately connected to the amount of empirical data available. Obviously, the inferred network will be less ambiguous the more data points are available. The amount of data needed to completely determine a unique network is known as the data requirement problem in network inference. The amount of data required depends on the sparseness of the underlying topology and the type of Boolean functions allowed. This can be understood intuitively. A network with few connections may be defined with few data points. In the worst case, the deterministic inference algorithms need on the order of $m=2^n$ transition pairs of data (experimental data points) to infer a densely connected Boolean network with general Boolean functions at the n vertices.

Since the advent of microarray experiments to quantify gene expression, a major problem has been the estimation of the **intergenic interaction matrix** M, see below. The matrix element m_{ij} of the interaction matrix M should be positive if gene G_j activates gene G_i, negative if gene G_j inhibits gene G_i, and equal to 0 if gene G_j and gene G_i have no interaction. The state of the Boolean variable G_i corresponding to gene i equals 1 if gene i is expressed and is zero otherwise. To calculate the m_{ij}s from experimental data points, one can determine the correlation $\rho_{ij}(s)$ between the state vector $\{x_j(t-s)\}_{t\in C}$ of gene j at time $t-s$ and the state vector $\{x_i(t)\}_{t\in C}$ of gene i at time t, t varying during the cell cycle C of length $K=|C|$ and corresponding to the observation time of the bio-array images:

$$\rho_{ij}(s) = \frac{\sum_{t\in C} x_j(t-s)x_i(t) - (1/K)\sum_{t\in C} x_j(t-s)\sum_{t\in C} x_i(t)}{\sigma_j(s)\sigma_i(s)},$$

where:

$$\sigma_j(s) = \left(\sum_{t\in C} x_j(t-s)^2 - \frac{1}{K}\left(\sum_{t\in C} x_j(t-s)\right)^2\right)^{1/2},$$

and then take:

$$\begin{aligned} m_{ij} &= \mathrm{sign}\left(\frac{1}{K}\sum_{s=1,\ldots,m} \rho_{jj}(s)\right) && \mathit{if}\,|m_{ij}|>\eta \\ m_{ij} &= 0 && \mathit{if}\,|m_{ij}|\le\eta \end{aligned}.$$

where η is a decorrelation threshold.

Given the interaction matrix M, the change of state x_i of gene G_i between t and $t+1$ obeys a threshold rule:

$$x_i(t+1) = H\left(\sum_{k=1,n} m_{ik}x_k(t) - b_i\right) \quad \text{or}$$
$$x_i(t+1) = H(Mx(t) - b),$$

where H is the step function with $H(y) = 1$ if $y \geq 0$ and $H(y) = 0$ if $y < 0$, and the b_is are threshold values. In the case of small regulatory genetic systems, knowing the matrix M makes it possible to know all feasible stationary behaviors of the organisms having the corresponding genome.

Let us look at a well-characterized example derived for the model plant organism *Arabidopsis*. As shown in Espinosa-Soto *et al.* (2004), the network shown in Figure 5.6 has about 140 000 possible initial conditions and it converges to only 10 fixed-point attractors or steady gene expression states that are listed in Table 5.1. Based on this model, one can now easily simulate the behavior of virtual loss-of-function mutants where the vertex corresponding to the knocked out gene is permanently turned off.

Boolean approaches suffer from their inability to describe intermediate levels of gene expression. Due to their discrete nature, they can sometimes give spurious

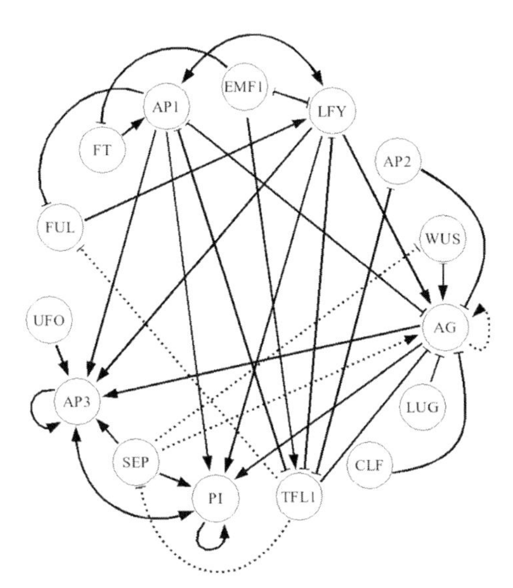

Figure 5.6 Gene network architecture for *Arabidopsis* floral organ fate determination. Each vertex corresponds to the concentration of active or functional protein encoded by each gene. The edges represent the regulatory interactions between vertices. Arrows represent positive (activating) interactions and blunt-end lines represent negative (repressing) interactions. AG, PI, AP3, WUS, etc., code for transcription factors, TFL1 and FT are likely membrane bound signaling molecules, EMF1 and LUG are positive or negative cofactors probably involved in transcription, and CLF is a chromatin remodeling protein. The four dashed lines constitute four novel predictions of the model that need to be tested experimentally at a later stage. Drawn after Espinosa-Soto *et al.* (2004).

Table 5.1 Gene expression in each of the 10 steady gene activation states in the wild-type *Arabidopsis* gene network (obviously, the example network discussed here is not truly Boolean as the variables may adopt three values 0, 1 and 2 rather than only 0 and 1).

FT	EMF1	TFL1	LFY	FUL	AP1	AP3	PI	AG	UFO	WUS	AP2	SEP	LUG	CLF	Cell type
0	1	2	0	0	0	0	0	0	0	0	0	0	1	1	Inf1
0	1	2	0	0	0	0	0	0	1	0	0	0	1	1	Inf2
0	1	2	0	0	0	0	0	0	1	1	0	0	1	1	Inf3
0	1	2	0	0	0	0	0	0	0	1	0	0	1	1	Inf4
1	0	0	2	0	2	0	0	0	0	0	1	1	1	1	Sep
1	0	0	2	0	2	2	2	0	1	0	1	1	1	1	Pe1
1	0	0	2	0	2	2	2	0	0	0	1	1	1	1	Pe2
1	0	0	2	2	0	2	2	2	1	0	1	1	1	1	St1
1	0	0	2	2	0	2	2	2	0	0	1	1	1	1	St2
1	0	0	2	2	0	0	1	2	0	0	1	1	1	1	Car

results. Potentially more accurate representations of gene networks use continuous functions, in which expression values are allowed to take on any positive value. These approaches are mathematically implemented by difference or differential equations, either linear or nonlinear.

5.4.3 Differential Equations Models

Differential equations are the starting point for the quantitative modeling of complex gene regulatory networks. They will be introduced more thoroughly in Chapter 7 in a different context and we will therefore keep the discussion short at this point. Differential equation models of gene networks are based on rate equations, quantifying the rate of change of gene expression as a function of the expressions of other genes (and possibly other factors as well). The general form of the equations, one for each of the n genes is:

$$\frac{\partial x_i}{\partial t} = f_i(x_{i1}, x_{i2}, \ldots, x_{il}),$$

where each x_j is a continuous function representing the expression of gene j. Each f_i describes the combined effect of its arguments (or regulators) on x_i, and subsumes all the biochemical effects of molecular interactions and degradation. $\{x_{i1}, x_{i2}, \ldots, x_{il}\}$ is a subset of all gene expression functions $\{x_1, x_2, \ldots, x_n\}$. In this way, only those gene expression levels are arguments for gene i that actually affect its expression (its parents, in the language of Boolean networks). In addition, there may appear additional factors such as constants. Identifying a gene network from observed data under this model means estimating (fitting) the parameters in the functions $f_i()$.

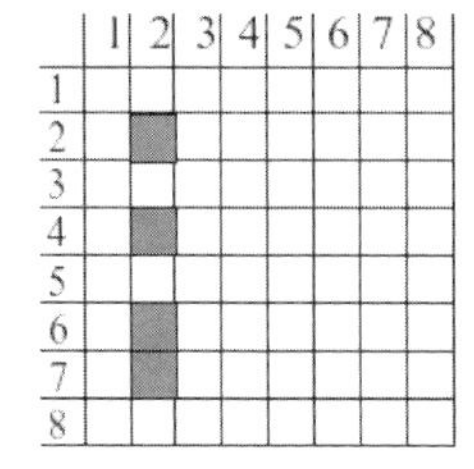

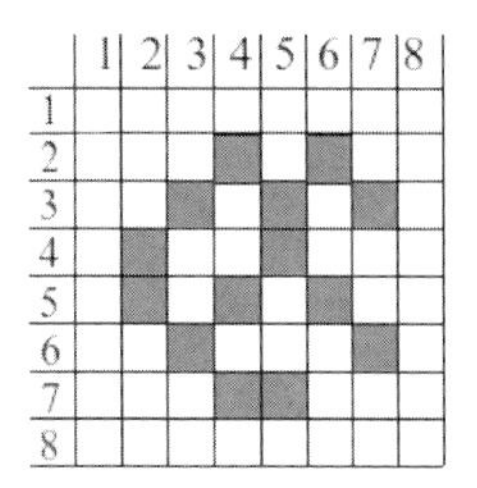

Figure 5.7 Connectivity matrix for causal regulation of transcription factor *j* (row) by transcription factor *i* (column). Dark fields indicate regulation. (Left) Feed-forward loop (FFL) motif. Transcription factor 2 regulates transcription factors 3 and 6, and transcription factor 3 again regulates transcription factor 6. (Middle) Single-input/multiple-output (SIM) motif, see text. (Right) Densely overlapping region (DOR).

5.5 Motifs

As mentioned in the introduction of this chapter, wiring diagrams of regulatory networks somehow resemble electrical circuits. To identify overrepresented motifs in these networks, Uri Alon and co-workers (Alon, 2007) tried to break down networks into basic building blocks. They searched for **network motifs** as patterns of interconnections that recur in many different parts of a network at frequencies much higher than those found in randomized networks. To do so, they again represented the transcriptional network as a connectivity matrix M such that $M_{ij} = 1$ if operon j encodes a transcription factor that transcriptionally regulates operon i and $M_{ij} = 0$ otherwise (Figure 5.7).

All $n \times n$ submatrices of M were generated by choosing n vertices that belong to a connected graph, for $n = 3$ and $n = 4$. Submatrices were enumerated by recursively searching for nonzero elements. P values representing the statistical significance were computed for submatrices representing each type of connected subgraph by comparing the number of times they appear in real network versus in a large number of random networks (Section 4.3). For $n = 3$, the only significant motif was the FFL, (see Figure 5.8). For $n = 4$, only the densely overlapping regulation motif was significant, (see Figure 5.10). SIM modules were identified by searching for isolated columns of M with many entries equal to 1.

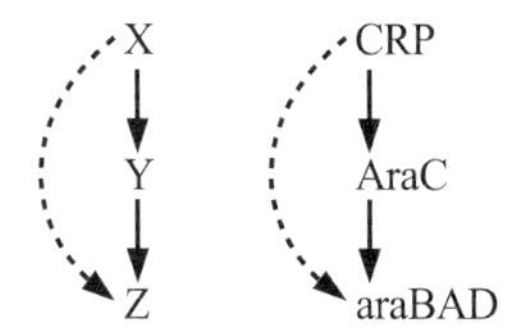

Figure 5.8 Example of a FFL (L-arabinose utilization in *E. coli*). The global transcription factor CRP is activated in the presence of cyclic AMP and de-activated in its absence.

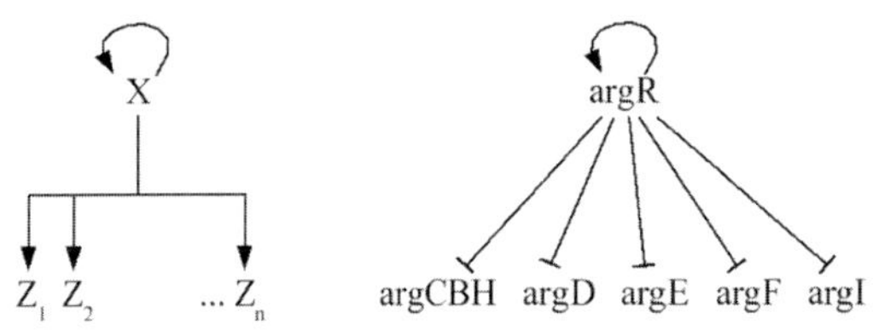

Figure 5.9 Example of a single input-multiple output SIM system (arginine biosynthesis in *E. coli*).

5.5.1
Feed-Forward Loop (FFL)

The term feed-forward loop (FFL) denotes a motif where a first transcription factor X regulates a second transcription factor Y, and both jointly regulate one or more operons $Z_1, \ldots, Z_n$. It appears in hundreds of gene systems in *E. coli* and yeast, as well as in other organisms (Figures 5.7, left and 5.8). As each of the three regulatory interactions in the FFL can be either activating or repressing, there are $2^3 = 8$ possible structures of FFLs.

5.5.2
SIM Motif

In a SIM motif (Figures 5.7, middle and 5.9), a single transcription factor X regulates a set of operons $Z_1, \ldots, Z_n$. In the purest form of a SIM motif, no other regulator regulates any of these genes, hence the name single-input motif (SIM). X is usually autoregulatory. All regulations are of the same sign. The main function of this motif is to allow coordinated expression of a group of genes with shared function. In addition, this motif has a more subtle dynamical property that is similar to that of the multi-output FFLs that are discussed above: it can generate a temporal expression programme, with a defined order of activation of each of the target promoters. X often has different activation thresholds for each gene, owing to variations in the sequence and context of its binding site in each promoter, So when X activity rises gradually with time, it crosses these thresholds in a defined order, first the lowest threshold, then the next lowest threshold, etc., resulting in a temporal order of expression. For example, the arginine biosynthesis system was found to adopt a SIM design in which the repressor ArgR regulates several operons that encode enzymes in the arginine biosynthesis pathway. When arginine is removed from the medium, these promoters are activated in a temporal order with minutes between the activations of different promoters. The order of activation matches the position of the enzymes in the arginine biosynthesis pathway.

5.5.3
Densely Overlapping Region (DOR)

A densely overlapping region (DOR) motif represents a set of operons $Z_1, \ldots, Z_m$ together with a set of input transcription factors, $X_1, \ldots, X_n$, by which they are regulated in a densely interconnected fashion (Figures 5.7, right and 5.10). The DOR algorithm (Shen-Orr *et al.*, 2002) detects dense regions of connections, with a high

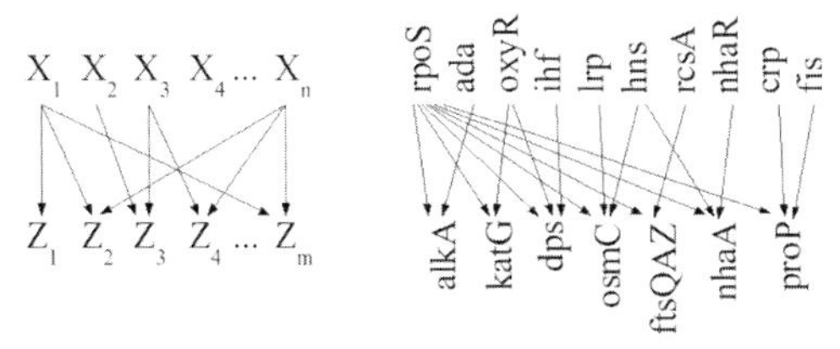

Figure 5.10 Example of a DOR. In this motif, many inputs regulate many outputs. (Right) This is an example of a DOR motif found in the stress–response system of *E. coli*.

ratio of connections to transcription factors. To identify DORs, all operons regulated by more than two transcription factors were considered. A (nonmetric) distance measure was defined between operons k and j, based on the number of transcription factors regulating both operons:

$$d(k,j) = \frac{1}{\left(\sum_n f_n M_{k,n} M_{j,n}\right)^2}.$$

where $f_n = 0.5$ for global transcription factors (to account for their unspecificity) and $f_n = 1$ otherwise. Operons were clustered with the average-linkage algorithm. DORs correspond to clusters with more than 10 connections, with a ratio of connections to transcription factors greater than two. *E. coli* has several DORs with hundreds of output genes, each responsible for a broad biological function. The DOR can be thought of as a gate array, carrying out a computation by which multiple inputs are translated into multiple outputs. Thus, to fully understand the function of the DOR, the connectivity arrows are not enough – the input functions in the promoter of each output gene must also be specified.

Exhaustive analysis of data from microarray experiments revealed that these three basic motifs are significantly overrepresented in natural gene regulatory networks. Comparing the regulation of homologous genes in different organisms revealed that they are often regulated by different classes of transcription factors. It is therefore hardly possible as in other areas of molecular and cell biology to transfer knowledge about particular gene regulation mechanisms from one organism to others. Remarkably, it turns out that if one gene is regulated by a FFL, the homologous gene is often regulated by another FFL in response to similar environmental stimuli. It seems therefore that the same network motifs have been "rediscovered" during evolution over and over again. The same applies to SIM or DOR networks motifs where similar output genes in different organisms are often regulated by unregulated transcription factors (Alon, 2007).

As networks become better characterized, new motifs and new motif functions will be discovered. Network motifs at the level of signaling networks and neuronal networks are only beginning to be investigated. If the current findings can be generalized, they suggest that complex biological networks have a degree of structural simplicity, in that they contain a limited set of network motifs. This raises the hope that the dynamics of large networks can be understood in terms of elementary circuit patterns. We will see a further example of how regulatory motifs are used in Chapter 10.

Summary

Gene regulatory networks are responsible for producing the right amount of protein in an intricately orchestrated timely fashion. When measured at the level of mRNA expression by microarrays, it is hard to deduce which gene is regulating which other gene as we only see the end result of the process. Biological cells are "robust" meaning that they tolerate significant amounts of stochastic fluctuations making it even more difficult to define precise threshold levels. Moreover, such fluctuations may also contain signals themselves. Recently, the discovery of siRNAs has added to the level of regulatory complexity. Fortunately, more and more crystal and nuclear magnetic resonance structures of protein:DNA complexes are becoming available. Researchers have started using molecular mechanics techniques as discussed in Chapter 9 to add to the structural understanding of the biomolecular association events underlying gene regulation.

Further Reading

Gene Networks

Brazhnik P, de la Fuente A, Mendes P (2002) Gene networks: how to put the function in genomics, *Trends in Biotechnology* **20**, 467–472.

Mathematical Modeling of Gene Regulatory Networks

Filkov V (2005) Identifying gene regulatory networks from gene expression data, in *Handbook of Computational Molecular Biology* (S. Aluru editor), Chapman & Hall/CRC Press, Boca Raton, FL.

Markowetz F, Spang R (2007) Inferring cellular networks – a review, *BMC Bioinformatics* **8**, S5.

Gene Regulatory Network of *E. coli*

Martínez-Antonio A, Collado-Vides J (2003) Identifying Global Regulators in Transcriptional Regulatory Networks in Bacteria. *Current Opinion in Microbiology* **6**, 482–489.

Resendis-Antonio O, Freyre-González JA, Menchaca-Méndez R, Guitérrez-Rios RM, Martínez-Antonio A, Ávila-Sánchez C, Collado-Vides J (2005) Modular analysis of the transcriptional regulatory network of *E. coli*, *Trends in Genetics* **21**, 16–20.

Boolean Network of *Arabidopsis* Floral Organ Fate

Espinosa-Soto C, Padilla-Longoria P, Alvarez-Buylla ER (2004) A gene regulatory model for cell-fate determination during *Arabidopsis thaliana* flower development that is robust and recovers experimental gene expression profiles, *The Plant Cell* **16**, 2923–2939.

Motifs in Networks

Shen-Orr SS, Milo R, Mangan S, Alon U (2002) Network motifs in the transcriptional regulation network of *Escherichia coli*, *Nature Genetics* **31**, 64–68.

Alon U (2007) Network motifs: theory and experimental approaches. *Nature Reviews Genetics* **8**, 450–461.

6
Metabolic Networks

In this chapter, we will introduce several mathematical techniques to quantitatively analyze some of the topological properties of metabolic networks.

6.1
Introduction

The division of cellular networks into metabolism, regulation and signaling has historical and life science curricular origins. As was mentioned already at various places throughout this textbook, these separations are now becoming obsolete because researchers have realized how tightly connected all those networks are. Often the same molecules participate in more than one of these networks and we should consider all networks at once. This, of course, is going to be more complex than studying each of them separately and may also require a mixture of different mathematical approaches or the development of new ones. In this chapter, we will therefore take a separate look at the mathematical modeling of metabolic cellular networks, which is one of the most mature areas of cellular networks.

Metabolites are the reactants and products of enzymatic reactions. To get a feeling for the most important players, Table 6.1 shows an overview of the most frequently used metabolites in the central metabolism of *Escherichia coli*. Table 6.1 illustrates that many substances participate in tens of different metabolic pathways. Measuring the cellular concentration of ATP, for example, will not allow us in any way to find out about the activity of a particular biochemical pathway involving ATP as, usually, many pathways using and producing ATP will be simultaneously active. In analogy to Chapters 3 and 4, it appears as if the highly connected metabolites will be the *hubs* – the important connectors of the cellular metabolic network.

As discussed in Section 1.3, multiple biochemical reactions are often chained together as **biochemical pathways**. Those substances occurring in multiple pathways will be the crossings of such linear biochemical pathways. This is the reason why we can no longer discuss these pathways separately when aiming at the quantitative

Principles of Computational Cell Biology – From Protein Complexes to Cellular Networks. Volkhard Helms

ISBN: 978-3-527-31555-0

Table 6.1 Metabolites most frequently found in the central metabolism of *E. coli* (after Ouzounis and Karp, 2000): occurrences refer to the number of reactions in which these most common substrates have been found.

Occurrence	Name of metabolite	Occurrence	Name of metabolite
205	H_2O	13	glyceraldehyde-3-phosphate
152	ATP	13	THF
101	ADP	13	acetate
100	phosphate	12	PRPP
89	pyrophosphate	12	(acyl carrier protein)
66	NAD	12	oxaloacetic acid
60	NADH	11	dihydroxy-acetone-phospate
54	CO_2	11	GDP
53	H^+	11	glucose-1-phosphate
49	AMP	11	UMP
48	NH_2	10	electron
48	NADP	10	phosphoenolpyruvate
45	NADPH	10	acceptor
44	coenzyme A	10	reduced acceptor
43	L-glutamate	10	GTP
41	pyruvate	10	L-serine
29	acetyl-CoA	10	fructose-6-phosphate
26	O_2	9	L-cysteine
24	2-oxoglutamate	9	reduced thioredoxin
23	*S*-adenosyl-L-methionine	9	oxidized thioredoxin
18	*S*-adenosyl-homocysteine	9	reduced glutathione
16	L-aspartate	8	acyl-ACP
16	L-glutamine	8	L-glycine
15	H_2O_2	8	GMP
14	glucose	8	formate

analysis of metabolic fluxes. Instead, we need to employ network approaches to describe the entirety of this **metabolic network** of interlinked biochemical pathways.

In this chapter, we will introduce two different levels for describing metabolic networks by computational methods. First, stoichiometric modeling (**flux balance analysis**) characterizes the feasible metabolic flux distributions of an integrated cellular network. Second, the automatic decomposition of metabolic networks into sets of generating vectors such as *elementary modes* and *extreme pathways* provides a general basis to discuss, for example, the effects of single gene deletions. Here, the definition of *minimal cut sets* will prove very helpful too.

We will not discuss kinetic modeling approaches of metabolic networks (one of the popular software packages in this area is the program package E-Cell) as these approaches are methodologically similar to the kinetic modeling of signaling transduction processes that is covered in Chapter 7. Often, these approaches are still facing lack of certain kinetic information on the dynamics and regulation of biologically or medically relevant aspects of cellular metabolism. Therefore,

Table 6.2 Information contained in EcoCyc, version 11.0.

Length of *E. coli* genome	**4.7 million DNA bases (600 times shorter than human)**
Number of predicted genes	4549
Of which code for proteins	4365
Coded enzymes	1323
Enzymatic reactions	1245
Transport reactions	238
Protein complexes	796
Compounds	1202

their range of applicability is either limited to well-understood model systems or comes at the price of a large experimental overhead to determine the required rate constants.

In one of the pioneering efforts in the field of metabolic networks, the database system EcoCyc (www.ecocyc.org) has compiled a comprehensive overview of the metabolic capabilities of the model system *E. coli*. This database has been developed to characterize the functional complement of *E. coli* and to allow comparisons of the biochemical networks of two organisms. EcoCyc provides to the user the metabolic map of *E. coli* defined as the set of all known pathways, reactions and enzymes of *E. coli* small-molecule metabolism. Some statistical data is listed in Table 6.2.

In Ouzounis and Karp (2000), each reaction on average contained 4.0 substrates and each distinct substrate occurred in an average of 2.1 reactions. EcoCyc describes 161 pathways involving energy metabolism, nucleotide and amino acid biosynthesis secondary metabolism, and 21 signaling pathways. Pathways vary in length from a single reaction step to 16 steps with an average of 5.4 steps. However, there is no precise biological definition of a pathway. The partitioning of the metabolic network into biochemical pathways [including the well-known examples from biochemistry textbooks such as glycolysis, pentose phosphate pathway and the tricarboxylic acid (TCA) cycle] is somehow arbitrary. These historical classifications, of course, then also affect the distribution of pathway lengths. EcoCyc also contains extensive information about the modulation of *E. coli* enzymes with respect to particular reactions, among which are activators and inhibitors of the enzyme, cofactors required by the enzyme, and alternative substrates that the enzyme will accept.

Whereas most *E. coli* enzymes catalyze only one reaction, about one-sixth of them are multifunctional, either having the same active site and different substrate specificities or having different active sites. The enzymes that catalyze the largest number of different reactions are purine nucleoside phosphorylase (seven reactions) and nucleoside diphosphate kinase (nine reactions) that phosphorylates many different nucleoside diphosphate substrates that each need to be counted as separate reactions. This relatively high proportion of multifunctional enzymes implies that the genome projects employing automatic functional annotation (where they usually only assign a single enzymatic function) may significantly underpredict multifunctional enzymes!

The 99 reactions belonging to multiple pathways appear to be the **intersection points** in the complex network of chemical processes in the cell. For example, one reaction present in six pathways is the reaction catalyzed by malate dehydrogenase, a central enzyme in cellular metabolism that participates in the glyoxylate cycle, gluconeogenesis, the TCA cycle, anaerobic respiration and in mixed acid fermentation.

Other recent database systems as BioCyc (www.biocyc.org) and the Kyoto Encyclopedia of Genes and Genomes (www.genome.jp/kegg) store analogous information for 260 and even more than 500 organisms. Based on this vast information, homology assignment based on sequence similarity may then even allow automatically reconstructing the network of biochemical reactions in an organism from a newly sequenced genome (Figure 6.1).

6.2 Stoichiometric Matrix

Any chemical reaction requires **mass conservation**. This fact may be exploited to quantitatively analyze the fluxomic capabilities of metabolic systems. The only knowledge required are the stoichiometries of all metabolic pathways involved and their metabolic demands. For each metabolite i we can write the time derivative dX_i/dt of its current concentration X_i as a balance equation:

$$v_i = \frac{dX_i}{dt} = V_{\text{synthesized}} - V_{\text{degraded}} - (V_{\text{used}} - V_{\text{transported}}). \quad (6.1)$$

where dX_i/dt indicates whether the concentration of X_i is increasing or decreasing. As written above, such changes over time can be derived simply from considering the net balance of all reactions involving X_i, synthesis reactions, degradation reactions and transport into the reaction compartment or out of the reaction compartment.

At steady-state conditions, all concentrations X_i will have reached an equilibrium meaning that the time derivatives of their concentrations dX_i/dt are equal to zero. This is exactly the definition of a **steady state**. If some time derivatives were still different from zero and the fluxes connecting the system with the outside were kept constant, the system would balance itself until eventually all time derivatives of the concentrations will have become zero. The above equation applies to metabolite i and there are m such equations for each metabolite in the network. One can conveniently combine all the balance equations into matrix equation 6.2 using the concept of the stoichiometric matrix **S** and the flux vector **v**. Under steady-state conditions, the mass balance constraints in a metabolic network can then be represented mathematically by the matrix equation:

$$\mathbf{S} \cdot \mathbf{v} = 0, \quad (6.2)$$

where **S** is the $m \times n$ **stoichiometric matrix**, with $m =$ the number of metabolites and $n =$ the number of reactions in the network, and **v** represents all n fluxes in the metabolic network, including the internal fluxes, transport fluxes and the growth flux.

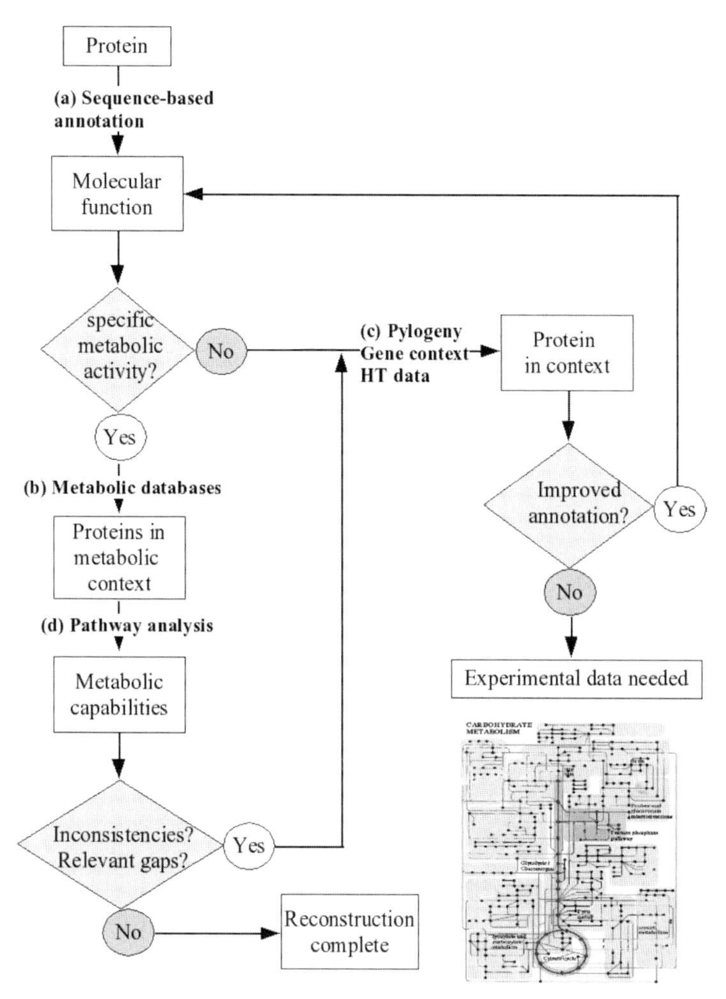

Figure 6.1 Flow chart of modern attempts to automatically reconstruct metabolic networks from a genome sequence alone. (a) Automatic assignment of molecular function for each protein is done on the basis of sequence homology and domain profiles. (b) When a specific metabolic activity is defined for the protein in the databases, an association is made with the corresponding reaction. (c) In case the molecular function is not specified, the assigned reaction is inconsistent with the metabolic context or when specific reactions are needed to close gaps in pathways, comparative genomics approaches are applied to put the protein into a context that might provide information on the molecular function. (d) Pathway analysis of the metabolic reconstruction leads to predictions of missing reactions in certain pathways and metabolic capacities that can be checked with experimental data. Drawn after Francke *et al.* (2005).

Deriving the stoichiometric matrix of a metabolic network is typically the first step of an *in silico* model of this network. Figure 6.2 shows a simple reaction network containing five metabolites A, B, . . . , E, six internal reactions $v_1, v_2, \ldots, v_6$, and two exchange fluxes with the environment b_1 and b_2.

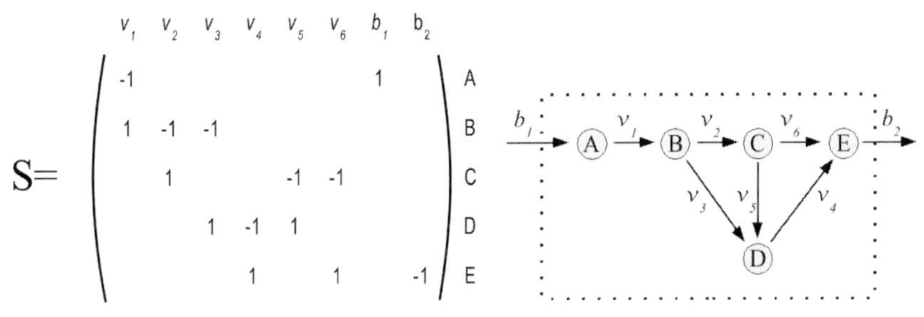

Figure 6.2 Simple network and the corresponding stoichiometric matrix. In this matrix, the nonzero entries of each column are the concentrations of those metabolites that are affected by a particular flux. This is why we have labeled all columns at the top by the various fluxes for visual help. These labels are not actually part of **S**. Similarly, the labels next to each row indicate by which reactions metabolite i is affected. For example, reaction ν_1 consumes one substance unit of metabolite A and produces one substance unit of metabolite B. The external flux b_1 produces one unit of A and the external flux b_2 consumes one unit of D. You may compare this matrix to the incidence matrix introduced in Section 2.3. The only differences are that the external fluxes in the stoichiometric matrix typically only affect a single metabolite and some internal reactions may couple more than two metabolites.

Starting from this matrix, the methods of extreme pathway and elementary mode analysis can be used to generate a set of generating vectors comparable to the three orthogonal axes of the Cartesian coordinate system (Section 6.6). For example, the set of three extreme pathways P1, P2 and P3 shown in Figure 6.14 suffices to construct all feasible steady-state fluxes **v** in the network shown in Figure 6.3 as a linear combination:

$$\mathrm{v} = \alpha_1 \cdot P_1 + \alpha_2 \cdot P_2 + \alpha_3 \cdot P_3,$$

with suitable parameters α_i. To be able to understand the mathematical concepts behind these algorithms, the next section will provide a short review of matrix algebra.

Although having only been developed since the mid-1990s, this research field of studying metabolic networks *in silico* has already found its way into biotechnological

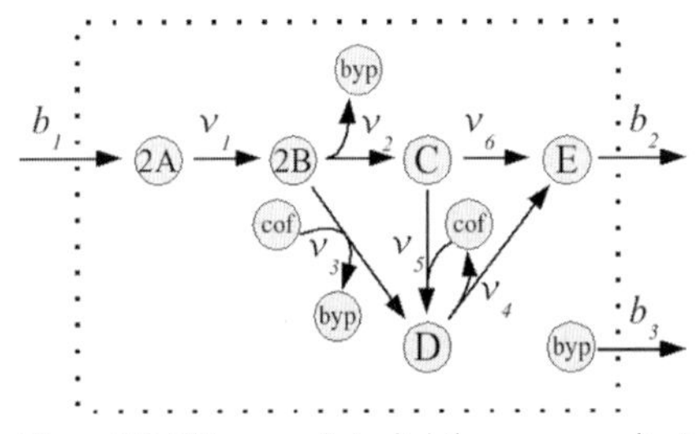

Figure 6.3 This example is slightly more complicated than the one in Figure 6.2. “cof” and “byp” are a cofactor needed in one reaction and a byproduct generated. “byp” is produced by two reactions. “cof” is produced by flux ν_4 and consumed by fluxes ν_3 and ν_5. Drawn after Papin *et al.* (2003).

applications and is even used in various companies. Some of the groups that are very active in this field (Bernard Palsson at UC San Diego, Naama Barkai at the Weizman Institute) have made available their reconstructions of *in silico* organisms including *E. coli, Helicobacter pylori, Haemophilus influenzae, Staphylococcus aureus, Methanosarcina barkeri, Saccharomyces cerevisiae,* the red blood cell, human cardiac mitochondria, cardiomyocytes and even an early model of *human sapiens,* including carbohydrate, lipid, amino acid, nucleotide, vitamin, cofactor, energy, glycan and secondary metabolite metabolism.

6.3 Linear Algebra Primer

6.3.1 Matrices: Definitions and Notations

Matrices are used in various mathematical disciplines. Here, we will focus on their usage in linear algebra to rotate vectors and to describe systems of linear equations. Concerning notation, we will use capital bold letters for matrices (**A**) and small bold letters for vectors (**v**).

A **matrix** is a rectangular table of numbers which will be assumed to be real numbers in this textbook. The horizontal lines in a matrix are called **rows** and the vertical lines are called **columns**. A matrix with m rows and n columns is called an m-by-n matrix (or $m \times n$ matrix) and m and n are called its **dimensions**. The entry of a matrix A that lies in the ith row and the jth column is called the (i,j)th entry of A. This is also written as $A_{i,j}$ or $A[i,j]$. We often write $A := (a_{i,j})_{m \times n}$ to define an $m \times n$ matrix A with each entry in the matrix $A[i,j]$ called a_{ij} for all $1 \leq i \leq m$ and $1 \leq j \leq n$.

- Example
- The matrix:
-

$$\begin{pmatrix} 1 & 3 & -8 \\ 1 & 3 & 4 \\ 3 & 9 & 2 \\ 5 & 0 & 5 \end{pmatrix},$$

- is a 4×3 matrix. The element $A[2,3]$ or $a_{2,3}$ is 4.

6.3.2 Adding, Subtracting and Multiplying Matrices

Given two $m \times n$ matrices A and B, their sum $A + B$ is defined as the $m \times n$ matrix computed by adding corresponding elements, i.e. $(A + B)[i,j] = A[i,j] + B[i,j]$. For example:

$$\begin{pmatrix} 1 & 4 & 2 \\ 1 & 0 & 0 \\ 3 & 2 & 2 \end{pmatrix} + \begin{pmatrix} 0 & 0 & 3 \\ 3 & -2 & 0 \\ 1 & 3 & 1 \end{pmatrix} = \begin{pmatrix} 1+0 & 4+0 & 2+3 \\ 1+3 & 0-2 & 0+0 \\ 3+1 & 2+3 & 2+1 \end{pmatrix}$$
$$= \begin{pmatrix} 1 & 4 & 5 \\ 4 & -2 & 0 \\ 4 & 5 & 3 \end{pmatrix}$$

Subtraction of two matrices is defined analogously. Given a matrix A and a number c, the **scalar multiplication** cA is defined by $(cA)[i,j] = cA[i,j]$. For example

$$2\begin{pmatrix} 1 & 4 \\ 2 & -2 \end{pmatrix} = \begin{pmatrix} 2\cdot 1 & 2\cdot\ 4 \\ 2\cdot 2 & 2\cdot(-2) \end{pmatrix} = \begin{pmatrix} 2 & 8 \\ 4 & -4 \end{pmatrix}.$$

Multiplication of two matrices is well-defined only if the number of columns of the first matrix is the same as the number of rows of the second matrix. If A is an $m\times n$ matrix (m rows, n columns) and B is an $n\times p$ matrix (n rows, p columns), then their product AB is the $m\times p$ matrix (m rows, p columns) given by $(AB)[i,j] = A[i,1] * B[1,j] + A[i,2] * B[2,j] + \ldots + A[i,n] * B[n,j]$ for each pair i and j.

For instance:

$$\begin{pmatrix} 2 & 0 & 4 \\ -1 & 3 & 2 \end{pmatrix} \cdot \begin{pmatrix} 3 & 1 \\ 1 & 0 \\ 2 & 1 \end{pmatrix} = \begin{pmatrix} 2\cdot 3+0\cdot 1+4\cdot 2 & 2\cdot 1+0\cdot 0+4\cdot \\ -1\cdot 3+3\cdot 1+2\cdot 2 & -1\cdot 1+3\cdot 0+2\cdot 1 \end{pmatrix}$$
$$= \begin{pmatrix} 14 & 6 \\ 4 & 1 \end{pmatrix}.$$

This multiplication has the following properties:

- $(AB)C = A(BC)$ for all $k\times m$ matrices A, $m\times n$ matrices B and $n\times p$ matrices C ("associativity").
- $(A+B)C = AC + BC$ for all $m\times n$ matrices A and B and $n\times k$ matrices C ("right distributivity").
- $C(A+B) = CA + CB$ for all $m\times n$ matrices A and B and $k\times m$ matrices C ("left distributivity").

It is important to note that commutativity does not generally hold, i.e. given matrices A and B and their product defined, then generally $AB \neq BA$.

6.3.3 Linear Transformations, Ranks and Transpose

Matrices can conveniently represent linear transformations because matrix multiplication neatly corresponds to the composition of maps. Let us identify the space of n-dimensional vectors with real-valued coefficients $\Re^n$ with the set of "rows" or $n\times 1$ matrices. For every linear map f: $\Re^n \rightarrow \Re^m$ there exists then a unique $m\times n$ matrix $\mathbf{A}$ such that $f(\mathbf{x}) = \mathbf{Ax}$ for all $\mathbf{x}$ in $\Re^n$. We say that the matrix $\mathbf{A}$ "represents" the linear map f. More generally, a linear map from an n-dimensional vector space to an m-dimensional vector space is represented by an $m \times n$ matrix, provided that bases (orthogonal set of unit vectors spanning the whole space) have been chosen for each.

The **rank** of a matrix **A** is the **dimension** of the **image** of the linear map represented by **A**. This is the same as the dimension of the space generated by the rows of **A** and also the same as the dimension of the space generated by the columns of **A**. The column rank (row rank, respectively) of a matrix **A** with entries in some field is defined to be the maximal number of columns (rows, respectively) of **A** which are **linearly independent**. The column rank and the row rank are indeed equal; this common number is simply called the rank of **A**. The easiest way to compute the rank of a matrix **A** is given by the Gauss elimination method. Consider for example the 4×4 matrix:

$$\mathbf{A} = \begin{pmatrix} 2 & 4 & 1 & 3 \\ -1 & -2 & 1 & 0 \\ 0 & 0 & 2 & 2 \\ 3 & 6 & 2 & 5 \end{pmatrix}.$$

In this simple example, we can easily verify that the second column is twice the first column, and that the fourth column equals the sum of the first and the third. Therefore, only the first and the third columns are linearly independent in this example, so the rank of **A** is two. (As a check, you can verify that row 1 + row 3 − row 2 gives row 4, so row 4 can be eliminated, too. Also, 2 × row 1 + 4 × row 2 gives three times row 3, so row 3 can be eliminated too. This shows that the column and row ranks are identical.)

The **transpose** of an $m \times n$ matrix A is the $n \times m$ matrix A^{T} formed by turning rows into columns and columns into rows, i.e. $A^{\mathrm{T}}[i,j] = A[j,i]$ for all indices i and j.

We have $(A+B)^{\mathrm{T}} = A^{\mathrm{T}} + B^{\mathrm{T}}$ and $(AB)^{\mathrm{T}} = B^{\mathrm{T}} A^{\mathrm{T}}$.

6.3.4
Square Matrices and Matrix Inversion

A **square matrix** is a matrix which has the same number of rows as columns. The unit matrix or **identity matrix** $\mathbf{I}_n$, with elements on the main diagonal set to 1 and all other elements set to 0, satisfies $\mathbf{MI}_n = \mathbf{M}$ and $\mathbf{I}_n\mathbf{N} = \mathbf{N}$ for any $m \times n$ matrix **M** and $n \times k$ matrix **N**. For example, if $n = 3$:

$$\mathbf{I}_3 = \begin{pmatrix} 1 & 0 & 0 \\ 0 & 1 & 0 \\ 0 & 0 & 1 \end{pmatrix}.$$

An n by n matrix **A** is **invertible** if and only if there exists a matrix **B** such that:

$$\mathbf{AB} = \mathbf{I}_n.$$

In this case, **B** is uniquely determined by **A** and is called the **inverse matrix** of **A**, denoted by $\mathbf{A}^{-1}$. A square matrix that is not invertible is called singular. As a rule of thumb, almost all matrices of maximal rank are invertible.

6.3.5
Eigenvalues of Matrices

An **eigenvector v** of a linear transformation matrix **A** is a vector whose direction is not changed by application of that transformation matrix. In other words, it is invariant upon application of the matrix up to a multiplication by a scalar number λ:

$$\mathbf{A} \cdot \mathbf{v} = \lambda \cdot \mathbf{I} \cdot \mathbf{v}.$$

For example, let us consider the rotation matrix:

$$\begin{pmatrix} \cos\alpha & -\sin\alpha & 0 \\ \sin\alpha & \cos\alpha & 0 \\ 0 & 0 & 1 \end{pmatrix},$$

that rotates every point in the xy-plane about an angle α, but leaves their z-coordinates unchanged. Obviously, the z-axis itself will be one of the eigenvectors of this matrix. The corresponding scalar value λ is called the **eigenvalue** of this eigenvector. Intuitively, it denotes "how large the vector appears" after application of the rotation matrix **A** to it.

An important tool for describing eigenvalues of square matrices is the characteristic polynomial. Saying that λ is an eigenvalue of **A** is equivalent to stating that the system of linear equations $(\mathbf{A} - \lambda\mathbf{I})\,\mathbf{v} = 0$ (where **I** is the identity matrix) has a nonzero solution **v** (an eigenvector) and so it is equivalent to the determinant $\det(\mathbf{A} - \lambda\mathbf{I})$ being zero. The function $p(\lambda) = \det(\mathbf{A} - \lambda\mathbf{I})$ is a polynomial in λ since determinants are defined as sums of products. This is the characteristic polynomial of **A**: the eigenvalues of a matrix are the zeros of its characteristic polynomial. If the matrix is small, the eigenvalues can easily be computed symbolically using the characteristic polynomial. However, this is often impossible for larger matrices, in which case we must use a numerical method.

All the eigenvalues of a matrix **A** can be computed by solving the equation $p_A(\lambda) = 0$. If **A** is an $n \times n$ matrix, then p_A has degree n and **A** can therefore have at most n eigenvalues. Conversely, the fundamental theorem of algebra says that this equation has exactly n roots (zeros), counted with multiplicity. An example of a matrix with no real eigenvalues is the 90° clockwise rotation in two dimensions:

$$\begin{pmatrix} 0 & 1 \\ -1 & 0 \end{pmatrix}$$

whose characteristic polynomial is $\lambda^2 + 1$ and so its eigenvalues are the pair of complex conjugates $i, -i$. The associated eigenvectors are also not real.

6.3.6
System of Linear Equations

A system of linear equations is a set of linear equations such as:

$$\begin{aligned} 3x_1 + 2x_2 - x_3 &= 1 \\ 2x_1 - 2x_2 + 4x_3 &= -2 \\ -x_1 + \tfrac{1}{2}x_2 - x_3 &= 0. \end{aligned}$$

The mathematical problem to solve is to find those values for the unknowns x_1, x_2 and x_3 which satisfy all three equations simultaneously. Systems of linear equations belong to the oldest problems in mathematics and they have many applications. An efficient way to solve systems of linear equations is given by the Gauss–Jordan elimination or by the Cholesky decomposition. In general, a system with m linear equations and n unknowns can be written as

$$\begin{aligned}
a_{11}x_1 + a_{12}x_2 + \ldots + a_1x_n &= b_1\\
a_{21}x_1 + a_{22}x_2 + \ldots + a_{2n}x_n &= b_2\\
&\vdots\\
&\vdots\\
a_{m1}x_1 + a_{m2}x_2 + \ldots + a_{mn}x_n &= b_m,
\end{aligned}$$

where $x_1, \ldots, x_n$ are the unknowns and the numbers a_{ij} are the coefficients of the system. We can separate the coefficients in a matrix as follows:

$$\begin{pmatrix} a_{11} & a_{12} & \ldots & a_{1n}\\ a_{21} & a_{22} & \ldots & a_{2n}\\ \vdots & \vdots & \ddots & \vdots\\ a_{m1} & a_{m2} & \ldots & a_{mn} \end{pmatrix}\begin{pmatrix} x_1\\ x_2\\ \vdots\\ x_n \end{pmatrix} = \begin{pmatrix} b_1\\ b_2\\ \vdots\\ b_m \end{pmatrix}.$$

If we represent each matrix by a single letter, this becomes

$$\mathbf{Ax} = \mathbf{b},$$

where $\mathbf{A}$ is the $m \times n$ matrix above, $\mathbf{x}$ is a column vector with n entries and $\mathbf{b}$ is a column vector with m entries. The above-mentioned Gauss–Jordan elimination applies to all these systems. If the field is infinite (as in the case of the real numbers), then only the following three cases are possible for any given system of linear equations:

- The system has no solution (in this case, we say that the system is overdetermined).
- The system has a single solution (the system is exactly determined).
- The system has infinitely many solutions (the system is underdetermined).

A system of the form:

$$\mathbf{Ax} = 0,$$

is called a *homogenous* system of linear equations. The set of all solutions of such a homogeneous system is called the **null space** of the matrix $\mathbf{A}$.

6.4 Flux Balance Analysis

After this short mathematical excursion, we will return to studying the solution space of Eq. (6.2). When including the external fluxes, this equation generalizes into:

$$\begin{pmatrix} \mathbf{S} \\ \text{external fluxes} \end{pmatrix} \cdot \mathbf{v} \quad \begin{matrix} = 0 \\ \geq 0 \end{matrix}, \tag{6.3}$$

where the internal fluxes satisfy the steady-state condition $\mathbf{S} \cdot \mathbf{v} = 0$, and the external fluxes are constrained to be nonnegative. Since the number of metabolites is generally smaller than the number of reactions ($m < n$) the **flux balance equation** (Eq. 6.3) is typically underdetermined. Recall from the end of Section 6.3 that only quadratic matrices ($n \times n$) with rank n have a unique solution. Therefore, there exists generally an infinite number of feasible flux distributions that satisfy the mass balance constraints. The set of solutions are confined to the null space of matrix **S**.

- Example
- Let us consider as a simple example in three dimensions the following set of two homogenous linear equations (Eq. 6.4):

$$\begin{pmatrix} 0 & 2 & 1 \\ 3 & -1 & 1 \end{pmatrix} \cdot \begin{pmatrix} x_1 \\ x_2 \\ x_3 \end{pmatrix} = \begin{pmatrix} 0 \\ 0 \\ 0 \end{pmatrix}. \tag{6.4}$$

- Carrying out the matrix multiplication gives the following two equations:
  - $2x_2 + x_3 = 0$
  - $3x_1 - x_2 + x_3 = 0.$
- This system can be simplified by subtracting x_3 on both sides of the first equation and by subtracting the first from the second equation:
  - $2x_2 = -x_3$
  - $3x_1 - 3x_2 = 0.$
- This means that we can freely choose the value of one variable, say x_3. If we set $x_3 = -2$, it follows that $x_2 = 1$ and $x_1 = 1$. You can verify the correctness of this solution by inserting (1, 1, −2) into Eq. (6.4).

In the general case, there will be a $(n - m)$-dimensional infinite space of solutions that satisfy Eq. (6.3). The intersection of the null space and the region defined by those linear inequalities defines a region in flux space that includes all feasible fluxes. The steady-state operation of the metabolic network is restricted to the region within a **cone**, termed the **feasible set** (Figure 6.4A). The feasible set contains all flux vectors that satisfy the physicochemical constraints. Thus, the feasible set defines the capabilities of the metabolic network.

To find the "true" biological flux in cells (that can be determined by proteomic experiments) one either needs additional (experimental) information or one may

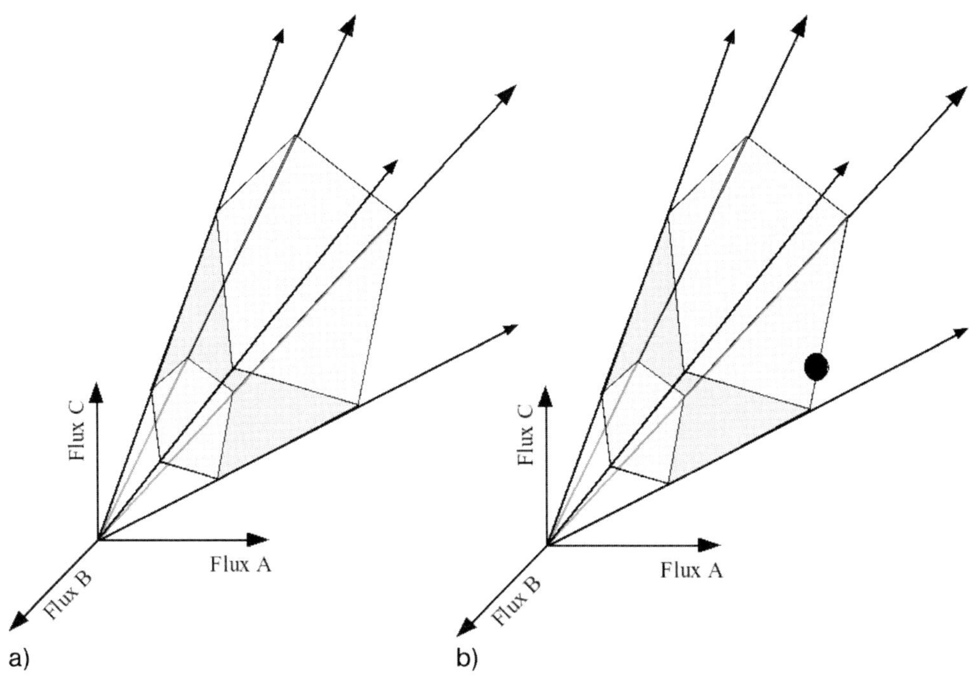

Figure 6.4 a) A "pointed" cone spanned by five generating vectors that intersect at the origin. All points inside the polygon represent feasible fluxes of the corresponding metabolic system. b) Additional constraints may allow to reduce the range of possible solutions to a single point.

impose constraints:

$$\alpha_i \leq v_i \leq \beta, \tag{6.5}$$

on the magnitudes of individual metabolic fluxes that may come from measurements of protein concentrations and enzymatic rates, or from regulatory constraints (Section 10.3). Such constraints will not reduce the dimensionality of the search space, but constrain the width of the solution space. In the limiting case, where all constraints on the metabolic network are known, such as the enzyme kinetics and gene regulation, the feasible set may be reduced to a single point (Figure 6.4B). This single point must, of course, still lie within the feasible set.

One mathematical method to solve systems of linear inequalities shown in Eq. (6.3) is linear programming (Figure 6.5) that is explained in introductory computer science textbooks.

Flux balance analysis constructs the **optimal network utilization** using the stoichiometries of metabolic reactions and capacity constraints. Many applications for *E. coli* have shown the *in silico* results to be **consistent** with experimental data. In particular, flux balance analysis showed that the *E. coli* metabolic network contains relatively few **critical gene products** in its central metabolism. However, the ability to adjust to different environments (growth conditions) may be diminished by gene

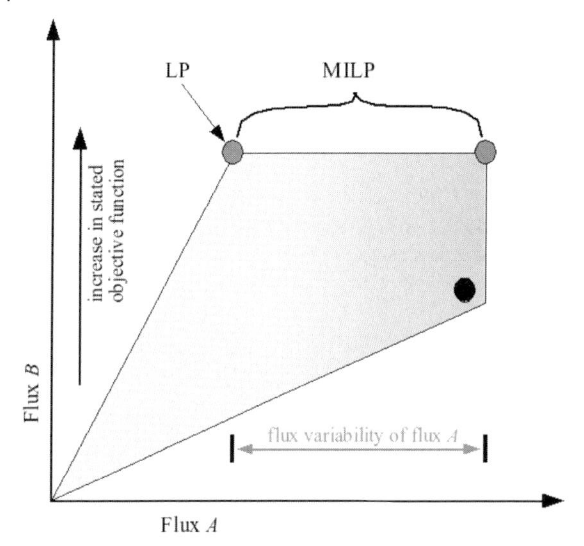

Figure 6.5 Strategy for determining optimal states of a biochemical network by flux balance analysis. If an objective is stated for the network, optimal solutions for this objective satisfying Eqs. (6.2) and (6.3) can be calculated. One objective may be optimal growth or biomass production. Linear programming (LP) will find one particular optimal solution, whereas mixed integer LP (MILP) can be used to find all of the basic optimal solutions. Flux variability analysis can be used to find ranges of values for all the fluxes in the set of alternate optima. Here, only flux *A* is variable across the alternate optima. The black point represents a solution that could have been obtained by application of an alternative objective function. Drawn after Price *et al.* (2004).

deletions. Flux balance analysis always identifies "the best" the cell can do, not how the cell actually behaves under a given set of conditions. So far, most studies have equated survival with growth although we do not know for sure whether prokaryotic organisms were optimized for optimal growth alone during the process of evolution.

6.5 Double Description Method

Note that this section is mathematically more ambitious than the remainder of this textbook.

We now arrive at the construction of a set of generating vectors to describe general metabolic flux distributions in a given metabolic network. Let us first consider the normal three-dimensional coordinate space. We are used to sloppily indicate the position of any point in this space by its three coordinates *x*, *y* and z. Is this information complete? No, we also need to indicate to which axis/vectors these coordinates belong. Normally we use the three orthogonal *x*-, *y*- and *z*-axes of Cartesian space without explicitly saying this because it is convention to use these three axes. However, this is only one particular choice and we could use an infinite number of different coordinate systems ("bases"). It is an appropriate choice, though, because these three axes are linearly independent. This means that the scalar product between each of them is equal to zero. The position **r** of a point can then be specified as a unique linear combination:

$$\mathbf{r} = x\mathbf{e}_x + y\mathbf{e}_y + z\mathbf{e}_z.$$

Here, $\mathbf{e}_x$, $\mathbf{e}_y$ and $\mathbf{e}_z$ are unit vectors (vectors of length one) parallel to the Cartesian x-, y- and z-axes. If we could generate a similar *basis* or *coordinate system* for the metabolic flux distributions, this would be extremely useful for understanding its underlying architecture. Using the available information about the multiplicity of feasible flux distributions that can be measured using some experimental technique, we need a robust method that can "re-engineer" the set of generating vectors forming a basis for these flux distributions.

As an example, let us assume that you are standing at night a few meters away from an open door and shine with a torch into a dark room behind this door (Figure 6.6). A certain portion of the room will be illuminated by the torch. How can we characterize this portion of the room?

In the formerly dark room, the border between dark and light is defined by the rays of the torch light that touch the door frame. To describe which points are in the illuminated area, we actually only need four **extreme rays**, i.e. those rays that go through the corners of the door. All points in the volume between these corner rays can be generated by a linear combination of those rays only allowing positive factors.

However, we could also choose a different representation to describe the illuminated area (Figure 6.7). For simplicity, we will only use one or two dimensions here. In the left picture showing a one-dimensional example, all points above the dividing line (the shaded area) fulfill the condition $x \geq 0$. In the middle showing a two-dimensional example, the points in the grey area fulfill the conditions $x_1 \geq 0$ and $x_2 \geq 0$.

However, how could we describe the points in the grey area on the right side in a correspondingly simple manner? Obviously, we could define a new coordinate system (r_1, r_2) meaning that we would construct a new set of generating vectors

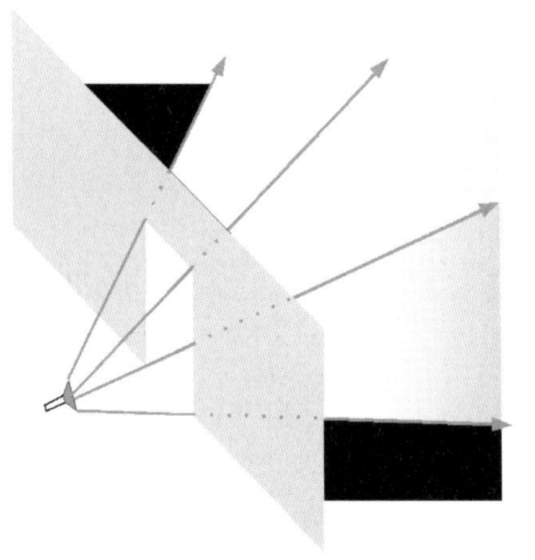

Figure 6.6 A torch light illuminates a dark room through an open door.

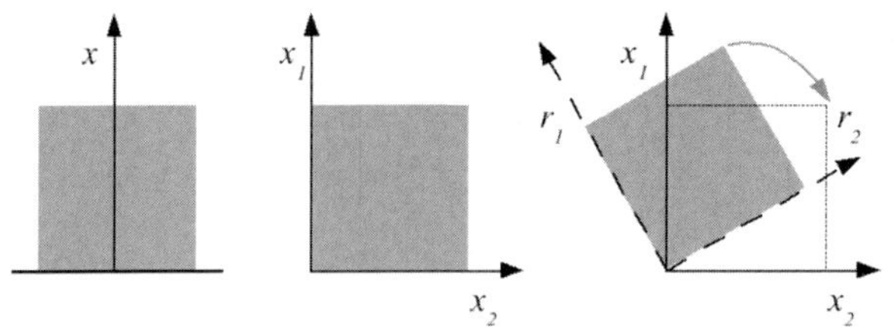

Figure 6.7 (Left) Points belonging to the grey shaded area fulfill the condition $x \geq 0$. (Middle) Points in the grey shaded area fulfill the two conditions $x_1 \geq 0$ and $x_2 \geq 0$. (Right) Here, the grey area is tilted by 30° with respect to the two axes. See text how one may define this area now.

as in Figure 6.6. But we could also try to transform this area back into the grey area of the middle panel. Then, the same description could be used as before using the old axes x_1 and x_2. In this simple example, this transformation can be obviously best performed by multiplying all vectors inside the grey area by a two-dimensional rotation matrix that will rotate them around the axis perpendicular to the x_1x_2 plane about a certain angle α (which is 30° here). Such rotation matrices read:

$$\begin{pmatrix} \cos\alpha & -\sin\alpha \\ \sin\alpha & \cos\alpha \end{pmatrix}.$$

Generalized to n dimensions, an n-dimensional rotation matrix may be applied to n-dimensional vectors to rotate them into the positive quadrant of a coordinate system constructed by a set of orthogonal basis vectors. Therefore, besides using the principles of extreme rays spanning a certain volume, we can also rotate this volume by applying a suitable rotation matrix and characterize it using an existing set of coordinate axes (here, x_1 and x_2).

This duality between a set of generating vectors and the space that is spanned from these vectors is a well-known problem in linear algebra that appears in many areas of applied mathematics. [Here, we will follow the presentation of Gagneur and Klamt (2004).] A pair (**A**,**R**) of real matrices **A** and **R** is said to be a **double description (DD) pair** if the relationship:

$$\mathbf{A}\,\mathbf{x} \geq 0 \text{ if and only if } \mathbf{x} = \mathbf{R}\,\lambda \text{ for some } \lambda \geq 0,$$

holds. This definition corresponds exactly to the examples discussed above. The vectors **x** in the feasible solution space (they can be constructed as linear combinations from the row vectors of **R**, see right side) are those that can be rotated by **A** into the positive n-dimensional quadrant (middle side):

$$\mathbf{R}\lambda = \begin{pmatrix} r_{11} & r_{12} & \ldots & r_{1n} \\ r_{21} & r_{22} & \ldots & r_{2n} \\ \ldots & \ldots & \ldots & \ldots \\ r_{m1} & r_{m2} & \ldots & r_{mn} \end{pmatrix} \begin{pmatrix} \lambda_1 \\ \lambda_2 \\ \ldots \\ \lambda_n \end{pmatrix} = \begin{pmatrix} r_{11}\lambda_1 + r_{12}\lambda_2 + \ldots + r_{1n}\lambda_n \\ r_{21}\lambda_1 + r_{22}\lambda_2 + \ldots + r_{2n}\lambda_n \\ \ldots \\ r_{m1}\lambda_1 + r_{m2}\lambda_2 + \ldots + r_{mn}\lambda_n \end{pmatrix}.$$

In Section 6.6, the columns of **R** will contain the set of generating vectors, and the elements of the vector $\mathbf{R} \cdot \lambda$ contain the λ-multiples of the flux vectors (see

Section 6.3), which are the feasible flux distributions. For a pair (**A**,**R**) to be a DD pair, the column size of **A** has to equal the row size of **R**, say d. For such a pair, the set $P(\mathbf{A})$ represented by **A** as $P(\mathbf{A}) = \{\mathbf{x} \in \Re^d : \mathbf{Ax} \geq 0\}$ is simultaneously represented by **R** as a linear combination of extreme rays.

A subset P of $\Re^d$ is called a **polyhedral cone** if $P = P(\mathbf{A})$ for some matrix **A** and **A** is called a **representation matrix** of the polyhedral cone $P(\mathbf{A})$. Then, we say **R** is a **generating matrix** for P. Each column vector of a generating matrix **R** lies in the cone P and every vector in P is a nonnegative combination of some columns of **R**.

The famous mathematician Hermann Minkowski proved the following theorem for polyhedral cones. For any $m \times n$ real matrix **A**, there exists some $d \times m$ real matrix **R** such that (**A**,**R**) is a DD pair, or in other words, the cone $P(\mathbf{A})$ is generated by **R**. The theorem states that every polyhedral cone admits a generating matrix. The non-triviality comes from the fact that the row size of **R** is finite. If we allow an infinite size, there is a trivial generating matrix consisting of all vectors in the cone. Also the converse is true as was shown by another famous mathematician, Hermann Weyl, for polyhedral cones. For any $d \times n$ real matrix **R**, there exists some $m \times d$ real matrix **A** such that (**A**,**R**) is a DD pair, or in other words, the set generated by **R** is the cone $P(\mathbf{A})$.

The **task** in a practical case is how to construct a matrix **R** from a given matrix **A**, and the converse? These two problems are computationally equivalent. Farkas' lemma shows that (**A**,**R**) is a DD pair if and only if ($\mathbf{R}^T$,$\mathbf{A}^T$) is a DD pair. An important modification of the problem is to require the minimality of **R**: find a matrix **R** such that no proper submatrix is generating $P(\mathbf{A})$. As is intuitively clear from Fig. 6.4, a minimal set of generators is unique up to positive scaling when we assume the regularity condition that the cone is **pointed**, i.e. the origin is an extreme point of $P(\mathbf{A})$. Geometrically, the columns of a minimal generating matrix are in one-to-one correspondence with the **extreme rays** of **P**. Thus the problem is also known as the **extreme ray enumeration problem**. No efficient (polynomial) algorithm is known so far for the general problem.

Suppose that the $m \times d$ matrix **A** is given and let $P(\mathbf{A}) = \{\mathbf{x}{:}\mathbf{Ax} \geq 0\}$. The DD method is an incremental algorithm to construct a $d \times m$ matrix **R** such that (**A**,**R**) is a DD pair. Let us assume for simplicity that the cone $P(\mathbf{A})$ is pointed. Let **K** be a subset of the row indices $\{1,2, \ldots, m\}$ of **A** and let $\mathbf{A_K}$ denote the submatrix of **A** consisting of rows indexed by **K**. Suppose we already found a generating matrix **R** for $\mathbf{A_K}$, or equivalently, ($\mathbf{A_K}$,**R**) is a DD pair. If $\mathbf{A} = \mathbf{A_K}$, we are done. Otherwise we select any row index i not in **K** and try to construct a DD pair ($\mathbf{A}_{\mathbf{K}+i}$,$\mathbf{R}'$) using the information of the DD pair ($\mathbf{A_K}$,**R**). Once we have found a way to iteratively construct larger and larger DD pairs, we have an algorithm to construct a generating matrix **R** for $P(\mathbf{A})$.

Let us use Figure 6.8 to geometrically understand the task and see how a possible solution may look like. Having a generating matrix **R** means that all extreme rays (i.e. extreme points of the cut-section) of the cone are represented by columns of **R**. Let the cube represent a cone of solutions satisfying $\mathbf{A_K x} \geq 0$ and let us assume that the cone is pointed and thus C is bounded. The cut-section plane connecting the points $bcij$ represents a new cut-section C' of the cone $P(\mathbf{A_K})$ with the hyperplane h in $\Re^d$ that represents the new condition $\mathbf{A}_i\mathbf{x} \geq 0$. It intersects every extreme ray of $P(\mathbf{A_K})$ at a single point.

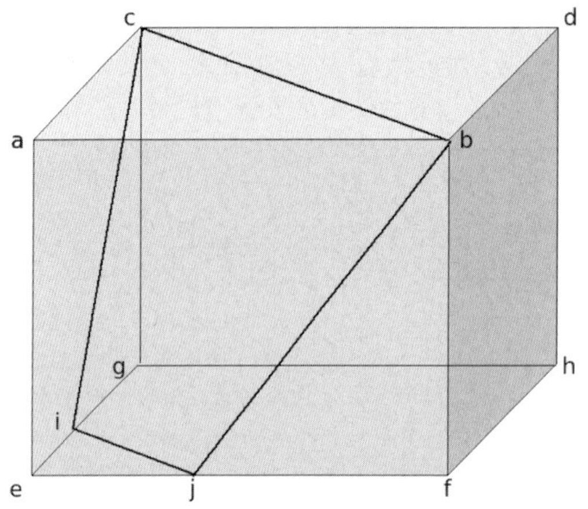

Figure 6.8 Cube *abcdefgh* represents a cone of solutions satisfying all inequality constraints up to iteration *K*. Adding a new constraint leads to cutting away a certain part of this solution space as represented by the hyperplane *acij*.

The newly introduced inequality $\mathbf{A}_i \cdot \mathbf{x} \geq 0$ partitions the space $\Re^d$ into three parts:

$$H_i^+ = \{\mathbf{x} \in \Re^d : \mathbf{A}_i \cdot \mathbf{x} > 0\}$$
$$H_i^0 = \{\mathbf{x} \in \Re^d : \mathbf{A}_i \cdot \mathbf{x} = 0\}$$
$$H_i^- = \{\mathbf{x} \in \Re^d : \mathbf{A}_i \cdot \mathbf{x} < 0\}.$$

The intersection of H_i^0 with P and the new extreme points i and j in the cut-section C are shown in Figure 6.8. Let J be the set of column indices of $\mathbf{R}$. The rays $\mathbf{r}_j$ $(j \in J)$ are then partitioned into three parts accordingly:

$$J^+ = \{j \in J : \mathbf{r}_j \in H_i^+\}$$
$$J^0 = \{j \in J : \mathbf{r}_j \in H_i^0\}$$
$$J^- = \{j \in J : \mathbf{r}_j \in H_i^-\}.$$

We call the rays indexed by J^+, J^0 and J^- the **positive**, **zero** and **negative** rays with respect to i, respectively. To construct a matrix $\mathbf{R}'$ from $\mathbf{R}$, we generate new $|J^+| \times |J^-|$ rays lying on the ith hyperplane H_i^0 by taking an appropriate positive combination of each positive ray $\mathbf{r}_j$ and each negative ray $\mathbf{r}_{j'}$ and by discarding all negative rays. The following lemma ensures that we have a DD pair $(\mathbf{A}_{\mathbf{K}+i}, \mathbf{R}')$, and provides the key procedure for the most primitive version of the DD method.

Lemma Let $(\mathbf{A}_\mathbf{K}, \mathbf{R})$ be a DD pair and let i be a row index of $\mathbf{A}$ not in $\mathbf{K}$. Then the pair $(\mathbf{A}_{\mathbf{K}+i}, \mathbf{R}')$ is a DD pair, where $\mathbf{R}'$ is the $d \times |J'|$ matrix with column vectors $\mathbf{r}_j$ $(j \in J')$ defined by:

$$J' = J^+ \cup J^0 \cup (J^+ \times J^-),$$

and:

$$\mathbf{r}_{jj'} = (\mathbf{A}_i \cdot \mathbf{r}_j) \cdot \mathbf{r}_{j'} - (\mathbf{A}_i \cdot \mathbf{r}_{j'}) \cdot \mathbf{r}_j \text{ for each } (j, j') \in J^+ \cdot J^-.$$

It is quite simple to find a DD pair ($\mathbf{A}_K$,$\mathbf{R}$) when $|\mathbf{K}| = 1$, which can serve as the initial DD pair. The strategy outlined here is very primitive, and the straightforward implementation will be quite useless, because the size of J increases very fast and goes beyond any tractable limit. This is because many vectors $\mathbf{r}_{jj'}$ generated by the algorithm defined in the Lemma are unnecessary. We need to avoid generating redundant vectors. This can be done, for example, by checking for each pair of extreme rays $\mathbf{r}$ and $\mathbf{r}'$ of $P(\mathbf{A}_K)$ with $\mathbf{A}_i\mathbf{r} > 0$ and $\mathbf{A}_i\mathbf{r}' < 0$ whether they are adjacent in $P(\mathbf{A}_K)$, which means that the minimal face of P containing both contains no other extreme rays. We will omit the further mathematical and practical details of the DD method and refer the interested reader to the according specialized literature.

6.6 Extreme Pathways and Elementary Modes

Metabolic pathway analysis searches for meaningful structural and functional units in metabolic networks. Today's most powerful methods are based on the concepts from convex analysis that were introduced near the end of Section 6.5. Two such approaches are the **elementary flux modes** (Schuster and Hilgetag, 1994) and **extreme pathways** (Schilling *et al.*, 2000). Both methods span the space of feasible steady-state flux distributions by nondecomposable routes, i.e. no subset of reactions involved in an elementary flux mode or extreme pathway can hold the network balanced using nontrivial fluxes. Both algorithms follow the principles of DD introduced in the previous section algorithm to generate from the set of observed fluxes ($\mathbf{S} \cdot \mathbf{v} \geq 0$) a set of extremal rays/generating vectors of a convex polyhedral cone. As both methods employ quite similar algorithms, we will introduce here only the extreme pathway method in order not to confuse the reader. Metabolic pathway analysis has been used, for example, to study routing and flexibility/redundancy of networks, functionality of networks, and the identification of futile cycles. It gives all (sub)optimal pathways with respect to product/biomass yield and it can be useful for calculability studies in metabolic flux analysis.

Seen from a practical point of view, we may say that the algorithms construct balanced combinations of multiple reaction fluxes that do not change the concentrations of metabolites when flux flows through them (input fluxes are channeled to products not to accumulation of intermediates). As the stochiometric matrix describes the coupling of each reaction i to the concentration of metabolites j, we need to construct combinations of reactions that leave their concentrations unchanged. When such combined pathways applied to metabolites do not change their concentrations, we will call the combined pathways "balanced". In algorithmic terms, we need to transform the stoichiometric matrix so that the matrix entries are brought to zero.

Looking back at the previous section, we will start with the system:

$$\begin{pmatrix} \mathbf{S} \\ \text{external fluxes} \end{pmatrix} \cdot \mathbf{v} \quad \begin{matrix} = 0 \\ \geq 0 \end{matrix} .$$

as in $\mathbf{A} \cdot \mathbf{x} \geq 0$ and derive the matrix $\mathbf{R}$ containing the basis vectors, the extreme rays.

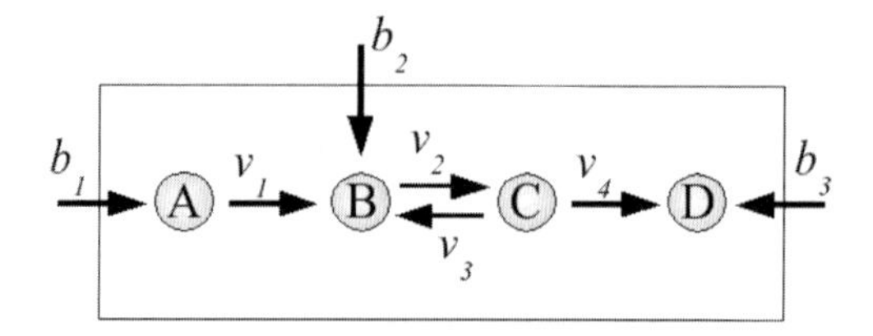

	v_1	v_2	v_3	v_4	b_1	b_2	b_3	
	-1				1			A
	1	-1	1			1		B
		1	-1	-1				C
				1			1	D

Figure 6.9 Simple metabolic network with four metabolites A–D connected by four internal flux reactions ν_1–ν_4 and three unconstrained external fluxes b_1–b_3 that connect the system with the environment. The lower part of the figure shows the stoichiometric matrix for this system.

- Steps of the extreme pathway algorithm
- Let us start with the very simple system shown in Figure 6.9.
- **(Step 0)** Construct the stoichiometric matrix of the metabolic system and take its transpose $\mathbf{S}^T$. Then attach an $n \times n$ identity matrix from the left side that will serve for bookkeeping purposes (Figure 6.10). Reconfigure the network if needed, which means splitting up all reversible internal reactions into two separate, irreversible reactions (forward and backward reaction).
- **(Step 1)** Identify all metabolites that do not have an unconstrained external flux associated with them. The total number of such metabolites is denoted by μ. The example network contains only one such metabolite C ($\mu = 1$). (The concentrations of these internal metabolites need to be balanced first. For the balancing of the other metabolites, we can later use the external fluxes as well.) Examine the constraints on each of the exchange fluxes as given by :
  - $\alpha_j \leq b_j \leq \beta_j$.
- There may be three cases:
- (a) If the exchange flux is constrained to be positive do nothing.
- (b) If the exchange flux is constrained to be negative multiply the corresponding row of the initial matrix by –1 (i.e. change the direction of flow).
- (c) If the exchange flux is unconstrained, move the entire row to a temporary matrix $\mathbf{T}^{(E)}$.
- Remove $\mathbf{T}^{(E)}$ containing all rows representing the unconstrained external fluxes for the moment (they will be added back later). Here, we are talking about the rows

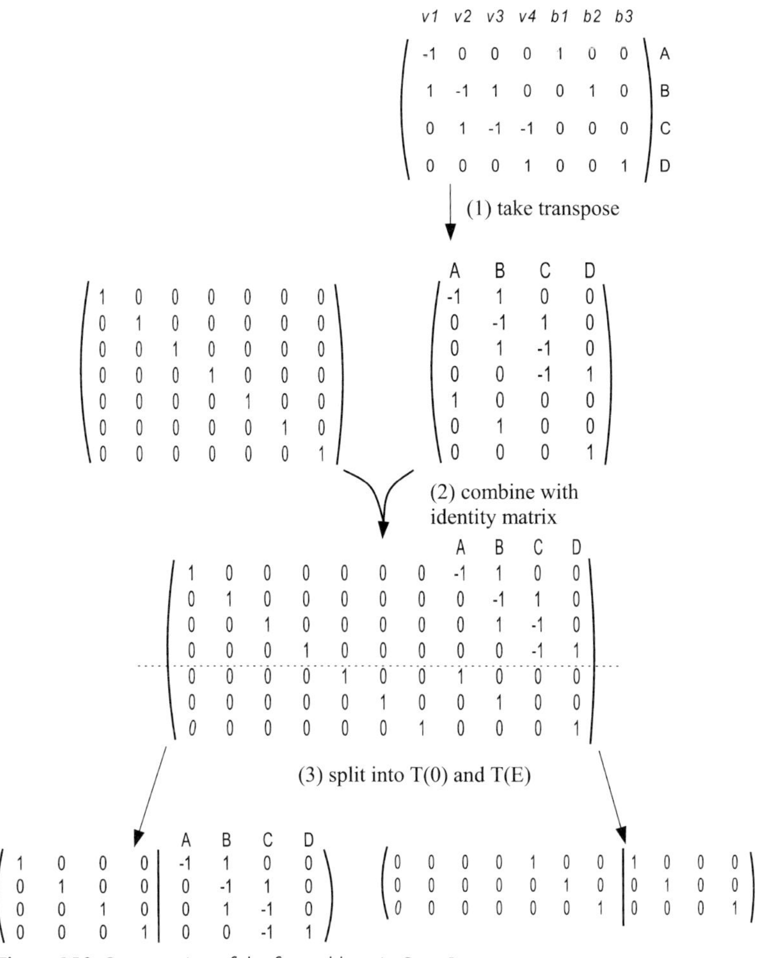

Figure 6.10 Construction of the first tableau in Step 1.

b_1–b_3. This completes the first tableau $\mathbf{T}^{(0)}$. Each element of this matrix will be designated T_{ij}. $\mathbf{T}^{(0)}$ for the example reaction system is shown in Figure. 6.10.

- Starting with $x = 1$ and $\mathrm{T}^{(0)} = \mathrm{T}^{(x-1)}$ the next tableau is generated in the following way:
- **(Step 2)** Begin forming the new matrix $\mathbf{T}^{(x)}$ by copying all rows from $\mathbf{T}^{(x-1)}$ which contain a zero in the column of $\mathbf{S}^{\mathrm{T}}$ that corresponds to the first metabolite identified in step 1, denoted by index c. (Here the third column of $\mathbf{S}^{\mathrm{T}}$.)
- **(Step 3)** Of the remaining rows in $\mathbf{T}^{(x-1)}$ add together all possible combinations of rows which contain values of the opposite sign in column C, such that the addition produces a zero in this column (Figure 6.11).
- **(Step 4)** For all of the rows added to $\mathbf{T}^{(x)}$ in Steps 2 and 3 check to make sure that no row exists that is a nonnegative combination of any other sets of rows in $\mathbf{T}^{(x)}$. This step is equivalent to removing superfluous rays during the DD method.

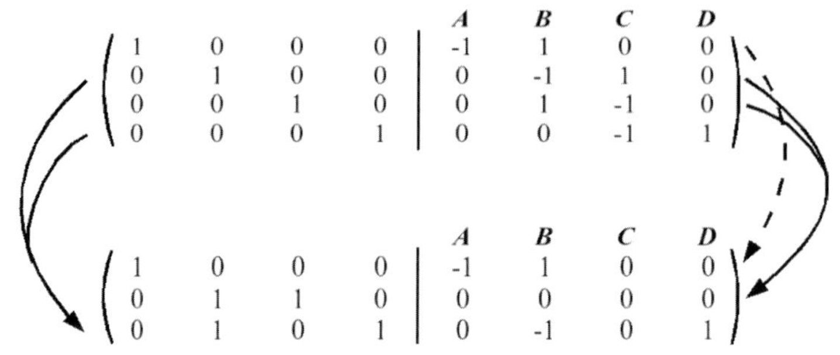

Figure 6.11 By going from the upper to the lower tableau, rows are being transferred if they already contain zero in the C column (dashed line – corresponding to Step 2) or appropriate combinations are formed (Step 3) to balance metabolite C. Here, row 2 can be added to row 3 (solid lines on the right) or to row 4 (solid lines on the left) to obtain zero in column C.

- One method that may be used for this check works as follows. Let $A(i)$ bet the set of column indices j for which the elements of row i equal zero. Here, $A(1) = \{2,3,4,7,8\}$, $A(2) = \{1,4,5,6,7,8\}$ and $A(3) = \{1,3,5,7\}$. Then check to determine if there exists another row (h) for which $A(i)$ is a subset of $A(h)$. Here, A(2) does not include columns 2 and 3, and A(3) does not include columns 2 and 4. Therefore, A(1) is not a subset of neither A(2) nor A(3). Moreover, it is easy to verify that neither A(2) nor A(3) is a subset of either one of the other two sets. Therefore, no row must be eliminated in this example.

- **(Step 5)** Complete Steps 2–4 for all of the metabolites that do not have an unconstrained exchange flux operating on the metabolite, incrementing x by one up to μ. (In the example here $\mu = 1$.) The final tableau will be $T^{(\mu)}$. Note that the number of rows in $T^{(\mu)}$ will be equal to k, the number of extreme pathways.

- **(Step 6)** Next we append $\mathbf{T}^{(E)}$ to the bottom of $\mathbf{T}^{(\mu)}$. This results in a new tableau (Figure 6.12).

- **(Step 7)** Starting in the $n+1$ column (or the first non-zero column on the right side), if $T_{i,(n+1)} \neq 0$ then add the corresponding nonzero row from $\mathbf{T}^{(E)}$ to row i so as to produce 0 in the $n+1$th column. In order to use the exchange flux in the appropriate direction, add the appropriate multiples of this nonzero row to row i if $T_{i,(n+1)} < 0$ or subtract them if $T_{i,(n+1)} > 0$.

- Repeat this procedure for each of the rows in the upper portion of the tableau so as to create zeros in the entire upper portion of the $(n+1)$ column. When finished, remove the row in $\mathbf{T}^{(E)}$ corresponding to the exchange flux for the metabolite just balanced.

- **(Step 8)** Follow the same procedure as in Step 7 for each of the columns on the right side of the tableau containing nonzero entries. (In this example we need to perform Step 7 for every column except the third column of the right side which corresponds to metabolite C.)

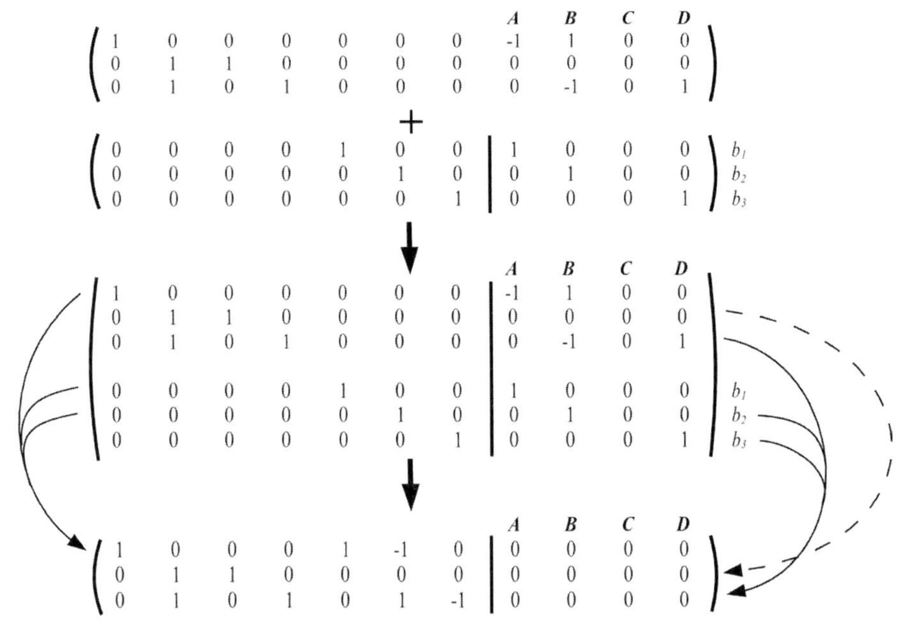

Figure 6.12 In the upper part of the picture, the tableau $\mathbf{T}^{(E)}$ with the exchange fluxes is combined with the previously formed tableau $\mathbf{T}^{(\mu)}$. By going from the combined to the bottom tableau, all rows are being transferred that already contain zero in all metabolite columns (dashed line) or appropriate combinations are formed (continuous lines left and right, see Step 3). The bottom tableau is the final tableau after Step 8.

- The final tableau $\mathbf{T}^{(\text{final})}$ will contain on the left side the transpose of the matrix **P** containing the extreme pathways in place of the original identity matrix.
- Note that this algorithm perfectly agrees with the principles of the DD method in that we generated from the set of observed fluxes (stoichiometric matrix $\mathbf{S} \equiv \mathbf{A}$) a set of generating vectors ($\mathbf{P} \equiv \mathbf{R}$). Figure 6.13 shows the pathways constructed for this simple network.

6.6.1 Analysis of Extreme Pathways

How do reactions appear in the pathway matrix? In the matrix **P** of extreme pathways, each column is an extreme pathway and each row corresponds to a reaction in the network. The numerical value of the *i,j*th element corresponds to the relative flux level through the *i*th reaction in the *j*th extreme pathway. Let us consider again the example network introduced at the beginning of this chapter (Figure 6.14).

The lengths of P_1, P_2 and P_3 are 6, 6 and 7, respectively, as they are composed from six, six or seven reactions. These values are also contained as diagonal elements of the

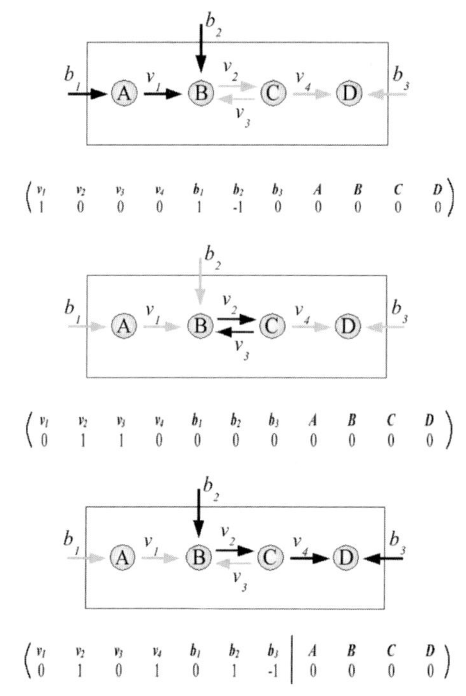

Figure 6.13 Three extreme pathways found that span the solution space of this network. Note that pathways P1 and P3 use the external fluxes b_2 and b_3 in the opposite direction. The cyclic pathway P2 formed by the opposing fluxes ν_2 and ν_3 has no net overall effect on the functional capabilities of the network. Such pathways are usually deleted at this point. You may wonder why the straight path $b_1 \to \nu_1 \to \nu_2 \to \nu_4 \to b_3$ is not among the solutions. It is indeed not an independent solution because it can be generated as a linear combination from the other three simply by adding P1 and P3.

symmetric pathway length matrix $\mathbf{P}_{\mathrm{LM}}$ that is calculated from the normalized pathway matrix $\tilde{\boldsymbol{P}}$ where all nonzero entries are replaced by 1:

$$\mathbf{P}_{\mathrm{LM}} = \tilde{\boldsymbol{P}}^T \cdot \tilde{\boldsymbol{P}}$$

$$\boldsymbol{P} = \begin{array}{c} P_1 \; P_2 \; P_3 \\ \left\{\begin{matrix} 2 & 2 & 2 \\ 1 & 0 & 1 \\ 0 & 1 & 0 \\ 0 & 1 & 1 \\ 0 & 0 & 1 \\ 1 & 0 & 0 \\ 2 & 2 & 2 \\ 1 & 1 & 1 \\ 1 & 1 & 1 \end{matrix}\right\} \end{array} \begin{matrix} \mathrm{v}_1 \\ \mathrm{v}_2 \\ \mathrm{v}_3 \\ \mathrm{v}_4 \\ \mathrm{v}_5 \\ \mathrm{v}_6 \\ b_1 \\ b_2 \\ b_3 \end{matrix} \quad \tilde{\boldsymbol{P}} = \begin{array}{c} P_1 \; P_2 \; P_3 \\ \left\{\begin{matrix} 1 & 1 & 1 \\ 1 & 0 & 1 \\ 0 & 1 & 0 \\ 0 & 1 & 1 \\ 0 & 0 & 1 \\ 1 & 0 & 0 \\ 1 & 1 & 1 \\ 1 & 1 & 1 \\ 1 & 1 & 1 \end{matrix}\right\} \end{array} \begin{matrix} \mathrm{v}_1 \\ \mathrm{v}_2 \\ \mathrm{v}_3 \\ \mathrm{v}_4 \\ \mathrm{v}_5 \\ \mathrm{v}_6 \\ b_1 \\ b_2 \\ b_3 \end{matrix} \quad \tilde{\boldsymbol{P}}^T \cdot \tilde{\boldsymbol{P}} = \begin{array}{c} P_1 \; P_2 \; P_3 \\ \left\{\begin{matrix} 6 & 4 & 5 \\ & 6 & 5 \\ & & 7 \end{matrix}\right\} \end{array} \begin{matrix} P_1 \\ P_2 \\ P_3 \end{matrix} .$$

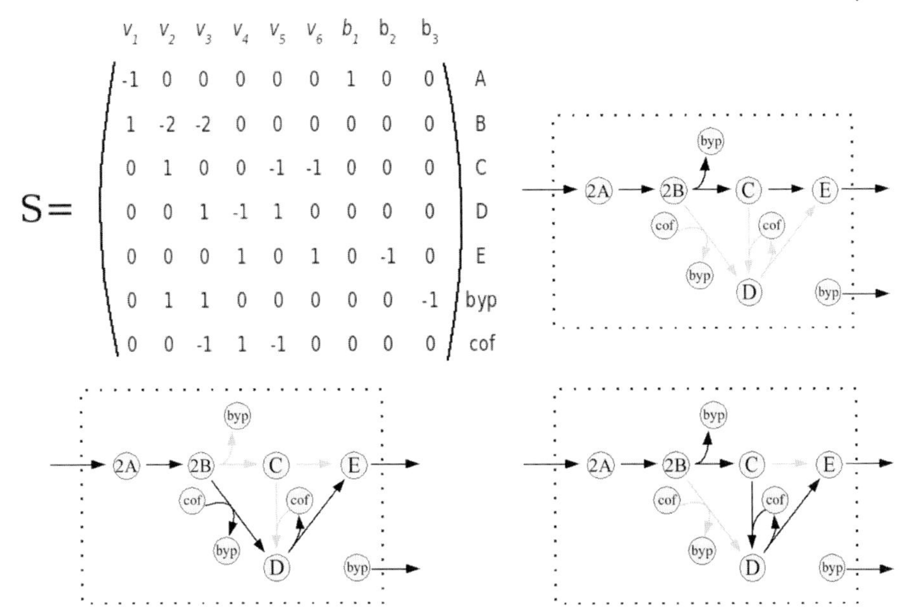

Figure 6.14 The same network as in Figure 6.3. (Top left) This is the stoichiometric matrix for the network shown in Figure 6.3. (Top right, bottom) Carrying out the same algorithm as before gives these three extreme pathways.

The off-diagonal terms of $\mathbf{P_{LM}}$ are the number of reactions that a pair of extreme pathways have in common. For example, the extreme pathways P_2 and P_3 have five reactions in common: v_1, v_4, and b_1, b_2 and b_3.

One can also compute a reaction participation matrix $\mathbf{P_{PM}}$ from $\mathbf{P}$:

$$\mathbf{P_{PM}} = \tilde{\boldsymbol{P}} \cdot \tilde{\boldsymbol{P}}^T.$$

where the diagonal correspond to the number of pathways in which the given reaction participates.

6.6.2 Elementary Flux Modes

The elementary flux mode technique was developed prior to that of the extreme pathway technique by Stephan Schuster, Reinhart Heinrich and coworkers. The method is very similar to the extreme pathway method to construct a basis for metabolic flux states based on methods from convex algebra. Extreme pathways are a subset of elementary modes and for many systems both methods coincide.

A pathway P(**v**) is an elementary flux mode **e** if it fulfills conditions (C1)–(C3).

- (C1) *Pseudo steady-state*: $\mathbf{S} \cdot \mathbf{v} = 0$. This ensures that none of the metabolites is consumed or produced in the overall stoichiometry.
- (C2) *Feasibility*: rate $e_i \geq 0$ if reaction i is irreversible. This demands that only thermodynamically realizable fluxes are contained in **e**.

- (C3) *Nondecomposability:* there is no vector **v** (unequal to the zero vector and to **e**) fulfilling (C1) and (C2), and that $P(\mathbf{v})$ is a proper subset of $P(\mathbf{e})$. This is the core characteristics for elementary flux modes and extreme pathways, and ensures the decomposition of the network into smallest units (that are still able to hold the network in steady state).

  The pathway $P(e)$ is furthermore an extreme pathway if it fulfills conditions (C1)–(C3) AND conditions (C4) and (C5).

- (C4) *Network reconfiguration.* Each reaction must be classified either as exchange flux or as internal reaction. All reversible internal reactions must be split up into two separate, irreversible reactions (forward and backward reaction).

- (C5) *Systemic independence.* The set of extreme pathways in a network is the **minimal** set of elementary flux modes that can describe all feasible steady-state flux distributions.

Elementary flux modes and extreme pathways are robust methods that offer great opportunities for studying functional and structural properties of complex metabolic networks such as enzyme subsets, essential reactions and optimal as well as redundant realizations of stoichiometric conversions. One practical problem is that the number of elementary flux modes or extreme pathways can easily reach into the thousands or even millions for medium-sized metabolic networks containing a few hundred reactions and metabolites. Some methods have been suggested, such as singular value decomposition, to select the most important of these vectors. More work is required in the future to figure out the most useful ways of analyzing millions of generating vectors. One way to go may be to subdivide a large metabolic system into modular components. Another significant reduction may come from combining metabolic models with models of cellular regulation which may significantly reduce the complexity of the solution space.

6.7 Minimal Cut Sets

Extreme pathways and elementary flux analysis address particular functional states in a metabolic network. For a direct characterization of structural failure modes, Steffen Klamt and Ernst-Dieter Gilles introduced the concept of **minimal cut sets**. They described the minimal cut set as the smallest "failure modes" in the network that render the correct functioning of a cellular reaction impossible and used the following example network for illustration (Figure 6.15).

The analysis of this network will focus on the synthesis of product *P*. To understand how the topology of the network affects the synthesis of *P*, we will use its computed elementary flux modes (Figure 6.16). Similarly, one could use its extreme pathways.

The production of *P* may be prevented by removing or inactivating one or more reactions. In mathematical graph theory, a cut set is defined as a set of vertices (or edges) whose removal increases the number of connected components of this graph. Here, Klamt and Gilles defined a cut set (with respect to a target reaction or

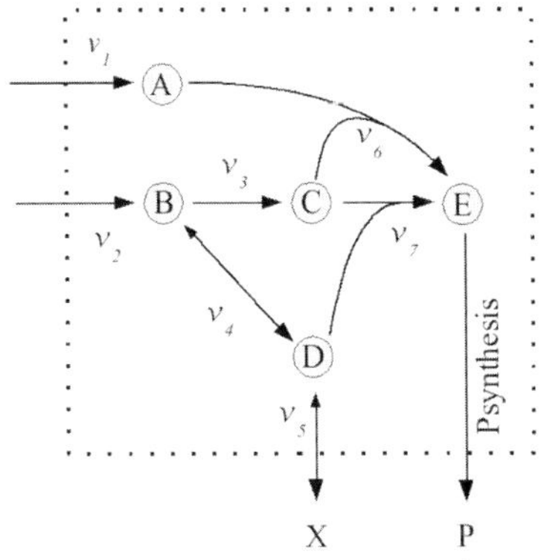

Figure 6.15 Example network with five internal metabolites and eight reactions. Two of its reactions, v_4 and v_5, are reversible. Drawn after Klamt (2006). Reactions crossing the system boundary are connected to buffered (external) substances.

product) as a set of reactions whose removal from the network will prevent any feasible steady state flux distribution involving the target.

Definition We call a set of reactions a **cut set** (with respect to a defined objective reaction) if after the removal of these reactions from the network no feasible balanced flux distribution involves the objective reaction.

For example, $C_1 = \{v_1, v_2, v_5\}$ is a cut set with respect to *Psynthesis*. However, inhibiting the three reactions of C_1 (by a gene knockout or by targeting the respective enzymes by small molecule inhibitors) would not be an efficient strategy since a subset of it, $C_2 = \{v_2, v_5\}$, is already a cut set that leads to the failure of the objective reaction *Psynthesis*. C_2 is optimal in the sense that no subset of C_2 is a cut set itself. Therefore, C_2 is called a minimal cut set.

Definition A **minimal cut set** must be a cut set itself and it is minimal in the sense that none of its elements can be excluded without losing the cut set property.

Figure 6.17 displays all eight minimal cut sets preventing synthesis of P in the model network.

Removing a minimal cut set always guarantees interruption of the objective function as long as the assumed network structure is correct. However, additional regulatory circuits or capacity restrictions may allow that even a proper subset of a minimal cut set may function as a cut set. This means that in a real cell, fewer gene deletions or small-molecule inhibitors may be required than found by the minimal cut set analysis. On the other hand, after removing a complete minimal cut set from the network, other pathways producing other metabolites may still be active.

A systematic computation of minimal cut sets must ensure (1) that the calculated minimal cut sets are cut sets (interrupting all possible balanced flux distributions

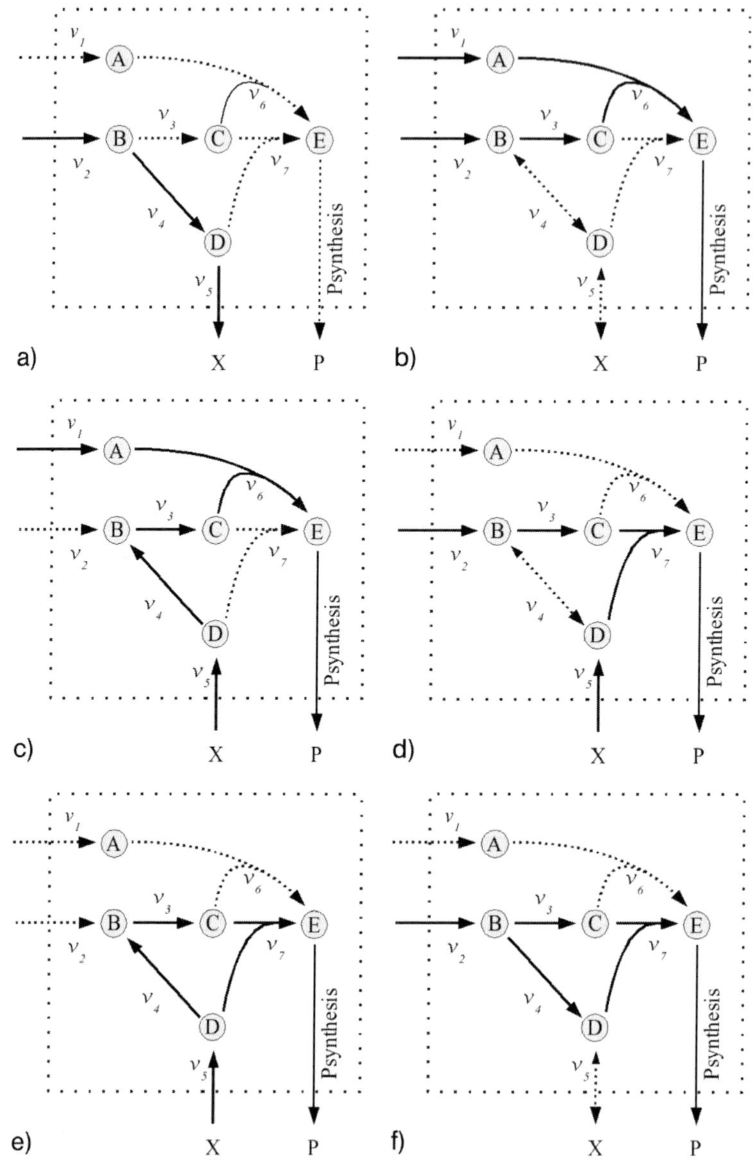

Figure 6.16 The example network contains six elementary flux modes. The latter five are coupled to product synthesis. Drawn after Klamt (2006).

involving the objective reaction), (2) that the minimal cut sets are really minimal and (3) that all minimal cut sets are found. When designing an algorithm to compute minimal cut sets, one may exploit the fact that any feasible steady-state flux distribution in a given network – expressed as vector **r** of the q net reaction rates – can be expressed

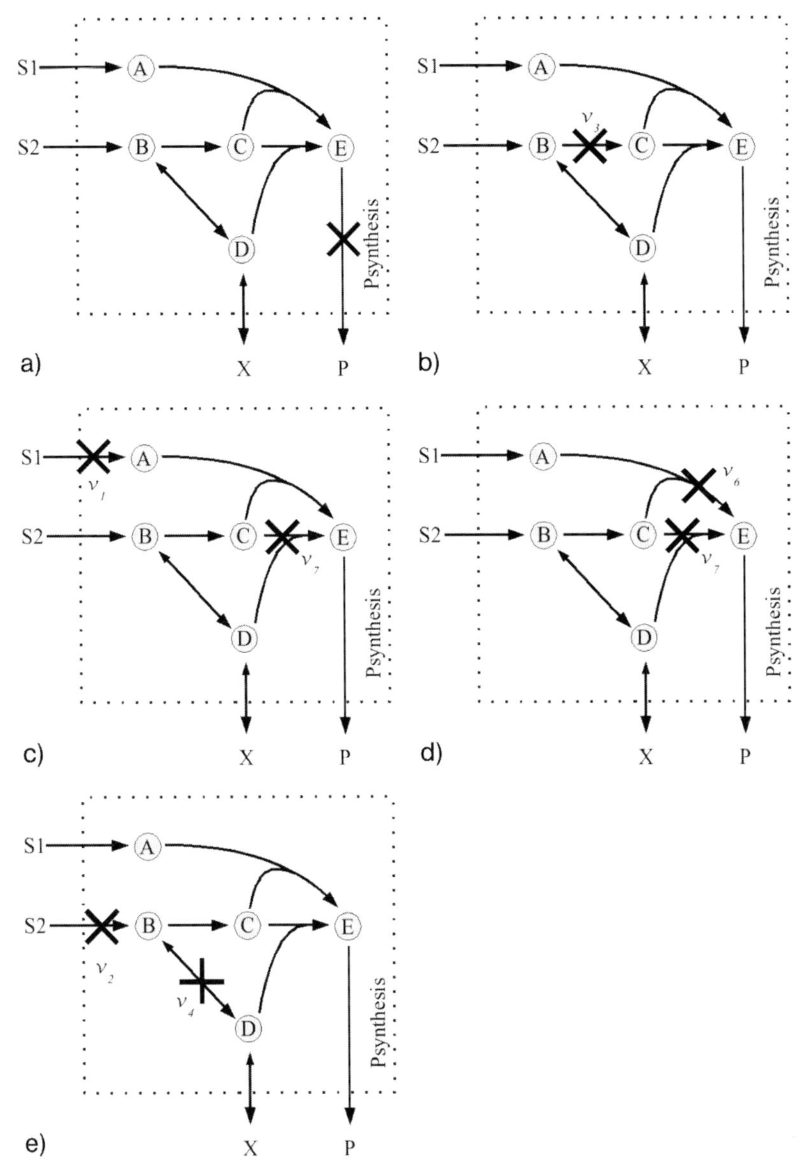

Figure 6.17 Minimal cut sets for repressing synthesis of P in the example network of Fig. 6.15. a) MCS 1, b) MCS 2, c) MCS 3, d) MCS 4, e) MCS 5, f) MCS 6, g) MCS 7, h) MCS 8. Drawn after Klamt (2006).

by a nonnegative linear combination of the N elementary modes:

$$\mathbf{r} = \sum_{i=1}^{N} \alpha_i EM_i, \quad \alpha_i \geq 0.$$

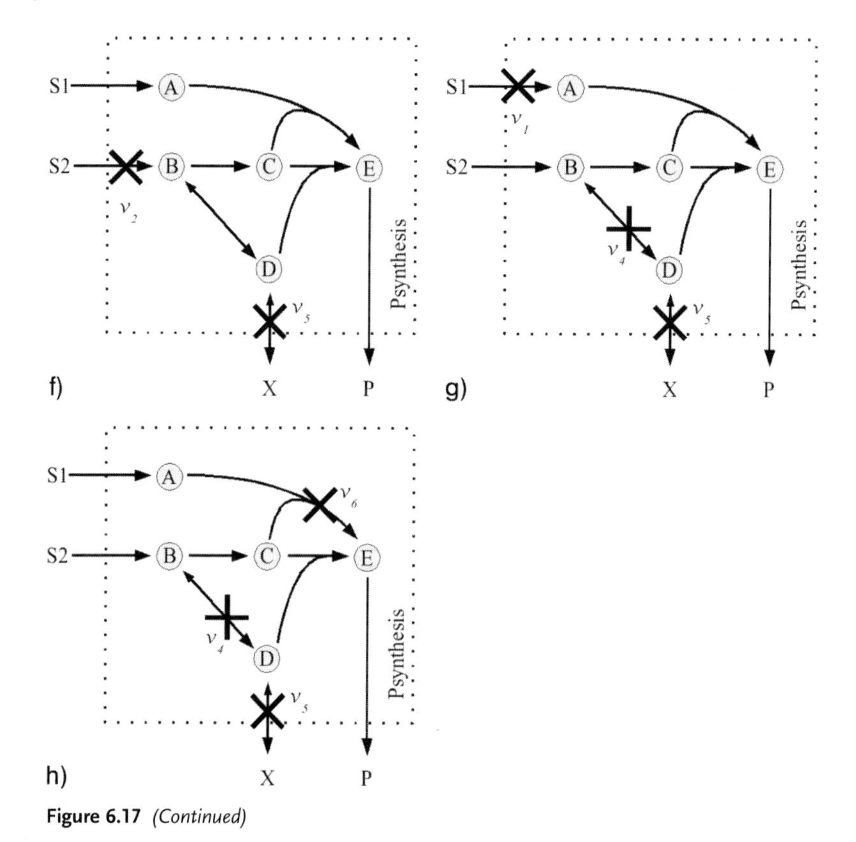

Figure 6.17 *(Continued)*

To ensure that the rate r_k of the objective reaction is 0 in all $\mathbf{r}$, each elementary mode must contain 0 at the kth place. If C is a proper cut set, for each elementary mode involving the objective reaction (with a nonzero value), there must be at least one reaction in C also involved in this elementary mode. This guarantees that all elementary modes in which the objective reaction participates will vanish when the reactions in the cut set are removed from the network. For a detailed discussion of algorithms to compute minimal cut set, we refer the reader to the original papers (Klamt and Gilles, 2004 and Klamt, 2006).

6.7.1 Applications of Minimal Cut Sets

1. **Target identification and repression of cellular functions.** Looking at all minimal cut sets facilitates identifying the best-suited manipulation of a cellular system for a desired target operation. For practical reasons, a small number of interventions is desirable (small size of minimal cut set) and if the cells are

to be kept viable it is preferable to choose minimal cut sets that affect other pathways in the network only weakly. Certainly, there may also be practical considerations for selecting a particular minimal cut set as some of the cellular functions might be difficult to shut down genetically or by inhibition, or when, for example, several other isozymes exist that catalyze a reaction that is part of a minimal cut set.

2. **Network verification and mutant phenotype predictions.** When targeting cellular reactions/processes that are essential for cell survival, shutting down a corresponding minimal cut set should definitely lead to cell death. Such predictions, derived purely from network structure, may be used to verify hypothetical or reconstructed networks. If the cell can survive the deletion of a set of genes in an experiment that were predicted to be fatal, the underlying network structure must be incomplete (a false-negative prediction). One may, however, as often in biology also face the opposite where a deletion was predicted to be still viable and the cells die in the experiment. This may be a hint that, for example, additional gene regulatory effects need to be taken into account.

3. **Structural fragility and robustness.** The concept of minimal cut sets is also known in completely different research fields such as risk analysis of industrial plants. There, a minimal cut set is also called failure mode and characterizes the minimal set of events that will together cause a breakdown of the system. In this regard, one desires to characterize the importance of individual reactions to lead to failure of the system. Intuitively, the most *fragile* reactions will be those whose removal directly leads to system failure (or shut down of the respective target). The next dangerous ones are those that, together with only one other reaction, lead to system failure. In this analysis, we will for the moment assume that each reaction in a metabolic network has the same probability to fail. Therefore, small minimal cut sets are most probable to be responsible for a failing objective function. We will define a **fragility coefficient F_i** as the reciprocal of the average size of all minimal cut sets in which reaction i participates:

$$F_i = \frac{1}{\text{avg}(\{|\text{MCS}_j| : i \in \text{MCS}_j\})}.$$

Figure 6.18 shows the fragility coefficients of the reactions in the example network of Figure 6.15 with respect to the production of P.

Apparently, reaction v_3 and *Psynthesis* itself are the most *fragile* reactions of the network with respect to synthesis of product P.

We conclude this chapter by summarizing that minimal cut sets are an irreducible combination of network elements whose simultaneous inactivation leads to a guaranteed dysfunction of certain cellular reactions or processes. Minimal cut sets are inherent and uniquely determined structural features of metabolic networks similar to elementary modes (or extreme pathways). Unfortunately, the computation of minimal cut sets and elementary modes becomes challenging in large networks.

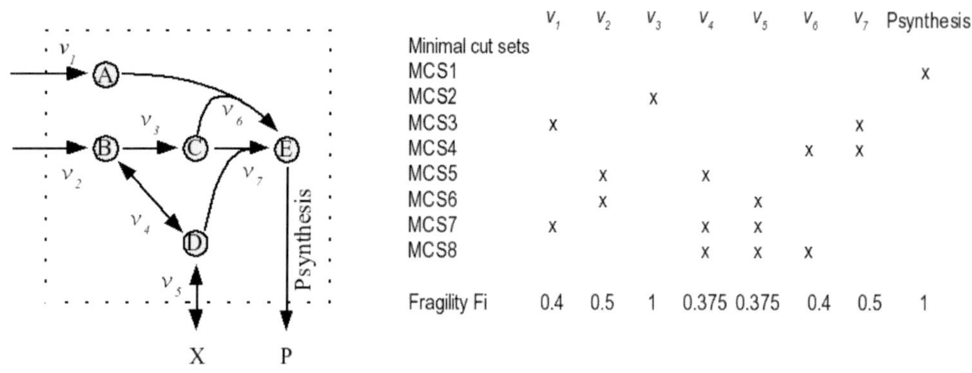

Minimal cut sets	v_1	v_2	v_3	v_4	v_5	v_6	v_7	Psynthesis
MCS1								X
MCS2			X					
MCS3	X						X	
MCS4						X	X	
MCS5		X		X				
MCS6		X			X			
MCS7	X			X	X			
MCS8				X	X	X		
Fragility Fi	0.4	0.5	1	0.375	0.375	0.4	0.5	1

Figure 6.18 (Left) Example network of Figure 6.15. (Right) Participation of individual reactions in the minimal cut sets shown in Figure 6.17 and computation of the fragility coefficient F_i. For example, reaction ν_2 is a member of MCS5 and MCS6 that both contain a total of two reactions. Therefore, the average size of the MCSs that ν_2 belongs to is 2, with the reciprocal 0.5.

Analyzing the minimal cut sets gives deeper insights in the structural fragility of a given metabolic network and is useful for identifying target sets for an intended repression of network functions.

6.8 High-Flux Backbone

As discussed before, recent work on the metabolic networks of model organisms required revising the picture of separate biochemical pathways into a densely woven metabolic network. Flux balance analysis as well as experimental data showed that the connectivity of substrates in this network follows a power law (see Section 4.4). Constraint-based modeling approaches (flux balance analysis) were successful in analyzing the capabilities of cellular metabolism including its capacity to predict deletion phenotypes, the ability to calculate the relative flux values of metabolic reactions and the capability to identify properties of alternate optimal growth states in a wide range of simulated environmental conditions. Open questions are which parts of metabolism are involved in adaptation to environmental conditions? Is there a central essential metabolic core? What role does transcriptional regulation play?

As an example of many recent successful studies employing flux balance analysis, we again use a study involving the Barabási group (Almaas *et al.*, 2004). This work utilized the stoichiometric matrix for *E. coli* strain MG1655 containing 537 metabolites and 739 reactions from the Palsson group. Flux balance analysis was applied to characterize the solution space (all possible flux states under a given condition) using linear programming and adapting constraints for each reaction flux ν_i of the form $\beta_{i,min} \le \nu_i \le \beta_{i,max}$. The flux states were calculated that optimize cell growth on various substrates. The authors restricted their analysis to the subspace of solutions

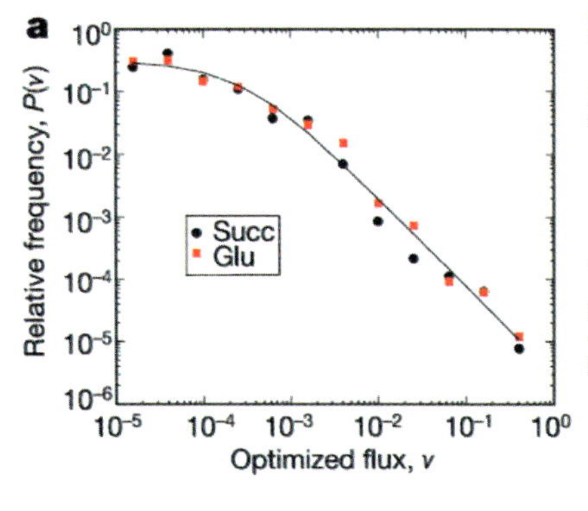

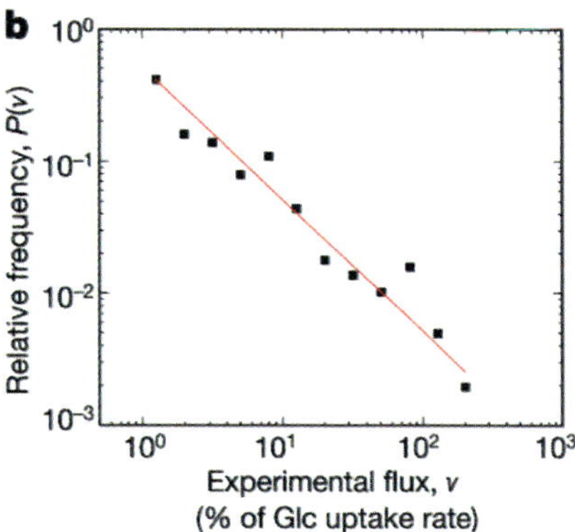

Figure 6.19 (a) Calculated flux distribution for optimized biomass production on succinate (black) and glutamate (red) substrates. The solid line corresponds to the power-law fit of the likelihood that a reaction has flux ν with $P(\nu) \propto (\nu + \nu_0)^{-\alpha}$, with $\nu_0 = 0.0003$ and $\alpha = 1.5$. (b) The distribution of experimentally determined fluxes from the central metabolism of *E. coli* shows a power-law behavior as well, with a best fit to $P(\nu) \propto \nu^{-\alpha}$ with $\alpha = 1$. Reprinted from Almaas *et al.* (2004) by permission from Macmillan publishers Ltd.

where all components ν_j are larger than zero. The **mass** carried by reaction j producing (consuming) metabolite i is denoted by:

$$\hat{\nu}_{ij} = |S_{ij}|\nu_j.$$

with the entries of the stoichiometric matrix S_{ij} and the fluxes ν_j.

Interestingly, the magnitudes of the individual fluxes were found to vary widely. For example, the dimensionless flux of succinyl coenzyme A synthetase reaction was 0.185, whereas the flux of the aspartate oxidase reaction was 10 000 times smaller, 2.2×10^{-5}. Figure 6.19 shows the calculated flux distribution for active (nonzero flux) reactions of *E. coli* grown in a glutamate- or succinate-rich substrate.

Remarkably, both computed and experimental flux distribution showed a wide spectrum of fluxes. The authors discussed that the observed flux distribution is compatible with two different potential local flux structures (Figure 6.20). (a) A **homogenous local organization** would imply that all reactions producing (consuming) a given metabolite have comparable fluxes, whereas (b) a more delocalized "**high-flux backbone** (HFB)" would be expected if the local flux organization was heterogeneous such that each metabolite has a dominant source (consuming) reaction.

To distinguish between these two schemes contrasted in Figure 6.20, they defined

$$Y(k,i) = \sum_{j=1}^{k} \left[\frac{\nu_{ij}}{\sum_{l=1}^{k} \nu_{il}} \right]$$

where ν_{ij} is the mass carried by reaction j which produces (consumes) metabolite i. If all reactions producing (consuming) metabolite i had comparable ν_{ij} values, $Y(k,i)$ would scale as $1/k$. If, however, the activity of a single reaction dominates, one would expect $Y(k,i)$ to scale as 1 (i.e. to be independent of k). The measured result for $k \cdot Y(k)$ is shown in Figure 6.21.

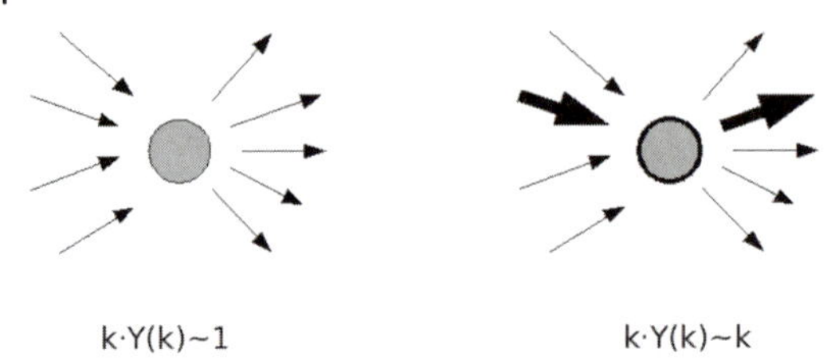

Figure 6.20 Schematic illustration of the hypothetical scenario in which (a) all fluxes have comparable activity, in which case one expects $kY(k) \sim 1$, and (b) the majority of the flux is carried by a single incoming or outgoing reaction, for which $kY(k) \sim k$.

Interestingly, an intermediate behavior is found between the two extreme cases. This proves that the large-scale inhomogeneity observed in the overall flux distribution is also partially valid at the level of the individual metabolites. The result for FAD suggests that the more reactions consume (produce) a given metabolite, the more likely it is that a single reaction carries most of the flux.

Encouraged by this finding, the authors went on to characterize the main flux backbone of the metabolic network of *E. coli*. From the large complex flux network, a simple algorithm removed for each metabolite systematically all reactions but the one providing the largest incoming (outgoing) flux distribution. In this way, the algorithm uncovered the high-flux-backbone of the metabolism, a distinct structure of linked reactions that form a giant component with a star-like topology (Figure 6.22). In Figure 6.22 only a few pathways appear disconnected, indicating that although these pathways are part of the HFB, their end product is only the second-most important source for another high-flux metabolite. The groups of individual HFB reactions largely overlap with traditional biochemical partitioning of cellular metabolism.

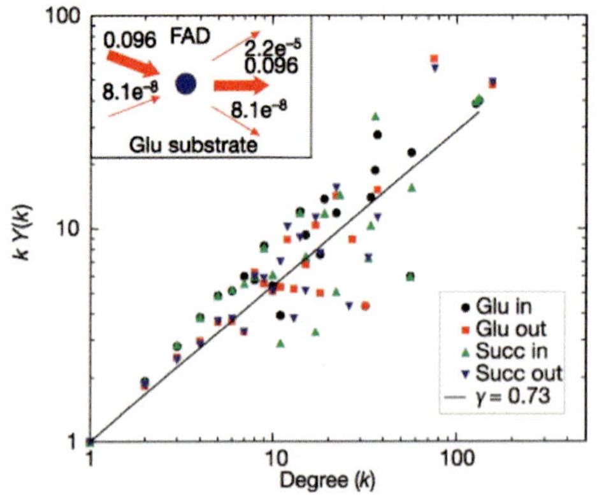

Figure 6.21 Measured $kY(k)$ shown as a function of k for incoming and outgoing reactions, averaged over all metabolites, indicates that $Y(k) \propto k^{-0.27}$. Inset shows nonzero mass flows, ν_{ij}, producing (consuming) FAD on a glutamate-rich substrate. Reprinted from Almaas *et al.* (2004) by permission from Macmillan publishers Ltd.

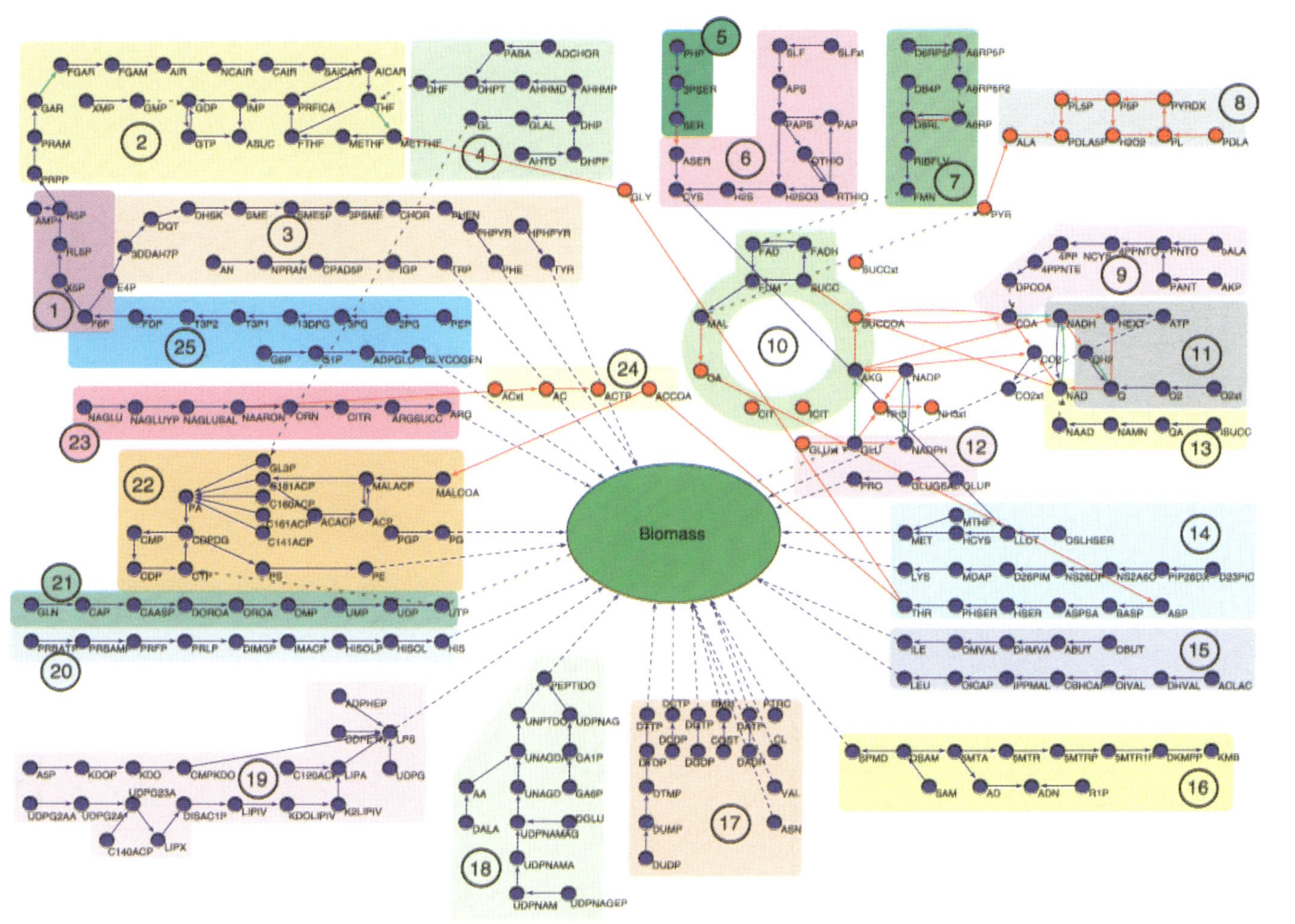

Figure 6.22 HFB in the metabolic network of *E. coli* as optimized by flux balance analysis for growth on glutamate-rich substrate. Two metabolites (e.g. A and B) are connected with a directed edge pointing from A to B only if the reaction with maximal flux consuming A is the reaction with maximal flux producing B. Shown are all the metabolites that have at least one neighbor after completing this procedure. The background colors indicate different known biochemical pathways. Reprinted from Almaas *et al.* (2004) by permission from Macmillan Publishers Ltd.

Summary

Compared to other cellular networks discussed in this textbook, our understanding of metabolic networks may be considered quite mature. This is helped by the almost complete characterization of central metabolism in most organisms and by the ability to perform direct fluxome measurement using, for example, ^{13}C-labeled substrate. The mathematical approaches of flux balance analysis, elementary modes and extreme pathways provide a robust toolbox to characterize topological properties of the networks and even make quantitative predictions. We pointed out that methodological improvements are still required to facilitate the analysis of millions of elementary modes.

Metabolic network use is highly uneven (power-law distribution) at the global level and at the level of the individual metabolites. Whereas most metabolic reactions have low fluxes, the overall activity of the metabolism is dominated by several reactions with very high fluxes. *E. coli* responds to changes in growth conditions by reorganizing the rates of selected fluxes predominantly within this high-flux backbone. Apart from minor changes, the use of the other pathways remains unaltered. These reorganizations result in large, discrete changes in the fluxes of the HFB reactions.

Problems

Static network properties: pathways
The first three assignments will introduce you to some aspects of networks under steady-state conditions. This includes taking apart a network into independent paths, identifying crucial metabolites and reactions, and simplifying networks.

(1) Identifying targets

The following hypothetical metabolic network produces "biomass" L from the substrates A and F. In various intermediate steps, accessory substances are produced or consumed. The metabolites are labeled with the letters A to L (Figure 6.23).

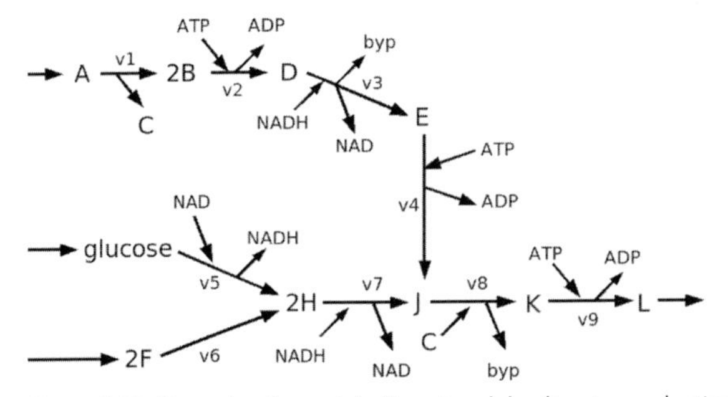

Figure 6.23 Example of a metabolic network leading to production of biomass L.

(a) Inspect the network visually and identify (without calculation) the important substrates that are essential for the functioning of the network, i.e. the production of L is stopped when these are missing. Explain your findings.

Now assume that this network was the central part of the metabolism of a dangerous bacterium and you want to develop an effective drug. On which enzymes (reactions) would you concentrate when searching for an inhibitor? Explain your answer.

Would you change your strategy, if you knew that high concentrations of "byp" slow down or even reverse reaction v_3? (Why?)

What if somebody discovers that high concentrations of D are lethal for the host?

(b) Simplify the above network of Figure 6.23 without changing its topology (structure). Do so by writing an equation for each reaction and solve for J and L.

Example: from v_1 you get $A \rightarrow 2B + C$, from v_2: $2B + ATP \rightarrow D + ADP$. This can be rewritten (for the moment) as $2B = D + ADP - ATP$. Insert this into $v1$ to get v_{12}: $A + ATP \rightarrow D + ADP + C$. Now try to eliminate D from this equation, etc.

Note: Some of the internal substrates can not be eliminated.

Once you are done with the elimination of the internal substrates (A, F, ..., are *not* internal) draw the simplified network in the same layout as the original network.

(2) Extreme pathways

Construct the stoichiometric matrix of the network in Figure 6.24. Do not forget to label rows and columns.

Hint: Split up v_4 into a forward and a backward reaction (reconfiguration).

Then calculate the extreme pathways from the stoichiometric matrix as explained in Section 6.6. Determine both the "pathway length matrix" and the "reaction participation matrix". What information do they provide? Which reaction(s) contribute(s) to the most pathways? Which are the shortest and the longest pathways?

Now assume that reaction b_2 is essential for the organism, i.e. it dies if there is no output via b_2. Determine from the extreme pathways which (combinations of) internal reactions are essential, i.e. if they are blocked then the output via b_2 is blocked, too.

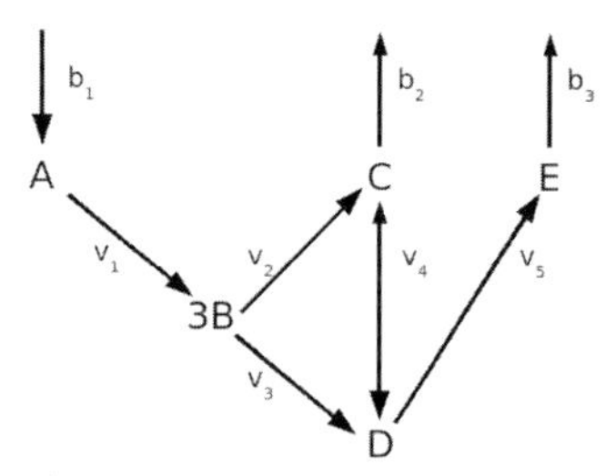

Figure 6.24 Simple metabolic network as in Figure 6.9.

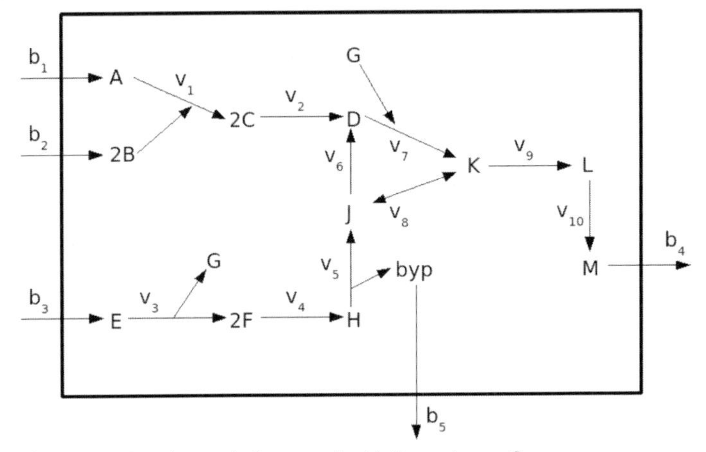

Figure 6.25 Simple metabolic network with five exchange fluxes.

(3) Extreme pathways

Characterize the steady-state properties of the network given in Figure 6.25 via its extreme pathways.

(a) **Simplify the network**
As a first step simplify the given network by grouping reactions and metabolites, respectively. Sketch the simplified network.

Hint: Write down the reactions as equations. Then you can proceed analogously as with a set of linear equations and eliminate (i.e. group) substances.

Example:

$$v_1 : A + 2B \rightarrow 2C$$

$$v_2 : 2C \rightarrow D$$

$$v_{12} : A + 2B \rightarrow D.$$

Comment: You can actually reduce the network to three internal substrates. "Internal" means that the metabolite is not connected to any of the b fluxes.

(b) **Stoichiometric matrix**
Construct the stoichiometric matrix for the simplified network and calculate from it the extreme pathways. Give the pathways as formulas and sketch them.

Determine both the "pathway length matrix" and the "reaction participation matrix" and interpret them.

Which reactions contribute to the most pathways; are there some reactions that do not contribute at all?

(c) **Extreme pathways of the of the original network**
Now reconstruct the extreme pathways of the original network from the simplified network. Again give the extreme pathways both as formulas and as sketches. Why does it work to determine the extreme pathways of the complete network via this "detour" of the simplified network?

From the pathway length matrices determine the lengths of the pathways for both the original and the simplified network.

The output ("biomass production") of our network corresponds to the flux through reaction b_4. Let us consider a reaction as "essential" for the network if blocking this reaction shuts down the network. List all essential reactions of the simplifying network.

Can you figure out how to determine these essential reactions from the extreme pathways?

(d) **Steady-state properties**
For the following step we will neglect the internal reactions of the pathways, i.e. only consider those reactins that are labeled with a letter b. Then we see how the (black box) network transforms input through b_1, b_2 and b_3 into output through b_4 and b_5.

Hint: Look at the pathways in their "formula" form.

Copy and complete the following table which relates the input through $b_1, \ldots, b_3$ to the output via b_4 and b_5. For each configuration give the contributions of each of the extreme pathways as, e.g. $n_1 * e_1 + n_2 * e_2 + n_3 * e_3$ with n_i as the contributions. Also determine the flux through the reactions v_2, v_5 and v_7 of the original network.

Configuration	I	II	III	IV	V	VI
b_1	0		1		1	
b_2	0		2		2	
b_3	1		2		5	
b_4		2		8		4
b_5		1		5		3

Hint: Extend the table so that it can hold all information asked for.

Further Reading

Metabolic Networks in General

Palsson BØ (2006) *Systems Biology: Properties of Reconstructed Networks* Cambridge University Press Cambridge.

Papin JA, Price ND, Wiback SJ, Fell DA, Palsson BO (2003) Metabolic pathways in the post-genome area, *Trends in Biochemical Sciences* **28**, 250–258.

Francke C, Siezen RJ, Teusink B (2005) Reconstructing the metabolic network of a bacterium from its genome, *Trends in Microbiology* **13**, 550–558.

Price ND, Reed JL, Palsson BØ (2004) Genome-scale models of microbial cells: evaluating the consequences of constraints, *Nature Reviews Microbiology* **2**, 886–897.

Metabolism of *E. coli*

Ouzounis CA, Karp PD (2000) Global properties of the metabolic map of *Escherichia coli*, *Genome Research* **10**, 568–576.

Flux Balance Analysis

Almaas E, Kovacs B, Vicsek T, Oltvai ZN, Barabási AL (2004) Global organization of metabolic fluxes in the bacterium *Escherichia coli*, *Nature* **427**, 839–843.

Extreme Pathway Method

Schilling CH, Letscher D, Palsson BO (2000) Theory for the systemic definition of metabolic pathways and their use in interpreting metabolic function from a pathway-oriented perspective, *Journal of Theoretical Biology* **203**, 229–248.

Elementary Modes

Schuster S, Hilgetag C, (1994) On elementary flux modes in biochemical reaction systems at steady state, *Journal of Biological Systems* **2**, 165–182.

Double Description Method

Gagneur J, Klamt S (2004) Computation of elementary modes: a unifying framework and the new binary approach, *BMC Bioinformatics* **5**, 175.

Minimal Cut Sets

Klamt S, Gilles ED (2004) Minimal cut sets in biochemical reaction networks, *Bioinformatics* **20**, 226–234.

Klamt S (2006) Generalized concept of minimal cut sets in biochemical networks, *BioSystems* **83**, 233–247.

7
Kinetic Modeling of Cellular Processes

So far, we have mostly used stationary, time-independent mathematical models to describe metabolic flux distributions, gene regulatory networks or protein–protein interactions. In cells, however, many processes undergo important temporal fluctuations, most notably the cell cycle itself. Therefore, we now introduce a new class of models that allow modeling of time-dependent cellular phenomena such as individual biochemical reactions, more complex processes such as signal transduction cascades or even the entire cell cycle.

7.1
Ordinary Differential Equation Models

When considering the time-evolution of a function f, a natural approach is considering the first time derivative of the function $f' = \partial f/\partial t$ and higher derivatives. An **ordinary differential equation (ODE)** is a relation that contains one or more functions of only one independent variable and one or more of their derivatives with respect to that variable.

A simple example is Newton's second law of motion describing the motion of a particle of mass m. It is an ODE containing the second derivative $\mathrm{d}^2x/\mathrm{d}t^2$ of its positional coordinate x with respect to time t and relates that to the force F acting on the particle at its current position x:

$$m\frac{\mathrm{d}^2x}{\mathrm{d}t^2} = F(x(t)) \tag{7.1}$$

The only independent variable here is the time t. Recall that the particle velocity v:

$$v = \frac{\mathrm{d}x}{\mathrm{d}t},$$

Principles of Computational Cell Biology – From Protein Complexes to Cellular Networks. Volkhard Helms

ISBN: 978-3-527-31555-0

expresses how the particle's position changes with time and the particle acceleration a:

$$a = \frac{dv}{dt} = \frac{d^2x}{dt^2},$$

expresses how the particle's velocity changes with time. Therefore, Newton's second law of motion describes whether the particle is accelerating or slowing down being subjected to the force $F(x)$.

The **order** of a differential equation is the order n of the highest derivative that appears. A **solution** of an ODE is a function $y(x)$ whose derivatives satisfy the equation. Such a function is not guaranteed to exist and, if it does exist, is usually not unique. A general solution of an nth-order equation is a solution containing n arbitrary variables, corresponding to n constants of integration. A particular solution is derived from the general solution by setting the constants to particular values.

When a differential equation of order n has the form:

$$f\left(x, y', y'', \ldots, y^{(n)}\right) = 0,$$

it is called an implicit differential equation, whereas the form:

$$f(x, y', y'', \ldots, y^{(n-1)}) = y^{(n)},$$

is called an explicit differential equation.

Ordinary differential equations are to be distinguished from **partial differential equations** where the function(s) depend on several independent variables, and the differential equation involves partial derivatives with respect to each of these variables (Section 7.3).

7.1.1 Examples for ODEs

The problem of solving a differential equation is to find the function y whose derivatives satisfy the equation. For example, the differential equation:

$$y'' + y = 0, \tag{7.2}$$

has the general solution $y = A\cos(x) + B\sin(x)$ where A and B are constants determined from boundary conditions. [Recall that the first derivative of the sin(x) function is cos(x) and the second derivative is –sin(x).]

Another illustrative example is the **mathematical pendulum** (Figure 7.1). For the sake of simplicity, one commonly assumes that the bob swinging is massless, the weight is a point mass m, there is no friction and the motion is restricted to the two-dimensional plane, i.e. the pendulum does not start to move out of the plane to make an ellipsoidal movement.

Assuming that the weight has been pushed out of its equilibrium position, the projection of the gravitational force mg onto the particle's plane of movement, $mg \sin\theta$, with the gravity constant g generates a restoring force on the weight that

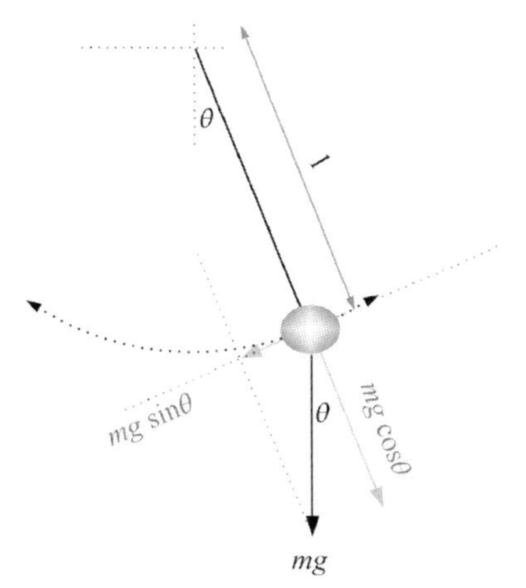

Figure 7.1 Force diagram for a mathematical pendulum consisting of a weight attached to a fixed point via a stiff connection of length l. The only acting force is gravitation.

directs the weight back to the central position (and actually beyond that into an infinite-length pendulum motion as we assume absence of friction). Let us use the coordinate of the arc length s to describe the weight's motion along its line of movement.

Newton's second law for the arc length coordinate s is:

$$F(s) = m\frac{d^2s}{dt^2}.$$

As force $F(s)$, we need to use the projection of the gravitational force $mg\sin\theta$ along the arc length (Figure 7.1):

$$\begin{aligned} -mg\sin\theta &= m\frac{d^2s}{dt^2} \\ -g\sin\theta &= \frac{d^2s}{dt^2}. \end{aligned} \tag{7.3}$$

The minus sign reflects that the gravitation force works against the particle's displacement from its equilibrium position. The coordinate s is related to the rod length l and the angle θ:

$$\begin{aligned} s &= l\theta \\ v &= \frac{ds}{dt} = l\frac{d\theta}{dt} \\ a &= \frac{d^2s}{dt^2} = l\frac{d^2\theta}{dt^2}. \end{aligned} \tag{7.4}$$

Combining Eqs. (7.3) and (7.4) gives:

$$-g\sin\theta = l\frac{d^2\theta}{dt^2},$$

or:

$$\frac{d^2\theta}{dt^2} + \frac{g}{l}\sin\theta = 0 \tag{7.5}$$

This is the differential equation which, when solved for $\theta(t)$, will give the motion of the pendulum. Unfortunately, it cannot be integrated directly. If we restrict the motion of the pendulum to relatively small amplitude, we can approximate:

$$\sin\theta \approx \theta.$$

This approximation is not too bad for small angles if you recall the shape of the *sinus* function that crosses the origin at a 45° angle. Then, Eq. (7.5) transforms into:

$$\frac{d^2\theta}{dt^2} + \frac{g}{l}\theta = 0, \tag{7.6}$$

which is almost exactly Eq. (7.2) above. The solution to this equation is readily obtained as:

$$\theta(t) = \theta_0 \cos\left(\sqrt{\frac{g}{l}}t + \varphi_0\right) \quad |\theta_0| \ll 1,$$

with the amplitude θ_0 and the initial phase φ_0. As expected, a pendulum undergoes an oscillatory motion. Its frequency decreases when the pendulum's length l increases.

There exist many analytical and numerical approaches to find solutions of differential equations. For example, in the case where the equations are **linear**, the original equation can be solved by breaking it down into smaller equations, solving those, and then adding the results back together. Unfortunately, many of the interesting differential equations are nonlinear, which means that they cannot be broken down in this way. There are also a number of techniques for solving differential equations using a computer. In the section, we will concentrate on finding stationary and oscillatory solutions of several model cases observed in cellular processes.

7.2 Modeling Cellular Feedback Loops by ODEs

As formulated by the late Reinhart Heinrich, a pioneer in the mathematical modeling of cellular processes, the mathematical description of signaling pathways helps answer questions like:

(1) How do the magnitudes of signal output and signal duration depend on the kinetic properties of pathway components?
(2) Can high signal amplification be coupled with fast signaling?
(3) How are signaling pathways designed to ensure that they are safely off in the absence of stimulation, yet display high signal amplification following receptor activation?

(4) How can different agonists stimulate the same pathway in distinct ways to elicit a sustained or a transient response, which can have dramatically different consequences?

We will see in the following how very simple signaling pathways can be embedded in networks using positive and negative feedback to generate quite complex behaviors – toggle switches and oscillators – which are the basic building complex of nonlinear control systems. The following discussion is largely based on the brilliant review paper by Tyson *et al.* (2003).

7.2.1 Protein Synthesis and Degradation: Linear Response

Let us start with the example of protein synthesis and degradation shown in Figure 7.2. Here, protein R is constantly produced at a base line rate k_0. R is termed the *response magnitude*. This synthesis rate may be further stimulated by the presence of a signal S. Here, S measures the *signal strength*, e.g. by the concentration of its corresponding mRNA. k_1 is the rate by which synthesis of R responds to the concentration of S. k_2 is the rate constant for the degradation of R. One may safely assume that degradation depends linearly on the concentration of R. This means that during each time interval, a constant fraction of R is degraded by digestion enzymes, etc.

We can set the rate equation as a simple balance equation:

$$\frac{\mathrm{d}R}{\mathrm{d}t} = k_0 + k_1 S - k_2 R, \qquad (7.7)$$

where R and S denote the concentrations of response and signal molecules, respectively. This equation describes that the concentration R increases over time with a constant rate k_0 plus the signal modulated rate $k_1 S$ and – at the same time – decreases with a constant rate $-k_2 R$ proportional to R itself.

Assuming a constant signal strength S, we expect that the response R will adjust to a steady state after a certain time interval. A **steady-state** solution of any differential equation, $\mathrm{d}R/\mathrm{d}t = f(R)$ is a constant R_{ss} that satisfies the algebraic equation $f(R_{ss}) = 0$. Therefore, we require:

$$0 = k_0 + k_1 S - k_2 R_{ss},$$

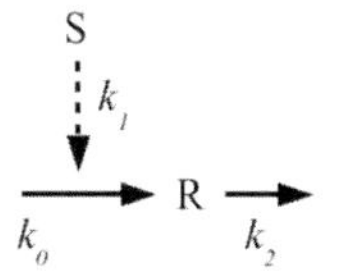

Figure 7.2 Simple model for synthesis of protein R ("response") under activation of a signal S and subsequent degradation. S could be a transcription factor that binds to the promoter region of the gene coding for R.

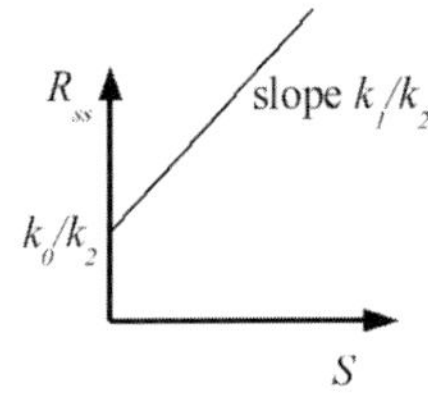

Figure 7.3 For the example of linear response, the steady-state response R_{ss} depends linearly on the signal strength S. At zero signal strength, there is a baseline response given by the ratio of the two rate constants k_0 and k_2. The magnitude of the response beyond this baseline depends on the ratio of the two constants k_1 and k_2.

or:

$$
\begin{aligned}
R_{ss} &= \frac{k_0 + k_1 S}{k_2} \\
&= \frac{k_0}{k_2} + \frac{k_1}{k_2} S.
\end{aligned}
$$

This means that the steady-state response R_{ss} depends linearly on the signal strength S with a proportionality constant given as the ratio of the two rate constants for input (k_1) and for degradation (k_2) (Figure 7.3).

7.2.2
Phosphorylation/Dephosphorylation – Hyperbolic Response

The following example models the equilibrium between a protein R ("response") and its phosphorylated form R_P (Figure 7.4). This example is a variation of the first example of protein synthesis because the degradation product of R_P is fed back to the starting substance R.

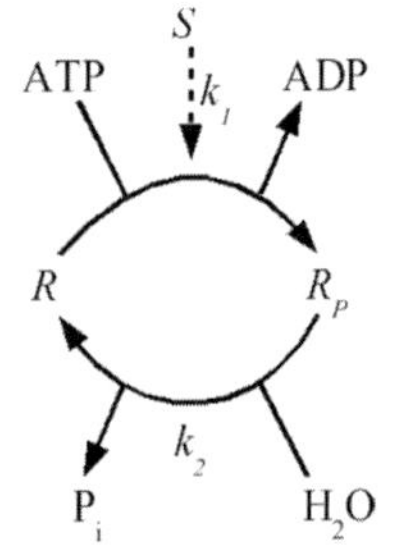

Figure 7.4 Simple model for the equilibrium between the phosphorylated form of a protein R_P and the dephosphorylated form R. In the back reaction, a water molecule is taken up from solution and a molecule of inorganic phosphate P_i is set free so that ATP can be regenerated from ADP.

The rate equation for the temporal change of the concentration of R_P is given by:

$$\frac{dR_P}{dt} = k_1 SR - k_2 R_P.$$

This example is almost identical to the previous case except for multiplication of S with the concentration of the nonphosphorylated form R. Here, we are not considering a basal activity k_0 as in the first example because the system is a closed cycle. As the total concentration of protein $R_T = R + R_P$ is constant, we can substitute:

$$R = R_T - R_P,$$

in the above equation and obtain:

$$\frac{dR_P}{dt} = k_1 S(R_T - R_P) - k_2 R_P. \quad (7.8)$$

The steady-state concentration of the phosphorylated form is again obtained by requiring:

$$0 = k_1 S(R_T - R_{P,ss}) - k_2 R_{P,ss},$$

leading to the stationary concentration:

$$\begin{aligned} R_{P,ss} &= \frac{k_1 S R_T}{k_2 + k_1 S} \\ &= \frac{S R_T}{(k_2/k_1) + S}. \end{aligned}$$

In the two limiting cases:

$$R_{P,ss} = \frac{S}{(k_2/k_1) + S} \cdot R_T \begin{cases} \xrightarrow{S \text{ small}} & \dfrac{k_1}{k_2} R_T \cdot S \\ \xrightarrow{S \text{ large}} & R_T. \end{cases}$$

In this case, the steady-state response is not linear, but hyperbolic (Figure 7.5). This is very easy to understand as the total concentration of protein on both sides is

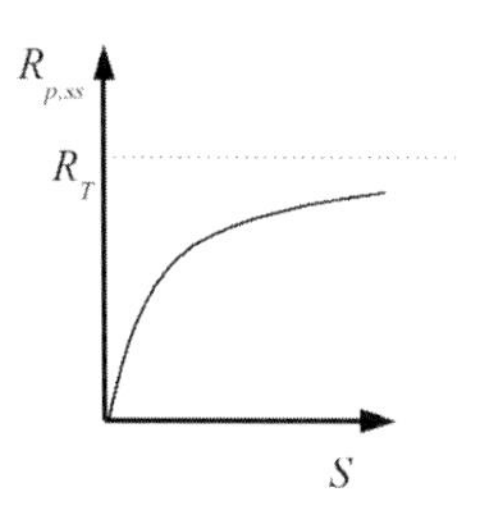

Figure 7.5 Steady-state concentration of phosphorylated protein R_P as a function of signal strength S. In the regime of small S, the rise of $R_{P,ss}$ is almost linear with S. The slope is determined by the ratio of the rates for synthesis and degradation. For larger S, the response saturates at the total concentration R_T of protein R.

constant. Increasing the signal strength S can shovel a large fraction of the protein into its phosphorylated form but the response must be limited to the total concentration of protein, R_T, of course.

7.2.3
Phosphorylation/Dephosphorylation – Buzzer

As a variation of the previous section, we will now assume that the phosphorylation and dephosphorylation reactions themselves are governed by Michaelis–Menten kinetics. Equation (7.7) then becomes:

$$\frac{dR_P}{dt} = \frac{k_1 S(R_T - R_P)}{K_{m,1} + R_T - R_P} - \frac{k_2 R_P}{K_{m,2} + R_P}.$$

The steady-state solution is obtained as:

$$k_2 R_P(K_{m,1} + R_T - R_P) = k_1 S(R_T - R_P)(K_{m,2} + R_P).$$

Here, the biophysically acceptable solutions must be in the range $0 < R_P < R_T$. By using the "Goldbeter–Koshland function" G (Goldbeter and Koshland, [2]) defined as:

$$G(u, v, J, K) = \frac{2uK}{\beta + \sqrt{\beta^2 - 4uK(v - u)}}$$

$$\beta(u, v, J, K) = v - u + u \cdot J + v \cdot K,$$

the steady-state solution of the above equation is:

$$\frac{R_{P,ss}}{R_T} = G\left(k_1 S, k_2, \frac{K_{m1}}{R_T}, \frac{K_{m2}}{R_T}\right).$$

When plotting $R_{P,ss}$ as a function of S (Figure 7.6) we obtain a sigmoidal response curve if K_{m1}/R_T and K_{m2}/R_T are both much smaller than 1.

This mechanism creates a switch-like signal–response curve which is called zero-order ultrasensitivity. All three examples considered so far give a "graded" and reversible behavior of R with respect to S. "Graded" means here that R increases continuously with S and "reversible" means that if S is changed from $S_{initial}$ to S_{final}, the response at S_{final} is the same whether the signal is being increased ($S_{initial} < S_{final}$) or decreased ($S_{initial} > S_{final}$). Although continuous and reversible, a sigmoidal response is abrupt. The element behaves like a **buzzer** where one must push hard

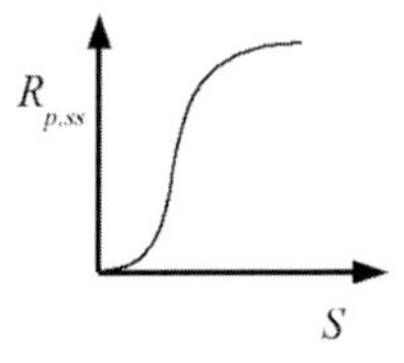

Figure 7.6 Response in a phosphorylation/dephosphorylation equilibrium with Michaelis–Menten kinetics.

enough on the button to activate the response. In terms of the phosphorylation signal, the signal S must be strong enough to create a noticeable change of the equilibrium.

7.2.4 Perfect Adaptation – Sniffer

Now, the simple linear response element of Figure 7.2 is supplemented with a second signaling pathway through species X (Figure 7.7). The signal influences the response via two parallel pathways that push the response in opposite directions (an example of feed-forward control similar to Section 5.5). The action of S on X corresponds exactly to our first example on protein synthesis. The steady-state response of X is linearly dependent on S. Here, however, we are plotting the relaxation process of adapting the response to new levels of the signal S that is being increased abruptly in discrete steps:

$$\frac{dR}{dt} = k_1 S - k_2 X \cdot R$$

$$\frac{dX}{dt} = k_3 S - k_4 X.$$

For the steady state, setting the second equation to zero gives:

$$X_{ss} = \frac{k_3 S}{k_4}.$$

Setting the first equation to zero and replacing X_{ss} by the expression just derived gives:

$$R_{ss} = \frac{k_1 S}{k_2 X} = \frac{k_1 k_4}{k_2 k_3}.$$

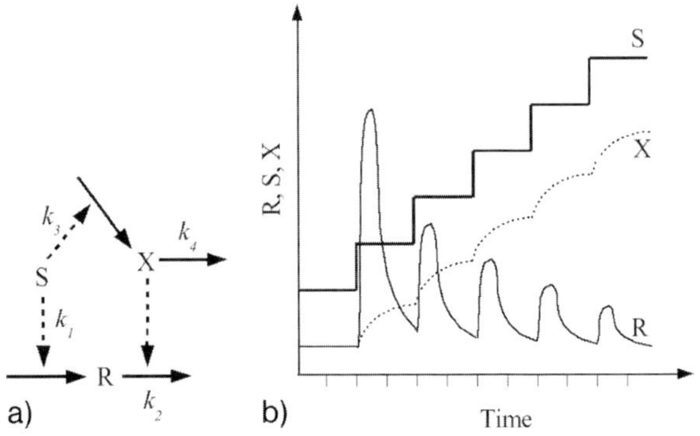

Figure 7.7 a) Coupling of the initial response pathway via R with a second signaling pathway (X). b) Transient response (R, thin solid line) as a function of stepwise increases in signal strength S (thick solid line) with concomitant changes in the indirect signaling pathway X (dashed line). The signal influences the response via two parallel pathways that push the response in opposite directions.

In this setup, the response mechanism exhibits perfect adaptation to the signal. Although the signaling pathway responds transiently to changes in signal strength R (Figure 7.7, right), its steady-state response R_{ss} is independent of S and is only controlled by the ratio of the four kinetic rates of the system! Such behavior is typical of chemotactic systems, which respond to an abrupt change in attractants or repellents, but then adapt to a constant level of the signal. Our own sense of smell operates this way. Therefore, this element is termed a **sniffer**.

7.2.5 Positive Feedback – One-Way Switch

In the previous example, the signal influenced the response via two parallel pathways that stimulated synthesis and degradation. This was an example for a feed-forward control system. Alternatively, some component of a response pathway may feed back to the signal. This will be the case in the example presented here, where the response element R activates enzyme E (by phosphorylation), and E_P enhances the synthesis of R (Figure 7.8):

$$\frac{dR}{dt} = k_4 E_P(R) + k_1 S - k_2 R.$$

$E_P(R)$ is again a Goldbeter–Koshland function depending on the rate constants $k_3 R$, k_4, and also the rate that describe the spontaneous back reaction from E_P to E. Solving this system gives the steady-state response shown in Figure 7.9.

The equilibrium between E_P and E is similar to the previous example of a phosphorylation/dephosphorylation equilibrium in that the total concentration of $E + E_P$ is constant. However, the previous example did not include the positive feedback of E_P into the production of P. In the response curve, the control system is found to be **bistable** between 0 and S_{crit}. In this regime, there are two stable steady-state response values (on the upper and lower branches, the solid lines) separated by an unstable steady state (on the intermediate branch, the dashed line). This is called a one-parameter **bifurcation**. Which value is taken depends on the history of the system. After the signal threshold S_{crit} has been crossed once, the system will remain on the upper curve. This is termed a **one-way switch**. Biological examples for this behavior include the maturation of frog oocytes in response to the hormone

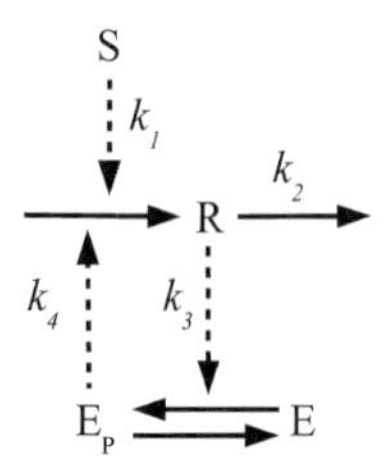

Figure 7.8 Example of a positive feedback system built from a protein E and the response R. E_P is the phosphorylated form of E. Here, R activates E by phosphorylation, and E_P enhances the synthesis of R.

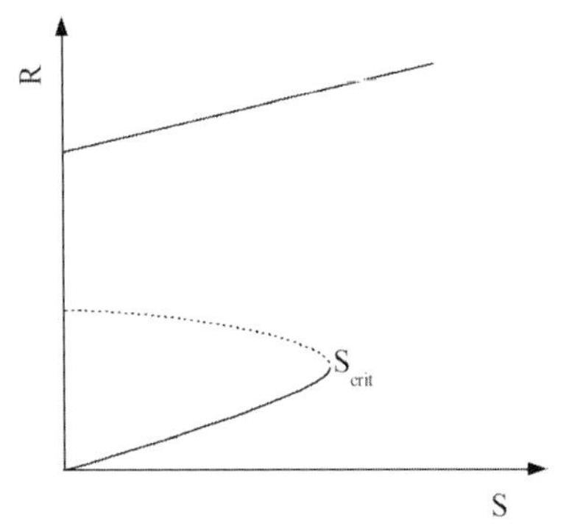

Figure 7.9 Positive feedback system. As S increases, the response is low until S exceeds a critical value S_{crit} at which point the response increases abruptly to a high value. Then, if S decreases, the response stays high.

progesterone, and apoptosis, where the decision to shut down the cell must be clearly a one-way switch.

7.2.6
Mutual Inhibition – Toggle Switch

The following example is a minute variation of the previous one in that the enzyme E now stimulates degradation of R (Figure 7.10, left).

This system again leads to a discontinuous behavior (Figure 7.10, right). This type of bifurcation is called a **toggle switch**. If S is decreased enough after starting from a high level, the switch will go back to the off-state on the lower curve meaning a small response R. For intermediate stimulus strength ($S_{crit1} < S < S_{crit2}$), the response of the system can be either small or large, depending on the history of $S(t)$. This is often called "**hysteresis**". Biological examples for such behavior include the lac operon in bacteria, activation of M-phase promoting factor in frog egg extracts, and the autocatalytic conversion of normal prion protein into its pathogenic form.

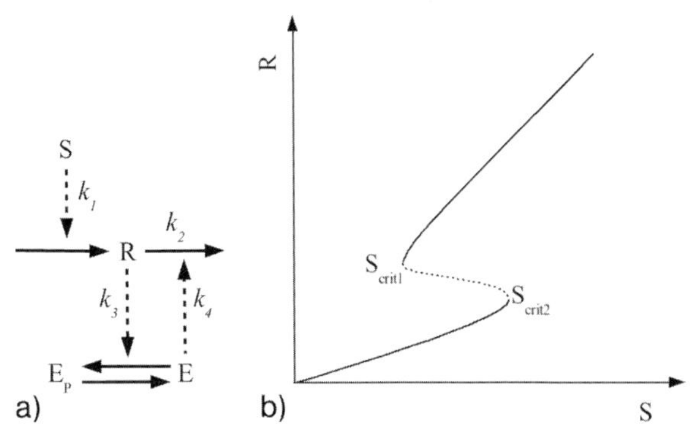

Figure 7.10 Positive feedback system.

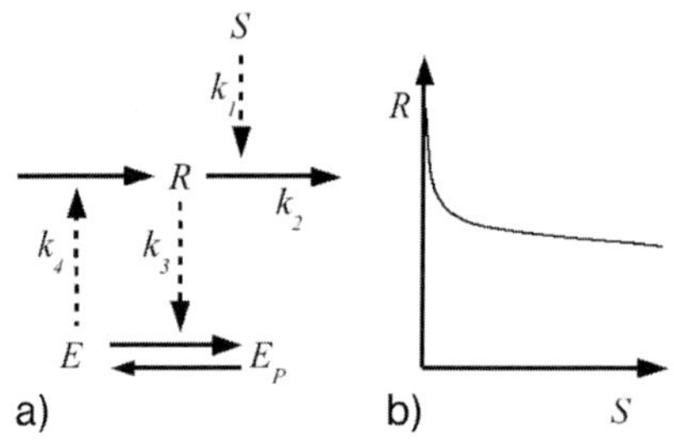

Figure 7.11 Negative feedback system.

7.2.7
Negative Feedback – Homeostasis

In negative feedback, the response counteracts the effect of the stimulus (Figure 7.11). Here, the response element R inhibits the enzyme E catalyzing its synthesis. Therefore, the rate of production of R is a sigmoidal decreasing function of R. The signal is in this case in demand for R. If there is not enough R present, increasing the signal S does not change the concentration of R much more:

$$\frac{dR}{dt} = k_4 E(R) - k_2 S \cdot R$$

$$E(R) = G(k_4, k_3 R, J_3, J_4).$$

This type of regulation, commonly employed in biosynthetic pathways, is called **homeostasis**. It is sort of an adaptation, but not a sniffer because stepwise increases in S do not generate transient changes in R.

7.2.8
Negative Feedback: Oscillatory Response

Negative feedback as in the two-component system $E \to R —| E$ can exhibit damped oscillations leading to a stable steady state but not sustained oscillations (—| indicates inhibition). Sustained oscillations typically require at least three components, $X \to Y \to R —| X$. The third component (Y) introduces a time delay in the feedback loop, causing the control system to repeatedly overshoot and undershoot its steady state. There are two ways to close the negative feedback loop: (1) R_P inhibits the synthesis of X and (2) R_P activates the degradation of X. Here, the second scenario is realized (Figure 7.12):

$$\frac{dX}{dt} = k_0 + k_1 S - k_2 X - k_7 R_P \cdot X$$

$$\frac{dY_P}{dt} = \frac{k_3 X (Y_T - Y_P)}{K_{m3} + (Y_T - Y_P)} - \frac{k_4 Y_P}{K_{m4} + Y_P}$$

$$\frac{dR_P}{dt} = \frac{k_5 Y_P (R_T - R_P)}{K_{m5} + (R_T - R_P)} - \frac{k_6 R_P}{K_{m6} + R_P}.$$

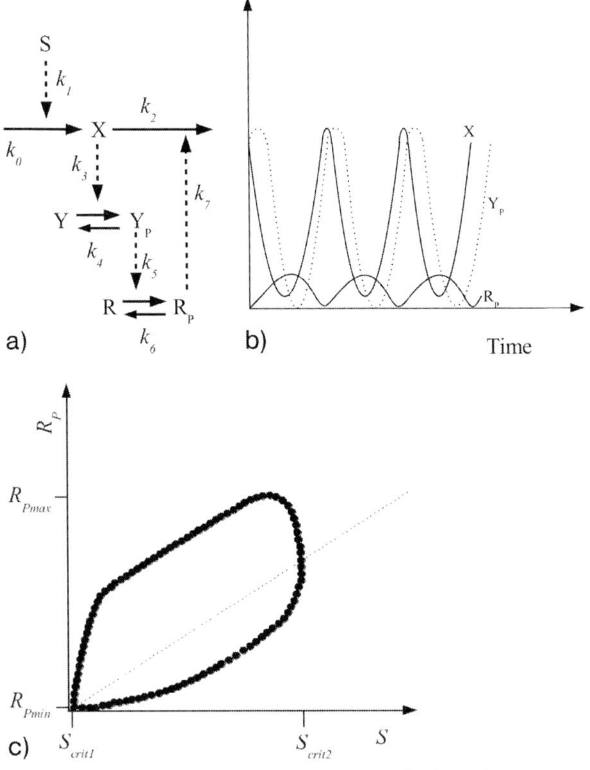

Figure 7.12 a) Three-component system with feedback loop. b) Feedback loop leads to oscillations of X (top solid line), Y_P (dashed line) and R_P (bottom solid line). c) Within the range $S_{crit1} < S < S_{crit2}$, the steady-state response $R_{P,ss}$ is unstable. Within this range, $R_P(t)$ oscillates between R_{Pmin} and R_{Pmax}.

Negative feedback has been proposed as a basis for oscillations in protein synthesis, activity of the mitosis-promoting factor, mitogen-activated protein kinase signaling pathways and circadian rhythms.

7.2.9
Cell Cycle Control System

As an example of a truly complex system that absolutely relies on the exact temporal control of all steps involved, we will now consider the wiring diagram for the cyclin-dependent kinase (Cdk) network regulating DNA synthesis and mitosis. Fortunately, this system can be partly decomposed into the motifs explained above. The network involving multiple proteins that regulate the activity of Cdk1–cyclin B heterodimers consists of three modules that oversee the G_1/S, G_2/M and M/G_1 transitions of the cell cycle (Figure 7.13). Let us first describe the individual components of this system. Cdk1 is a two-domain catalytic protein kinase domain whose function is to phosphorylate

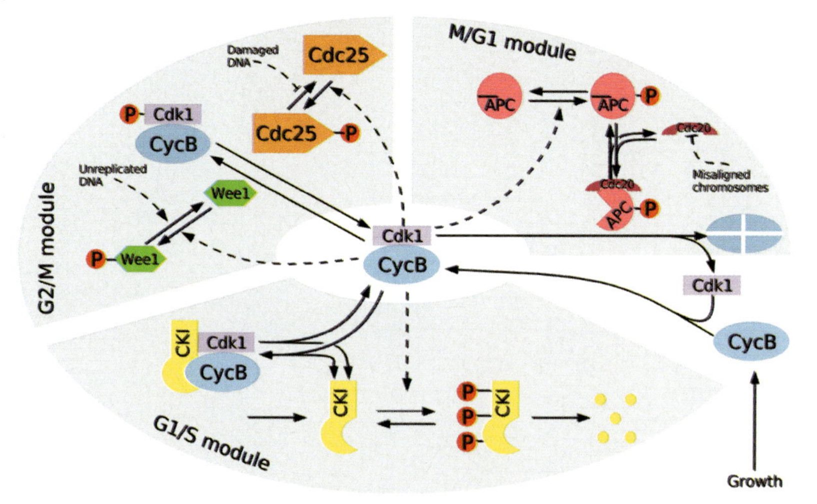

Figure 7.13 Wiring diagram of the cell cycle regulation in eukaryotes. Major events of the cell cycle are triggered by a Cdk1 in combination with cyclin B (CycB). The active dimer (see center) can be inactivated in the G_1/S module by combination with an inhibitor (CKI) or in the G_2/M module by phosphorylation of the kinase subunit by a kinase called Wee1. The inhibitory phosphate group is removed by a phosphatase (Cdc25). Cdk1 activity can also be destroyed by proteolysis of its cyclin partner, mediated by the anaphase-promoting complex (APC) in combination with Cdc20.

other proteins. It is activated by the binding of cyclin B and inhibited by the binding of the cyclin kinase inhibitor (CKI). Wee1 is another kinase that phosphorylates and thereby de-activates the Cdk1 bound to cyclin B. This phosphorylation can be eliminated by phosphatase Cdc25, thus re-activating Cdk1. Cdk1 itself activates Cdc25 by phosphorylation, meaning that this is a positive feed-forward mechanism. Cdk1 also phosphorylates Wee1, preventing its own phosphorylation by Wee1.

The G_1/S module is a toggle switch, based on mutual inhibition between Cdk1–cyclin B and CKI – a stoichiometric Cdk inhibitor (Figure 7.14). The G_2/M module is a second toggle switch, based on mutual activation between Cdk1–cyclin B and Cdc25 (a phosphatase that activates the dimer) and mutual inhibition between Cdk1–cyclin B and Wee1 (a kinase that inactivates the dimer) (Figure 7.15). The M/G_1

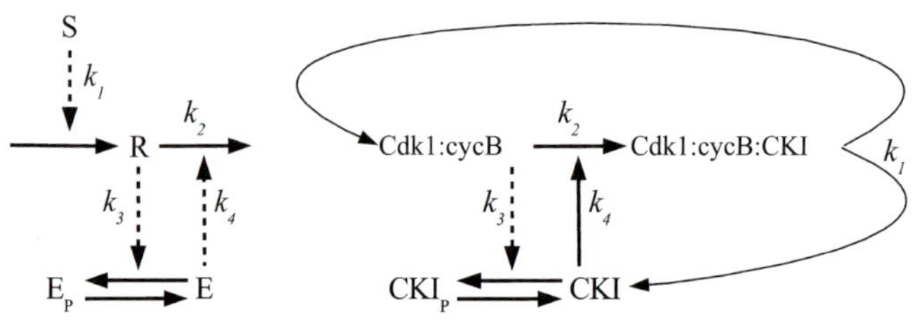

Figure 7.14 (Left) Three-component system with feedback loop (cf. Figure 7.10). (Right) Toggle-switch in the G_1/S module of the cell cycle.

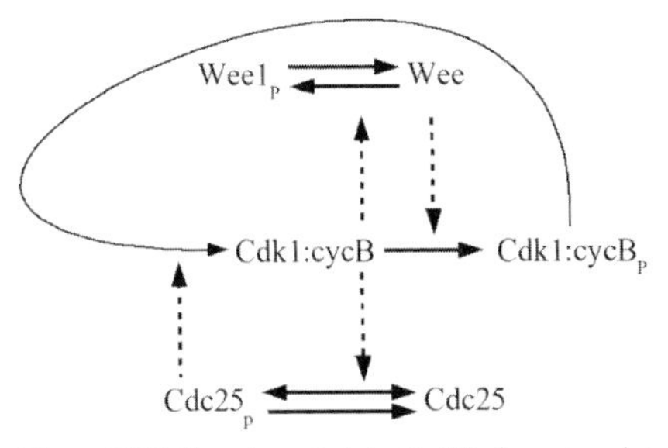

Figure 7.15 Toggle-switch in G_2/M phase involving the activating interaction of Cdk1:cyclin B on Cdc25 (cf. Figure 7.8) and the inhibitory interaction of Cdk1:cyclin B on Wee (cf. Figure 7.10).

module is an oscillator, based on a negative-feedback loop. Cdk1–cyclin B activates the APC, which activates Cdc20, which degrades cyclin B. The "signal" that drives cell proliferation is cell growth. A newborn cell cannot leave G_1 and enter the DNA synthesis/division process (S/G_2/M) until it grows to a critical size.

7.3 Partial Differential Equations

A **partial differential equation (PDE)** is a differential equation involving an unknown function of several independent variables and its partial derivatives with respect to those variables. PDEs are used to model many different types of processes that are distributed in space, or distributed in space and time, involving the propagation of sound and heat, electrostatics, electrodynamics, fluid flow or elasticity. Solving PDEs is an advanced field of mathematics.

We will restrict ourselves here to discussing the **diffusion equation** which describes density fluctuations in a material undergoing diffusion. This problem is closely related to many cellular phenomena. The equation is usually written as:

$$\frac{\partial \rho(\mathbf{x}, t)}{\partial t} = \nabla \cdot (D(\rho, \mathbf{x}) \nabla \rho(\mathbf{x}, t)), \tag{7.9}$$

with the density of the diffusing material ρ, the macroscopic diffusion coefficient D, the spatial coordinate $\mathbf{x}$ and the time t. We have already encountered the gradient operator:

$$\nabla := \begin{pmatrix} \partial/\partial x \\ \partial/\partial y \\ \partial/\partial z \end{pmatrix},$$

in Section 2.6. The diffusion equation above can be derived from (a) the *continuity equation* that states that a change in density in any part of the system is due to inflow and outflow of material into and out of that part of the system:

$$\frac{\partial \rho}{\partial t} + \nabla \cdot \mathbf{j} = 0,$$

with the flux **j** and (b) from the phenomenological *Fick's first law* that assumes that the flux of the diffusing material in any part of the system is proportional to the local density gradient:

$$\mathbf{j} = -D(\rho)\nabla\rho(\mathbf{x}, t).$$

This equation says that material will flow from a region of higher concentration to regions of lower concentrations. This is why **j** points in the direction of the negative gradient (Section 2.6). Inserting Fick's law into the continuity equation gives the diffusion equation (eq. 7.9).

For ease of understanding, let us look at the one-dimensional form of the diffusion equation, where we additionally assume that the diffusion coefficient is independent of ρ and t:

$$\frac{\partial\rho(\mathbf{x}, t)}{\partial t} = D\frac{\partial^2\rho(\mathbf{x}, t)}{\partial x^2}.$$

This equation describes a one-dimensional process of diffusion. One solution of this equation is the Gaussian function:

$$\rho(\mathbf{x}, t) = \frac{\rho_0}{\sqrt{2\pi Dt}}e^{-x^2/2Dt}$$

as can be verified by twice taking the derivative with respect to x times D and once with respect to time t. This solution describes how a certain amount of material that was initially all localized in one point at the coordinate origin, spreads out into the surrounding medium for $t > 0$. From this form of the solution, you can immediately imagine how diffusion proceeds. Initially, the Gaussian function is quite well focused. With increasing time, its width increases proportional to $\sqrt{t}$, but its amplitude decreases.

In computational cell biology, solving the diffusion equation by numerical methods allows modeling the diffusive motion of particles in cells. Some time ago, this may have sounded like a pure exercise to theoreticians. Nowadays, modern optical techniques allow tracking of fluorescent tracer molecules or Green Fluorescent Protein-labeled proteins. Therefore, comparing spatially resolved simulations with experiments has become more important than ever. An important program package in this area is the "Virtual Cell" initiative by the National Resource for Cell Analysis and Modeling at the University of Connecticut Health Center. The Virtual Cell environment (www.vcell.org) allows users from the experimental biologist's community to utilize a very sophisticated simulation package without having to deal with all the mathematical background. The package itself will model diffusional and transport processes by appropriate ODE and PDE approaches.

7.3.1
Spatial Gradients of Signaling Activities

In cells, proteins are not equally distributed throughout the cell volume, but are often localized to specific sites. Spatial gradients (differences of local concentrations) of protein activities may then organize signaling processes around cellular structures, such as membranes, chromosomes and scaffolds. The basic requisite for signaling gradients is the spatial segregation of opposing reactions (e.g. kinases and phosphatases) in a universal protein modification cycle (Figure 7.16). For a protein that is phosphorylated by a membrane-bound kinase and dephosphorylated by a cytosolic

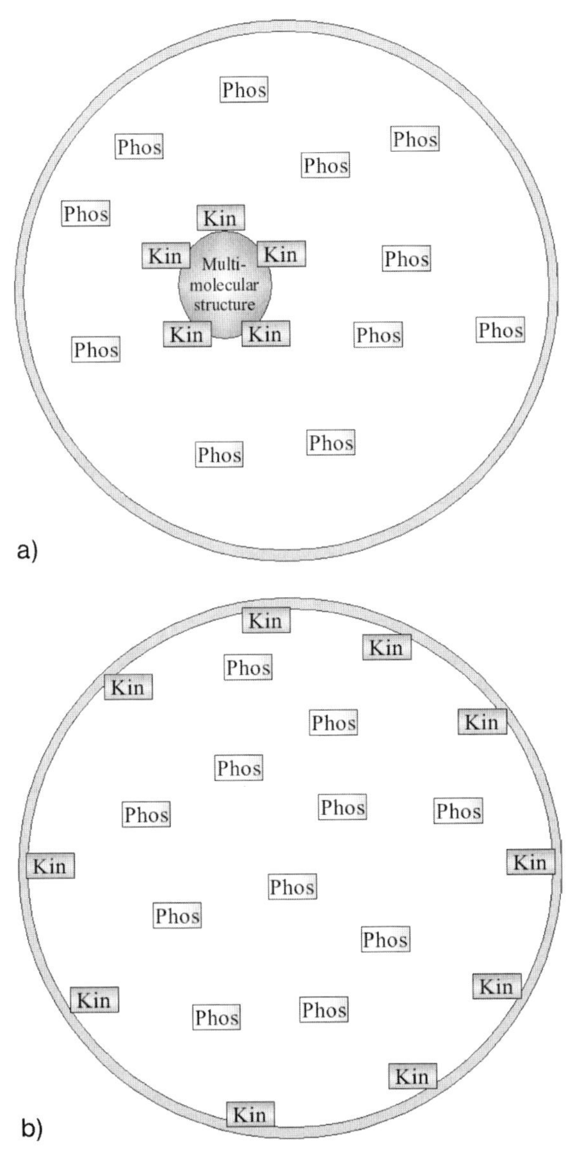

Figure 7.16 Spatial segregation of two opposing enzymes in a protein-modification cycle generates intracellular gradients. (Top) Kinases (Kin) localize to a supramolecular structure and the counteracting phosphates (Phos) are distributed elsewhere in the cytosol. (Bottom) Kinases localize to the membrane. Drawn after Kholodenko (2007).

phosphatase (see bottom part of Figure 7.16) a gradient of the phosphorylated protein will establish with high concentration close to the membrane and low concentration within the cell. The magnitude of the gradient will depend on the ratio of diffusion constants versus reaction rates of kinase and phosphatase.

7.4
Dynamic Monte Carlo (Gillespie Algorithm)

In the traditional computational approach to chemical/biochemical kinetics, one starts with a set of coupled ODEs that describe the time-dependent concentration of chemical species, and uses some **integrator** to calculate the concentrations as a function of time given the rate constants and a set of initial concentrations. Successful **applications** involve studies of yeast cell cycle, metabolic engineering, whole-cell scale models of metabolic pathways using the package E-cell and others.

This approach assumes that particle concentrations are constant in the reaction volume. On the other hand, cellular processes occur in very small volumes and frequently involve very small number of molecules. For example, one *Escherichia coli* cell contains on average only 10 molecules of the important Lac repressor. As a consequence, the modeling of reactions as continuous fluxes of matter is no longer correct. In such cases, one has to account for the fact that significant **stochastic fluctuations** occur. Popular approaches to study such scenarios have been stochastic formulations of chemical kinetics or Monte Carlo computer simulations.

In 1976, **Daniel Gillespie** introduced the exact **Dynamic Monte Carlo** (DMC) method (Gillespie, 1976,1977) that connects traditional chemical kinetics and stochastic approaches. In the usual implementation of DMC for kinetic simulations, each reaction is considered as an event and each event has an associated probability of occurring. Assuming that the system is well mixed, the rate constants appearing in the DMC and the ODE methods are related.

The probability $P(E_i)$ that a certain chemical reaction E_i takes place in a given time interval Δt is proportional to an effective rate constant $\tilde{k} = N/V$ with the volume V and to the number of molecules from the chemical species that can take part in that event, e.g. the probability of the first-order reaction $X \rightarrow Y + Z$ would be $\tilde{k}_1 N_X$ with N_x as the number of molecules of species X and $\tilde{k}_1$ as the effective rate constant of the reaction. Similarly, the probability of the reverse second-order reaction $Y + Z \rightarrow X$ would be $\tilde{k}_2 N_Y N_Z$.

As the method is a probabilistic approach based on "events", "reactions" included in the DMC simulations do not have to be solely chemical reactions. Any process that can be associated with a probability can be included as an event in the DMC simulations, e.g. a substrate attaching to a solid surface can initiate a series of chemical reactions. One can split the modeling into the physical events of substrate arrival, of attaching the substrate, followed by the chemical reaction steps.

7.4.1
Basic Outline of the Gillespie Method

(Step i) A list of the components/species is generated and the initial distributions at time $t=0$ are defined.

(Step ii) A list of possible events E_i (chemical reactions as well as physical processes) is generated.

(Step iii) Using the current component/species distribution, a probability table $P(E_i)$ is prepared of all the events that can take place. The total probability is computed of all events:

$$P_{\text{tot}} = \sum P(E_i),$$

where $P(E_i)$ is the probability of event E_i in a given time interval.

(Step iv) Two random numbers r_1 and $r_2 \in [0, \ldots, 1]$ are generated to decide which event E_μ will occur next and the amount of time τ by which E_μ occurs later since the most recent event.

Using the random number r_1 and the probability table, the event E_μ is determined by finding the event that satisfies the relation:

$$\sum_{i=1}^{\mu-1} P(E_i) < r_1 P_{\text{tot}} \leq \sum_{i=1}^{\mu} P(E_i),$$

The second random number r_2 is used to obtain the amount of time τ between the reactions:

$$\tau = 1/P_{\text{tot}} \ln(r_2).$$

As the total probability of the events changes in time, the time step between occurring steps varies. To complete one run of the simulation, steps (iii) and (iv) are repeated at each step of the simulation until some final time is reached. The necessary number of independent runs depends on the inherent noise of the system and on the desired statistical accuracy.

In this way, every process is attempted as frequent as its probability contributes to the total probability. We will see an example application of the Gillespie algorithm in Section 7.5.

7.5
Stochastic Modeling of a Small Molecular Network

As discussed before, the bioinformatic aspects of the new field of systems biology may include the graph description of protein–protein interaction networks, differential equation systems for signal transduction networks, matrix equations for metabolic networks or heuristic reverse engineering techniques for gene expression networks. All these were covered in previous chapters. However, typically, these descriptions do not consider the molecular nature of the individual gene and protein components. Here, we will now encounter a particular system where it is of essential importance to account for the effects of low particle numbers in a stochastic modeling approach utilizing variants of the Gillespie algorithm.

7.5.1
Model System: Bacterial Photosynthesis

The photosynthetic apparatus of purple bacteria is a very attractive model system for developing *in silico* approaches at the molecular level. In these evolutionarily old species, the photosynthetic machinery consists of only four different, relatively simple and well-known transmembrane proteins, i.e. the reaction center (RC), the light-harvesting complexes (LHC), the cytochrome bc_1 complex and the F_0F_1ATPase, as well as two transporter molecules, the soluble electron carrier protein cytochrome c_2 and the electron/proton carrier ubiquinone/ubiquinol that diffuses in the hydrophobic membrane. In the following, we will abbreviate ubiquinone and ubiquinol as quinone and quinol. Except for some mechanistic details, the biological function of each macromolecule is known precisely (Figure 7.17). Moreover, the three-dimensional structures of all components could be determined in recent years at atomic resolution (except for the membrane part of the F_0F_1ATPase).

In some species, such as in *Rhodobacter sphaeroides*, this system is spatially confined to small vesicles of 30–60 nm in diameter, which consequently contain a manageably small number of some 100 proteins in total, most of which are the simple light harvesting complexes. Figure 7.18 shows a cartoon of such a vesicle derived from stoichiometric and mechanistic considerations.

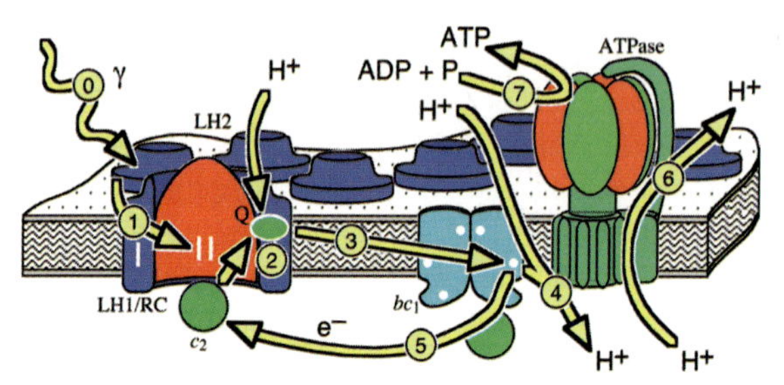

Figure 7.17 Artistic textbook-style rendering of the photosynthetic apparatus of purple bacteria. The inside of the chromatophore vesicle is below the membrane. The light harvesting complexes labeled LH2 and LH1 collect the incident photons (process 0) and hand their energy on to the reaction centers (RCs) in the form of electronic excitations (process 1). The RC passes this energy on onto a waiting quinone (Q) in the form of an electron–proton pair, where the proton (H^+) is taken up from the cytoplasm (process 2). Later, this quinone, which has become a quinol (QH2) by the uptake of a second of these pairs, unbinds from the RC and diffuses inside the membrane (process 3) to deliver its freight to the cytochrome bc_1 complex (bc1). The bc1 complex releases the protons to the inside of the vesicle and the stored energy is used to pump two further protons across the membrane (process 4). The electrons are then shuttled back to the RC by the water soluble electron carrier protein cytochrome c_2 (c2) (process 5), while the proton gradient is the driving force for the synthesis of ATP from ADP and inorganic phosphate in the F_0F_1–ATP synthase (ATPase) (processes 6 and 7). Drawn after Geyer and Helms (2006).

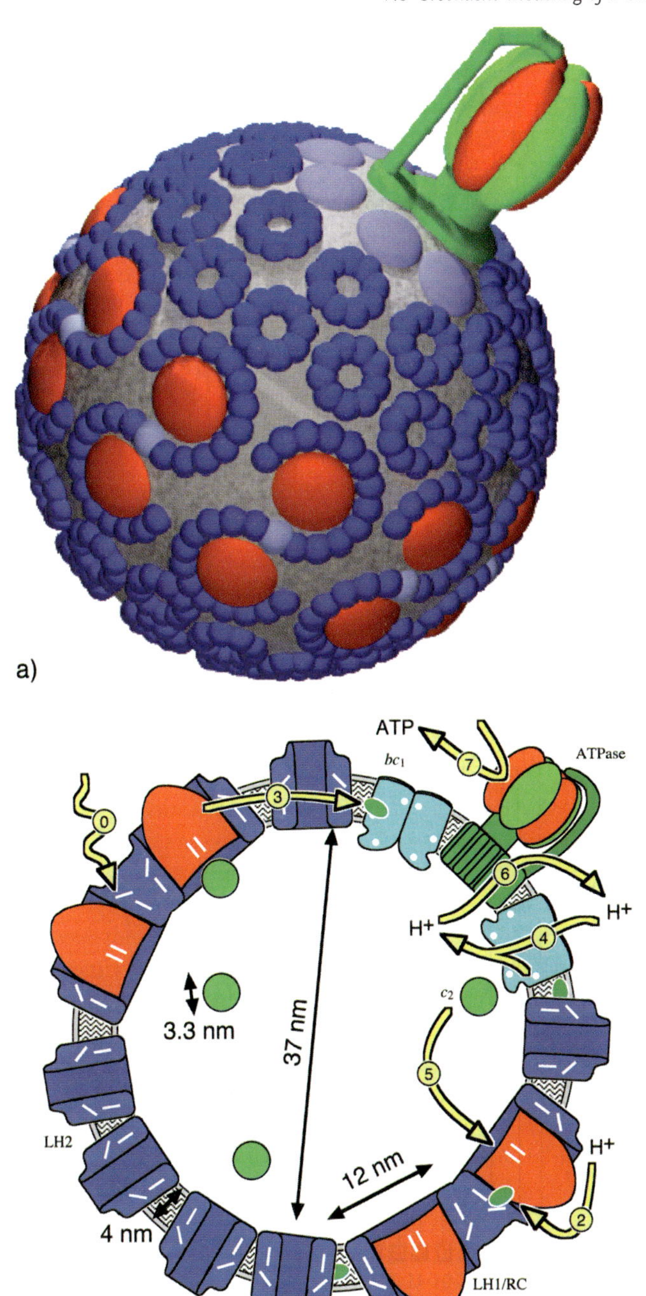
a)
ATP
bc_1
ATPase
0
3
7
6
H+
H+
4
3.3 nm
37 nm
c_2
5
LH2
12 nm
H+
2
4 nm
LH1/RC
b)

Consequently, these vesicles are already large and complex enough to represent a nontrivial metabolic subunit of a cell, but are still small enough to be considered at the molecular level. Also, they are easily accessible in experiments and, as light is the central 'metabolite' for them, they can be probed and monitored in dynamic experiments on timescales ranging from the picosecond scale for electronic transitions over the millisecond range for association and dissociation dynamics up to quasi-steady-state conditions.

We will now introduce a kinetic model of photosynthetic chromatophore vesicles from the purple bacterium *R. sphaeroides* that we compiled from a number of experimental references reported by various research groups up to the year 2006. As the experimental data was measured using a variety of different spectroscopic and physicochemical techniques either for individual reactions or for the entire vesicle, it was not clear at all whether it could all be combined into one general picture. Fortunately, we found the data to be overall consistent, so that the entire system could be assembled into a model of a conversion chain working at steady state conditions. Considering the stoichiometric relationships and the geometric dimensions of the individual proteins determined in electron microscopy and atomic force microscopy experiments also allowed us also to suggest a detailed spatial model of a chromatophore vesicle at molecular resolution as shown in Figure 7.18(a).

7.5.2
Pools-and-Proteins Model

A metabolic network, such as the one for bacterial photosynthesis, can be looked at from two different sides; from the network side with a focus on the various metabolites, and from the protein side where the proteins with their internal reactions form the building blocks. The protein based view corresponds to the classical microbiological approach, where the functions and structures of individual proteins are figured out first, while the network view reflects the experimentally accessible concentrations of the metabolites. The "pools-and-proteins" model shown in Figure 7.19 combines the details of the individual proteins, each with its own discrete internal states and reactions, and the network view of uniform concentrations of a discrete number of indistinguishable metabolite molecules.

Figure 7.18 (Top) A reconstructed chromatophore model vesicle of 45 nm diameter. The surface of this particle is 6300 nm^2 leaving just enough room to position one ATPase molecule (shown in green at the top), 11 LH1/RC rings (blue/red) around the circumference and the smaller LH2 rings (blue). The cytochrome bc_1 complexes are shown in light blue. The Z-shaped LH1/RC dimers form a linear array around the "equator" of the vesicle, determining the vesicle's diameter by their intrinsic curvature. The yellow arrows suggest diffusion of the protons out of the vesicle via the ATPase and to the RCs and bc_1s. (Bottom) Shown is a cut-section of the vesicle detailing some of the dimensions and illustrating the diffusion of the enclosed soluble electron carrier proteins cytochrome c_2. Drawn after Geyer and Helms (2006).

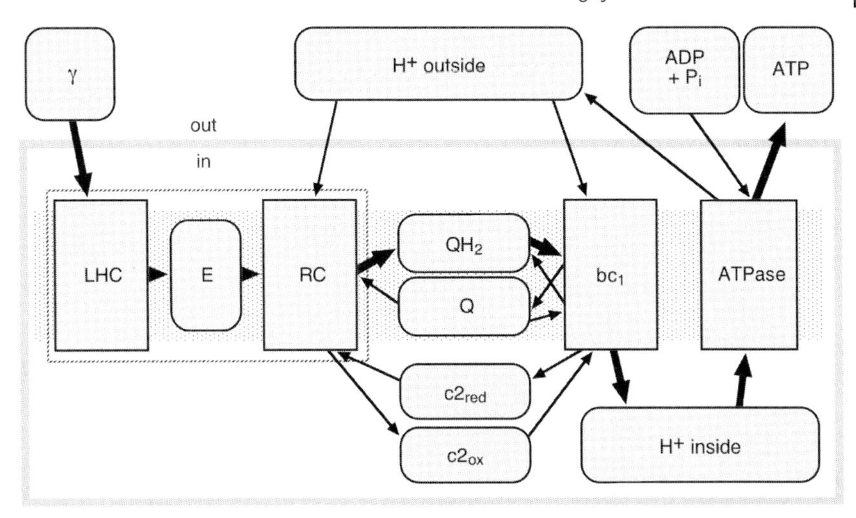

Figure 7.19 "Pools-and-proteins" view of bacterial photosynthesis. The network is formed from the proteins (rectangular boxes), which each exist in multiple copies, and one pool for each of the metabolites (rounded boxes), which are connected by the reactions (arrows). The path of the photon energy through the system into its chemical form of ATP is shown by the thick arrows. The core complex of RC and LHC with their own private exciton pool is enclosed by the dashed line. The membrane, indicated by the gray area, and the placement of the proteins and pools with respect to the membrane is not relevant for the simulation. Drawn after Geyer *et al.* (2007).

In the actual computational implementation (see below), the proteins are treated as the "machines" where the conversions between the metabolites take place. For these to take place as *in vivo*, certain metabolites that otherwise diffuse freely in the cytosolic or membrane compartments first have to bind to the proteins. Then the proteins act on the bound metabolites and modify them into some product forms, which are then released. Correspondingly, in this model each protein has a set of input connectors, where individual metabolite molecules are taken up from the corresponding pools and a set of outputs that finally release the product molecules into the respective pools. Thus, the network is built up from the respective numbers of the different proteins present in the system and a pool for each of the metabolites.

7.5.3
Evaluating the Binding and Unbinding Kinetics

In the stochastic molecular simulation each reaction is modeled as the binding and unbinding of discrete molecules according to the respective rate constants k_{on} or k_{off}. Importantly, these reactions often can only take place conditionally. A binding event, for example, can only occur if there is an empty binding site available. For an electron transfer to a bound substrate, on the other hand, this electron has to be available to the binding site. The objective is therefore to first check whether such a reaction can take place and then conditionally determine the probability for the reaction to occur during a time step, given the rate constant. For example, the binding reaction:

$$A + B \rightarrow A:B,$$

where a metabolite B binds to a protein A to form the complex A:B can only take place at a given protein A if its binding site for B is empty. Then the rate of binding events R_{on} at this protein A is proportional to the concentration of B, [B], and the association rate:

$$R_{on} = k_{on}[B].$$

When the time step Δt is small enough, i.e. $\Delta t \ll (k_{on} \times [B])^{-1}$, the probability P_{on} for this reaction to take place during Δt is well approximated by:

$$P_{on} = \Delta t \cdot k_{on}[B].$$

At each time step, the occupation state of each binding site of every protein is determined. P_{on} is then evaluated only for empty binding sites and compared to a random number r in the interval [0,1]. If $r < P_{on}$, one molecule B is placed in the respective binding site. Then the binding reaction needs not to be considered anymore until B unbinds again, often after some charge transfer has occurred.

To model the corresponding unbinding reaction:

$$A:B \rightarrow A + B,$$

with its rate k_{off}, a more efficient technique may be used than probing the reaction at every single time step. Note that the inverse of the rate constant k_{off} gives the lifetime T_{off} of the complex A:B. Correspondingly, a random unbinding duration t_{off} is chosen from the (nonnormalized) exponential probability distribution $P(t_{off}) = \exp[-t_{off}/T_{off}]$ of the unbinding times. A timer is then initialized with t_{off} that triggers the unbinding event. This timer approach can of course only be applied to unbinding reactions which are independent of the concentration of B. Otherwise, as for all association reactions, the reaction probability has to be determined at each time step in order to account for a possible change of [B] after the timer was initialized.

7.5.4
Pools of the Chromatophore Vesicle

A pool is characterized by the type of metabolite it contains and by its volume. In practice, a pool counts the number of particles and determines the respective concentration only when necessary. To be able to handle infinitely large pools such as the reservoir of protons outside of a vesicle or the incident light, a pool can be fixed to a given constant particle number. Pools do not contain any geometric information except for their volumes. However, special cellular geometries can be modeled by spatial discretization where each volume element ("voxel") is represented by a small pool which is connected to its direct neighbors.

According to the overview shown in Figure 7.19, a simulation of a chromatophore vesicle requires eight pools for biochemical metabolites (both cytochrome c_2 and

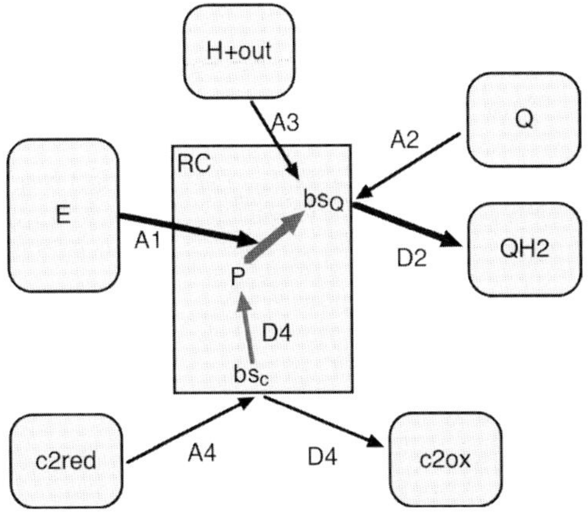

Figure 7.20 Reactions modeled in the RC. The respective rate constants are: A1 (exciton-initiated electron transfer from P to Q_b) = $10^{10}\,s^{-1}$; A2 (quinone binding) = $1.25 \times 10^5\,nm^2/s$; D2 (quinol unbinding) = $40\,s^{-1}$; A3 (proton uptake and transfer to Q_b-) = $10^{10}\,nm^3/s$; A4 (binding of reduced cytochrome c_2) = $1.4 \times 10^9\,nm^3/s$; D4 (reduction of P and unbinding of oxidized c_2) = $10^3\,s^{-1}$.

quinone in their oxidized and reduced forms, protons inside and outside of the vesicle, and one each for ADP and for ATP), one pool for the incident light and a number of exciton pools. The number of exciton pools depends on the number and connectivity of LHCs and RCs. An average-size vesicle of 45 nm diameter has an inner volume of $2.65 \times 10^4\,nm^3$, determining the volumes of the two pools of the reduced and oxidized cytochrome c_2. The quinones diffuse in the effectively two-dimensional plane between the two lipid layers of the vesicle membrane which has an area of $5.28 \times 10^3\,nm^2$. Correspondingly, this two-dimensional volume was used for the two pools of the membrane-bound quinones and quinols. Otherwise, the exact thickness of the quinone layer had to be known. As examples, Figures 7.20 and 7.21 describe the individual reactions modeled that are connected to the reaction center and to the cytochrome bc_1 complex, respectively.

7.5.5 Results for the Steady-State Regimes of the Vesicle

Even such a very simple biological system can already show biologically interesting behaviors under steady-state conditions. For low light intensities the supply of photons limits the throughput of the whole photosynthetic conversion chain, whereas for high intensities the total throughput of the cytochrome bc_1 complexes is the bottleneck. This is illustrated in Figure 7.22 that shows the steady-state rate of ATP production R_{ATP} versus the light intensity I from stochastic dynamic simulations of a vesicle consisting of 20 monomeric LHC/RC complexes, five cytochrome bc_1 dimers, one ATPase and 20 cytochrome c_2 molecules. For small light intensities,

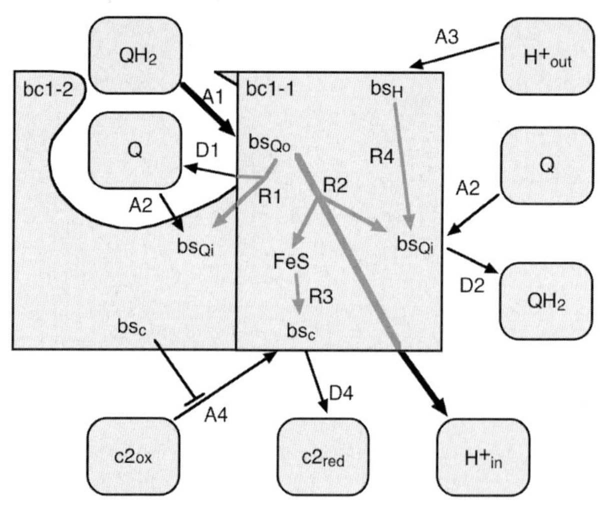

Figure 7.21 Reactions modeled in the bc_1 complex. The respective rate constants are: A1 (QH2 binding at Q_O) = 1.5×10^5 nm^2/s; D1 (Q unbinding from Q_O, if Q was not transferred directly to Q_i of the other dimer half in reaction R1) = 180 s^{-1}; A2 (Q binding at Q_i) = 1.5×10^5 nm^2/s; D2 (QH2 unbinding from Q_i) = 180 s^{-1}; A3 (binding of a proton from the outside) = 10^{10} nm^3/s; A4 (binding of oxidized c_2 with a mutual inhibition of the two binding sites of the dimmer) = 1.4×10^7 nm^3/s; D4 (unbinding of reduced c_2) = 1000 s^{-1}; R1 (swap of the Q unbinding from Q_O directly to Q_i of the other dimer half, otherwise reaction D1 may occur) = 10^4 s^{-1}; R2 (transfer of the electrons from Q_O to Q_i and FeS plus release of the protons into the vesicle; rate is exponentially reduced with increasing transmembrane voltage 300 s^{-1}) = $\Phi 0$ = 200 mV; R3 (movement of the FeS unit and electron transfer onto the bound c_2) = 120 s^{-1}; R4 (transfer of a bound proton to Q_i) = 1000 s^{-1}.

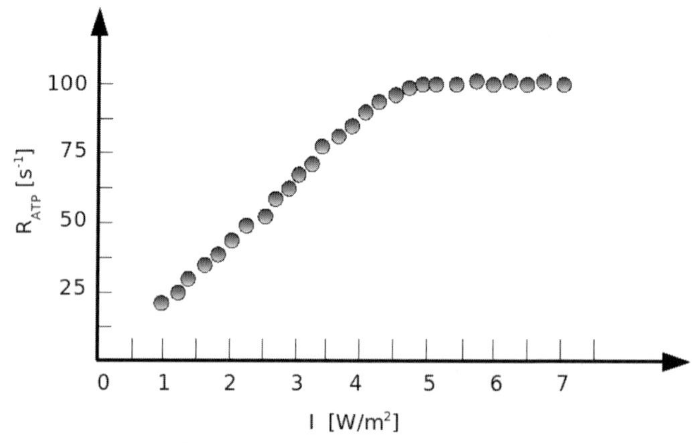

Figure 7.22 Rate of ATP production per second as a function of the light intensity and for different numbers of bc_1 complexes. At high light intensity, the production saturates because the bc_1 complexes are working at full speed and cannot pump more protons inside the vesicle.

R_{ATP} increases linearly with I before saturating at a light intensity above around 4 W/m^2. You may wonder about these small light intensities that are far smaller than that at a bright sunny day (around 1000 W/m^2). Interestingly, these bacteria live in a quite muddy environment at a light intensity of only about 18 W/m^2.

Which of the system components responds to the transition between the two different operation modes of the photosynthetic chain shown in Figure 7.22? This is apparent from Figure 7.23 that plots the redox state of the electron carrier cytochrome c_2 inside the vesicle. For small light intensities most of the 20 c_2 particles in the vesicle are reduced, while for high intensities they are essentially all oxidized. The "titration intensity" of a half-reduced c_2 pool corresponds to the transition intensity between the linear increase of R_{ATP} and the saturation regime. For little light, the RCs are slower than the bc_1s and, consequently, the c_2 are oxidized slower at the RCs than they can be reduced at the bc_1s. For high light intensities the situation is reversed. Then the RCs are faster and the oxidized c_2s are queuing up at the bc_1s, waiting to be reduced again. The transition between the two regimes takes place within a small intensity interval, as it results from the dynamic balance between the light-dependent turnover of the RCs and the light-independent rate of the bc_1s. Without the statistical fluctuations of the turnovers of the RCs and the bc_1s the transition would be a sharp step.

There are many more interesting biological phenomena that can be studied with this computational tool. Some are discussed in Geyer *et al.* (2007). However, as it is more instructive to perform such simulations yourself and to observe the reactions of the system to modifications of the parameters, the stochastic simulation of a chromatophore vesicle introduced here, termed Vesiweb, is available online at our website http://gepard.bioinformatik.uni-saarland.de/services/vesiweb.

As mentioned before, chromatophore vesicles are ideal systems to learn how to do computational molecular systems biology, because they are small, well understood and experimentally well characterized. In the scope of dynamical systemic

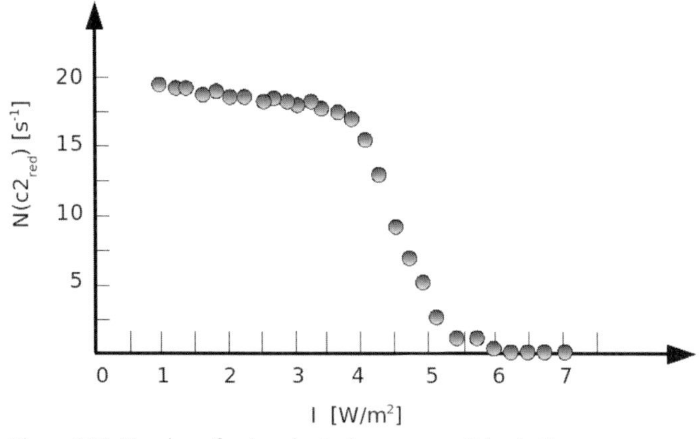

Figure 7.23 Number of reduced cytochrome c_2 particles in the stochastic simulation model at different light intensities.

simulations, they are a nearly unique system, because they can be naturally probed and monitored by light as today's technology allows to easily generate illumination signals of arbitrary shape and intensity.

7.6 Parameter Optimization with Genetic Algorithms

Although biological cells often only contain relatively small numbers of particular proteins and metabolites, biological cells can utilize them in an amazing variety of ways through different metabolic pathways or signaling cascades that often utilize the same proteins and metabolites. This frequent re-usage presents a huge challenge to computer simulations of metabolic systems. Even though we may only be interested in the temporal change of the concentration of a single metabolite, we often need to take into account large pieces of the full metabolic environment. The same problem naturally applies to experimental determinations of rate constants where some rates can be measured more easily, while others are masked by neighboring reactions. Instead of trying to understand the dynamic behavior of a metabolic system in a bottom-up approach from the kinetics of its individual reactions, one may also follow a systemic top-down approach and probe and analyze the whole network as one entity.

The following section will be based on the same simulation model of a chromatophore vesicle as introduced in the previous section. We will keep the concentrations of pools and proteins fixed at the values found to be consistent with experimental data, and we also fix the reaction rules of which system components interact with each other. However, we will now consider a set of 17 system properties such as the kinetic constants for association and dissociation reactions, the dielectric constant of the vesicle membrane, the cross-section of the LHCs for photon absorption, etc., as free parameters of an optimization problem. For the moment, we will ignore the fact that we actually do know quite a bit about the exact values from experiments and try to optimize the behavior of the system to best reproduce nine different sets of time-dependent experimental data. These are, for example, the change of the transmembrane voltage after a single light flash (measured as the change of light absorption at 503 nm minus that at 487 nm), the change of cytochrome *c* oxidation after a single light flash (measured as the change of absorption at 551 nm minus that at 542 nm), the change of the average oxidation state of the special pair subjected to a series of 16 fast light flashes, etc. To steer the kinetic parameters of the simulation system towards the correct part of the solution space, we need to score the behavior of the system by extracting global parameters from the simulations and compare those to the analogous experimental data. The correlation between the sets of experiments and simulations then allows judging both the consistency of the experimental data and the validity of the reconstructed *in silico* model.

Now we explain how the system parameters will be altered in an iterative process to arrive at one or multiple sets of optimal values that best reproduce the experimental behavior of the system. One iteration step (generation) consists of n simulations that are conducted in parallel for different system parameters and are scored afterwards. In the next step (generation), n new simulations will be performed using new, altered system parameters. To drive the system parameters toward the desired direction, one may use the concept of a genetic algorithm that is a well-established method used in optimization problems in high-dimensional spaces (Figure 7.24). For example, the new parameter generation could be a combination of: (a) the best parameter sets of the previous generation based on the scoring result, (b) random variations of some parameters of the best members of the previous generation, (c) crossings between the best members and (d) randomly generated new parameter sets. For simplicity, we used the same number of members from each of these four classes, although these proportions could be optimized too. To prevent rerunning parameter sets that were evaluated before, all previously executed runs are collected in a data base as n-dimensional vectors. Only vectors that are sufficiently different from all previous runs will be accepted as members of subsequent generations.

First, we idealize each experimental data set i by a suitable mathematical fit function $fit_i(t)$ that suppresses the stochastic noise of the various experimental data points (Figure 7.25). Δ_i^0 measures the deviation (e.g. the sum of quadratic errors) of the fit function $fit_i(t)$ from the corresponding experiment. We then extract from the simulation runs a property $sim_i(t)$ that corresponds to the property probed in

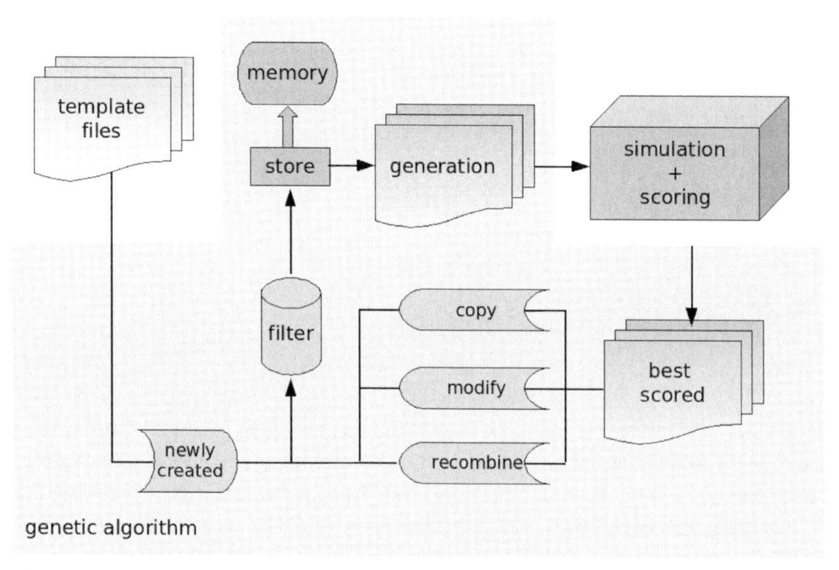

Figure 7.24 Optimization of kinetic system parameters by an evolutionary algorithm. A set of stochastic simulations (top, right) is run for a first "generation" of parameter sets. The results are scored and compared against suitable experiments (see text). The best parameter sets are kept unchanged and new combinations are generated from the promising parameter sets. This results in a new "generation" of parameter sets as input for a new set of simulations. This iterative cycle is repeated a given number of steps or until convergence is reached.

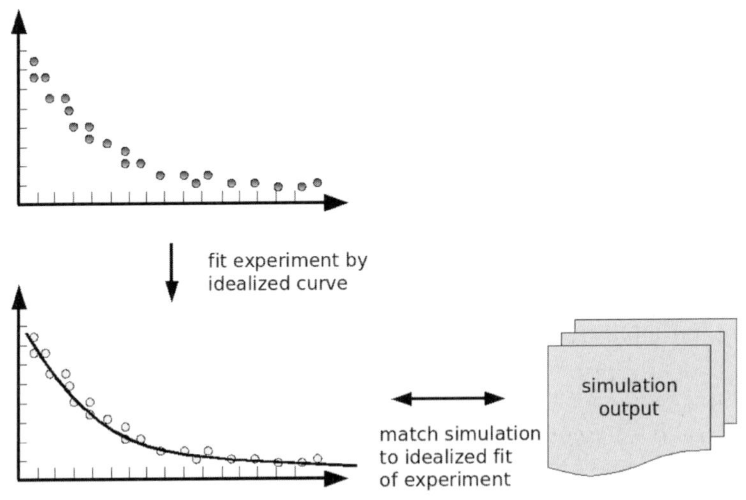

Figure 7.25 Fit of experimental results (top, left) by analytical curve (bottom, left) and comparison to simulation output using different parameter sets.

experiment i. This simulation property is then compared to the idealized experiment resulting in the following deviation (Figure 7.25):

$$\Delta_i = \frac{1}{T}\sum_{t=1}^{T}(fit_i(t) - sim_i(t))^2.$$

This score is normalized by Δ_i^0 as defined before:

$$score_i = \frac{\Delta_i}{\Delta_i^0}.$$

This prevents individual scores adopting very large values. Individual scores for reproducing individual experiments may be combined into a master score for the set of simulations using a particular parameter set:

$$masterscore = \prod_{i=1}^{N} score_i.$$

In this way, a failure to reproduce one single experiment will necessarily result in a quite bad score.

In principle, searching a 17-dimensional parameter space is a horrendous undertaking and one may be skeptical whether this procedure may converge anywhere near to the experimental values known beforehand. However, we found that the simultaneous optimization of 17 system parameters worked surprisingly well. Some parameters may be varied over a large range without affecting too much the kinetic aspects of the system that were probed in these simulation experiments. Others such as the k_{off} constant for unbinding of cytochrome bc_1 or the absorption cross section of

LHC complexes have a well-defined maximum value. Those are likely the parameters that are probed most sensitively by the experiments used as reference here. It is possible that the values of the broad-ranged parameters could be narrowed down further by having suitable experiments available. For the moment, we note that the computational optimization scheme could successfully generate a parameterized model simulation that yields similar time-dependent behavior as found in nine different time-dependent experiments on real vesicles. In retrospective, it appears quite reasonable that kinetic experimental data are useful fitting targets to optimize the kinetic parameters of individual system reactions. We note that such spectroscopic data is usually not very hard to generate for cellular systems. The problem so far was mostly the interpretation of this data because the spectroscopically "active" system components may sometimes belong to parts of the system that are considered "not interesting". However, we argue here that when combined with systemic simulations, such data may become quite valuable as is shown here for calibrating simulations against experiment.

In the future, similar techniques as introduced here for the chromatophore vesicles will be applied to more complex metabolic systems and signaling pathways thus helping to shed light onto processes and reactions that are not directly accessible to experiments.

Summary

Compared to other areas, kinetic modeling of cellular processes has a long history. This branch belongs to the fields of theoretical biology or complex systems that have strong roots in the mathematics community of the former Soviet Union. Remarkably, a few connections of signal and response substances suffice to give rise to very rich dynamic phenomena. Difficulties with modeling larger networks arise from lacking knowledge about enzymatic rate constants and those for biomolecular association and dissociation reactions. Apart from these "practical problems" this area is a well-established research field that has already made many important contributions to our understanding of cellular systems.

Problems

Dynamic simulations of networks

A static analysis of a (metabolic) network can reveal its steady-state properties such as the most important flux modes or identify seemingly redundant reactions. However, a network can exhibit a different or unexpected behavior, when subjected to time dependent concentration changes of the metabolites.

This is where dynamic network simulations come into play. For these dynamic simulations two major approaches exist: for large densities of the relevant molecules the network can be treated by a set of differential equations that describe the continuous time evolution of the densities, while for small densities, where the dynamics are governed by the binding and unbinding events of individual molecules, stochastic approaches are more appropriate.

Problem (1) first introduces you to the basic simulation techniques with a simple four-species network, before the second part exemplifies how the stochastic nature of individual binding and unbinding events can make signaling cascades sensitive to a few molecules.

(1) A simple reaction network

For this part consider the network displayed in Figure 7.26: two molecules of A associate to create one molecule of B, which is converted into substance D, when it encounters one molecule of C.

(a) Setting up the differential equations

A convenient recipe to compile the (sometimes complicated) set of differential equations that describe a system is to start from the stoichiometric matrix (Section 6.2). To do so first set up the stoichiometric matrix **S** for the above example network.

$$\frac{dR_1}{dt} = k_1 A^2$$

$$\frac{dA}{dt} = -2\frac{dR_1}{dt}$$

(i) Then walk through the columns of **S** to derive the rates dR_1/dt and dR_2/dt for the reactions R_1 and R_2, respectively, from the entries that have a negative sign (these are the educts for the corresponding reaction) (Figure 7.27).

(ii) Via the columns you can then figure out, which reactions contribute to the time evolution of a given molecule. This recipe is explained for R_1 and dA/dt in the figure above. Note that the stoichiometric matrix is not complete.

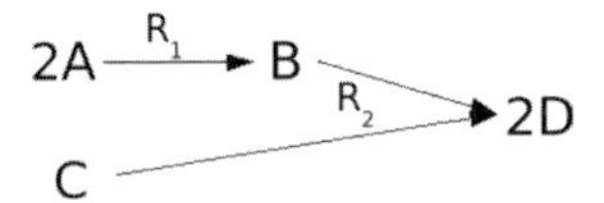

Figure 7.26 A simple reaction network (problem 1).

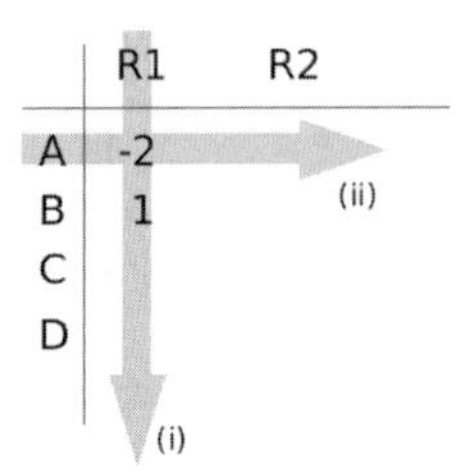

Figure 7.27 Construct stoichiometric matrix for reaction network of Figure 7.26.

From the complete stoichiometric matrix, give dR_1/dt and dR_2/dt explicitly, and list the rates for the changes of A, B, C and D in terms of the rates for R_1 and R_2.

Note: The amounts of metabolites – A, B, C and D – are given as densities with units of particles per volume (e.g. $1/nm^3$, mol/l, etc.). Consequently, the actual size of the system (test tube) is neglected in this description.

(b) Implementation with rate equations

The simplest way of solving a differential equation is based on the Taylor expansion truncated after the linear term:

$$A(t+\Delta t) \approx A(t) + \Delta t^* \, dA(t)/dt = A(t) + \Delta A(t).$$

This simple approximation requires small time steps to be fairly accurate.

Note that the increments ΔA, ΔB, ..., are calculated at the beginning of the time step of size Δt. To use this so-called "Forward Euler" integrator to simulate the time evolution of the above reaction system, the densities A, B, C and D are initialized to their values at $t=0$ and then with these values the increments ΔA, ΔB, ..., are determined and added to A, B, ... Then the time is advanced to $t+dt$. These steps are repeated until the final time is reached.

Note: All increments are evaluated at the beginning of the time step, therefore take care to first determine all values of ΔA, ΔB, ..., and only then to add them to their respective variable.

With the differences per time step ΔA, ΔB, ΔC and ΔD implement a differential equation model of the above network. Use a time step Δt of 0.1 s and a final time of 250 s. At $t=0$ start from $A=20\,\mu m^{-3}$, $C=10\,\mu m^{-3}$ and $B=D=0$. Set the reaction constants to $k_1=10^{-3}\,\mu m^3/s$ and $k_2=3\times10^{-3}\,\mu m^3/s$.

Plot the time traces of $A(t)$, $B(t)$, $C(t)$ and $D(t)$ into a single plot, describe them, and explain from their behavior the dynamics of the network. For comparison, the traces of A and D should look as sketched in Figure 7.28.

Then, run the simulation until $t=100$ s and give the final values of A, B, C and D.

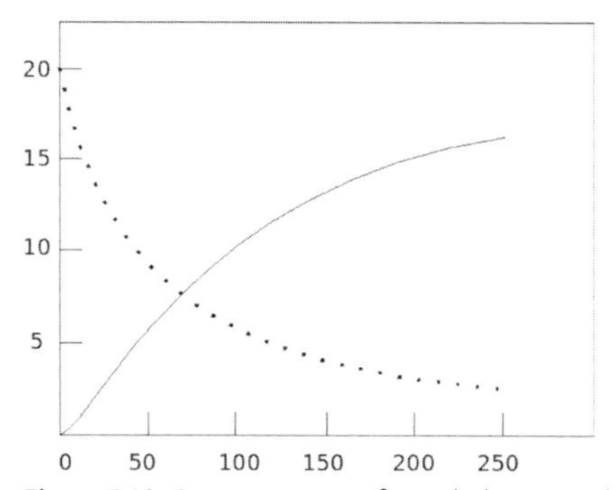

Figure 7.28 Concentrations of metabolites A and D at various time points.

(c) Stochastic implementation

Section 7.4 introduced the stochastic Gillespie method to solve a set of reactions. However, a more simple and direct method will be used here, where for each of the considered reactions the reaction probability P for a time step Δt is compared to a random number. If this random number is smaller than P, the event takes place during the actual time step and forms the corresponding product molecules.

Hint. How to calculate the reaction probabilities is explained here for the binary reaction $A + B \rightarrow AB$. The density of molecules of, for example, type A, [A], is determined by their number N_A and the corresponding volume V as $[A] = N_A/V$. With this, the rate for the change of the density of A becomes (the square brackets indicating the density are omitted in the following):

$$\frac{d[A]}{dt} = \frac{dA}{dt} = -kAB$$

$$\frac{1}{V}\frac{dN_A}{dt} = -k\frac{N_A}{V}\frac{N_B}{V}$$

$$\Delta A = \Delta t\frac{dN_A}{dt} = -\Delta t N_A N_B \frac{\Delta t}{V}.$$

With the change of the number of molecules of type A during Δt of $\Delta A = -P\, N_A N_B$, we can identify the probability for this bimolecular reaction between any of the As and any of the Bs to take place during Δt as $P = k\Delta t/V$.

Correspondingly, for the above network the parts of the inner loop that serve to integrate the number of molecules A look as follows:

```
while (t<= tEnd):
   dR1 = 0
if(NA >1): why this test???
       for i1 in range(NA):
           for i2 in range(NA):
               if random.random() <=P1:
                      dR1 += 1
                      dR2 =...
:
     NA  += (-2 * dR1)
     NB += ...
:
     t += dt
```

After each time step print out the densities of the molecules, i.e. N_A/V, etc., even though the simulation uses the particle numbers. Set the volume to 5 and 2 μm^3, respectively, and use the same rate constants and initial densities as above.

Hint: How many particles A do you need to have an initial density of 5 μm^{-3} at the given volumes?

Hint: Note that the volume enters into some of the probabilities.

For each of the volumes create a plot of the time traces $A(t)$, $B(t)$, ..., as from the continuous model and compare the three plots. Which differences do you observe? Explain your observations.

Hint: What do you observe, when you repeat the stochastic simulations a few times?

(d) Stochastic uncertainties

As a rule of thumb one can expect statistical fluctuations of $N^{1/2}$ for N particles. To check this rule, run the stochastic simulations 50 times until $t = 100$ s for both of the volumes of 5 and 2 μm^3. Determine the average numbers of A, B, C and D and their standard deviations σ_A, σ_B, σ_C and σ_D. For each molecule check whether $\sigma N^{-1/2}$ yields the same number at both volumes.

(2) Signaling

In a greatly simplified signaling cascade a time-dependent signal is given via the molecules S to the receptor R, which then switches into its activated state R* (Figure 7.29).

After some time, as determined by the rate constant k_2, the signaling molecule is degraded and unbinds as molecule T. While the receptor is activated, the soluble kinases K are phosphorylated. When the concentration of the phosphorylated form KP is above a certain threshold KP_0, the response reaction R_5 is enabled. The total amount of B, when all signaling molecules are used up, is used as a measure for the characteristic of this signaling reaction. In our model, A is assumed to be available in such quantities that its concentration is unchanged during a signaling event, i.e. $dA/dt = 0$.

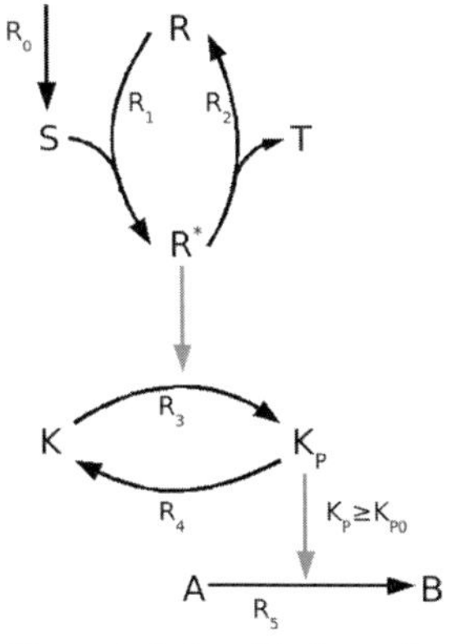

Figure 7.29 Simplified signaling cascade (problem 2).

(a) Setup

Proceed as in problem (1): set up the stoichiometric matrix, then determine the rates of the individual reactions and finally the rates of change of the metabolite concentrations. Note that dR_3/dt is determined by the concentration of R*, but that R* is not altered by R_3.

To model the time-dependent signal use:

$$R_0(t) = \frac{S_0}{V(t_{off} - t_{on})},$$

where S_0 is the total number of signaling molecules fed into the system during the interval $t_{on}, \ldots, t_{off}$. Instead of explicitly modeling a bistable switch for the response reaction (Section 7.2.3) use:

$$\frac{dR_5}{dt} = k_5 A \Theta(KP - KP_0),$$

with the step function $\Theta(x) = 1$ for $x \geq 0$ and 0 otherwise.

(b) Deterministic model: sensitivity threshold

Implement a deterministic model with the continuous densities and plot the time traces of the densities S, KP, R* and B into a single plot. Create two plots, one with $S_0 = 11$ molecules and one with $S_0 = 30$.

Run the simulation until $t = 2000$ s with a time step of $\Delta t = 0.1$ s. The signal is given between $t_{on} = 200$ s and $t_{off} = 220$ s. Use initial values of $R = 10\,\mu m^{-3}$, $K = 100\,\mu m^{-3}$ and $A = 10\,\mu m^{-3}$; all other metabolites have zero density initially. Set the volume to $V = 5\,\mu m^3$ and $KP_0 = 50\,\mu m^{-3}$. Set the rates $k_1 = k_3 = 0.01\,\mu m^3/s$ and $k_2 = k_4 = k_5 = 0.01\,s^{-1}$.

Now repeat the simulation for different signal strengths of $S_0 = 0, 1, 2, \ldots, 30$ molecules and plot the response N_B at the end of the simulation run, i.e. at $t = 2000$ s versus the signal strength S_0. Describe and explain the observed response characteristic.

Note that the step function of R_5 is fully visible in the response curve.

(c) Stochastic model: sensitivity by averaging

Before implementing a stochastic model of the signaling network, determine the probabilities $P_0, \ldots, P_5$. Again create two plots with the time traces of S, KP, R* and B for $S_0 = 11$ and 30 molecules and compare them to the results from the continuous model. Describe and explain your observations.

To get reproducible results for $N_B(S_0)$, run the stochastic simulation 30 times for every value of $S_0 = 0, 1, \ldots, 30$ molecules and plot the averaged response versus S_0. For better comparison also plot the result from the continuous model into this figure. What do you observe? Explain your findings.

Hint: For a biological interpretation consider that many discrete "macroscopic" events, e.g. an electrical spike on a neuron or the emission of a trafficking vesicle, need some kind of triggering with a threshold. The strength of the signal is finally encoded in the frequency of neuronal spikes or the number of vesicles. The above model can give you a clue about the interplay of stochastic kinetics, thresholds for macroscopic events and the enormous sensitivity of certain signaling cascades.

Further Reading

Differential Equation Models of Cellular Processes

Tyson JJ, Chen KC, Novak B (2003) Sniffers, buzzers, toggles and blinkers: dynamics of regulatory and signaling pathways in the cell, *Current Opinion in Cell Biology* **15**, 221–231.

Goldbeter A, Koshland DE (1981) An amplified sensitivity arising from covalent modification in biological systems, *Proceedings of the National Academy of Sciences USA* **78**, 6840–6844.

Kholodenko BN (2007) Untangling the signalling wires, *Nature Cell Biology* **9**, 247–249.

Stochastic Dynamics

Gillespie DT (1976) A general method for numerically simulating the stochastic time evolution of coupled chemical reactions, *Journal of Computational Physics* **22**, 403–434.

Gillespie DT (1977) Exact stochastic simulation of coupled chemical reactions, *Journal of Physical Chemistry* **81**, 2340–2361.

Geyer T, Helms V (2006) Reconstruction of a Kinetic Model of the Chromatophore Vesicles from Rhodobacter Sphaeroides, *Biophysical Journal*, **91**, 927–937.

Geyer T, Helms V (2006) A Spatial Model of the Chromatophore Vesicles of Rhodobacter Sphaeroides and the Position of the Cytochrome bc1 Complex, *Biophysical Journal*, **91**, 921–926.

Geyer T, Lauck F, Helms V (2007) Molecular Stochastic Simulations of Chromatophore Vesicles from Rhodobacter Sphaeroides, *Journal of Biotechnology*, **129**, 212–228.

8
Structures of Protein Complexes and Subcellular Structures

Chapter 1 introduced proteins as some of the key players of biological cells. Many of them are enzymes; others are membrane transporters or structural proteins that provide cell stability and motility. Enzymes are typically highly specialized "experts" that catalyze one particular biochemical reaction with high catalytic rate and specificity. The functional dependencies between them lead to the formation of biochemical pathways as one way of cellular organization where several enzymes are chained one after the other catalyzing several reactions on one molecule. A biochemical pathway may therefore involve many association and dissociation steps of this metabolite and its reaction products to various cellular enzymes interrupted by extended diffusional periods of the molecules before they find a representative of the next step of the pathway.

On the other hand, it is now well accepted that another important hierarchical element is the formation of protein complexes involving two up to 100 proteins. Many cellular functions are mediated by the structural association of separate proteins and their coordinated activities, not by random diffusion and transient associations. Protein complexes are thought to assemble in a particular order (Section 9.6) and often require energy-driven conformational changes, specific posttranslational modifications or chaperone assistance for proper formation. Furthermore, their composition is known to vary according to cellular requirements. To explain why complexes have evolved as an important form of cellular organization, we must resort to speculating that organizing proteins into large complexes must have functional advantages for the cell.

We have already introduced several such complexes, such as the ribosome formed by 80 proteins and rRNA and polymerase I formed by 10 proteins. We will now re-encounter these and a few other well-known examples also involving complexes where multiple enzymes of a biochemical pathway are permanently arranged into a large complex.

8.1
Examples of Protein Complexes

The first three examples show three protein complexes that execute the central processing from DNA over RNA to protein. The **RNA polymerase** shown in Figure 8.1

Principles of Computational Cell Biology – From Protein Complexes to Cellular Networks. Volkhard Helms

ISBN: 978-3-527-31555-0

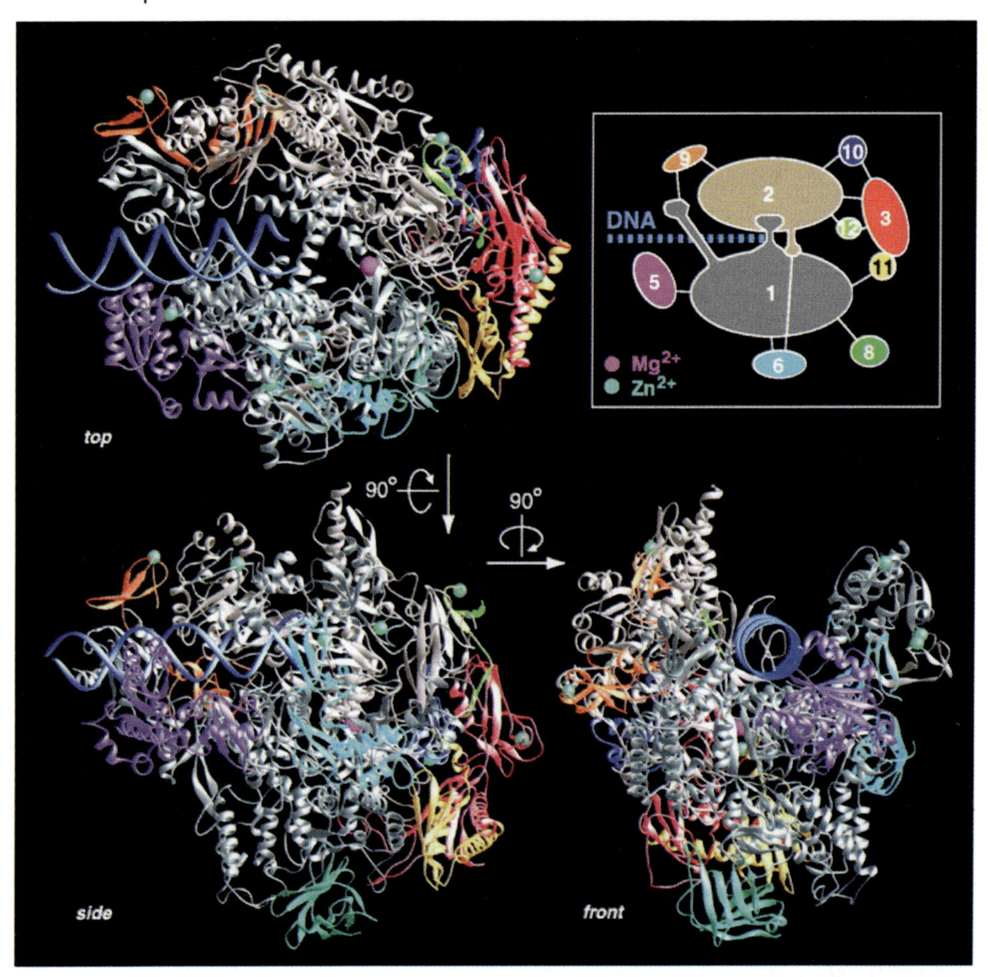

Figure 8.1 RNA polymerase II is the central enzyme of gene expression. It synthesizes all mRNA in eukaryotes. Shown is the high-resolution crystal structure taken from Cramer *et al.* (2000). Reprinted by permission from AAAS.

is the central enzyme of gene transcription and synthesizes a copy of mRNA from a DNA template. For determining the three-dimensional atomic structure of RNA polymerase, Roger Kornberg was awarded the 2006 Nobel Prize in Chemistry.

The **spliceosome** is a multicomponent macromolecular machine that is responsible for the splicing of pre-mRNA. Figure 8.2 shows a three-dimensional image reconstruction from cryo-electron microscopy images showing the spliceosome at a resolution of about 2 nm. The complex is composed of two subunits connected to each other with a tunnel in between that could accommodate the pre-mRNA component of the spliceosome.

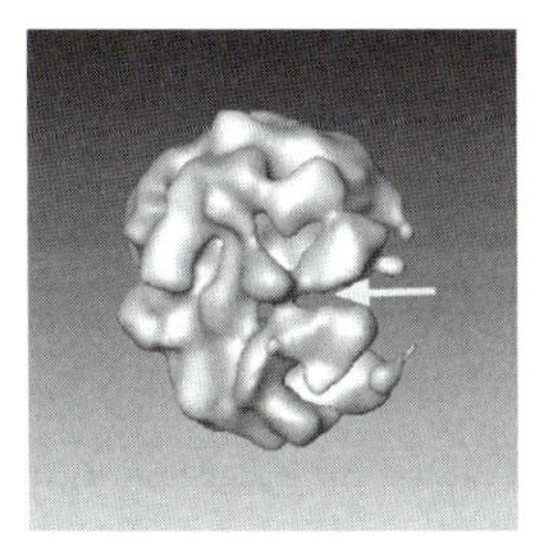

Figure 8.2 Spliceosome. Shown is a low-resolution structure of this cellular "editor" that "cuts and pastes" the first draft of RNA straight after it is formed from its DNA template. It has two distinct, unequal halves surrounding a tunnel. The larger part appears to contain proteins and the short segments of RNA, while the smaller half is made up of proteins alone. On one side, the tunnel opens up into a cavity, which is believed to function as a holding space for the fragile RNA waiting to be processed in the tunnel itself. Reprinted from Azubel *et al.* (2004) by permission from Elsevier.

The ribonucleoprotein called the **ribosome** is responsible for catalyzing the last step of the gene expression pathway where the genomic information encoded in mRNAs is translated into protein sequence. The prokaryotic ribosome is composed about two-thirds of RNA and one-third protein. It consists of two subunits, with the larger one (sedimentation at 50S in prokaryotes) being about twice as large as the small subunit. The large subunit shown in Figure 8.3 catalyzes the peptide bond formation.

The next example in Figure 8.4 shows the Arp 2/3 complex formed by structural proteins binding to the cytoskeleton. We will re-encounter this macromolecular complex in Section 8.8.

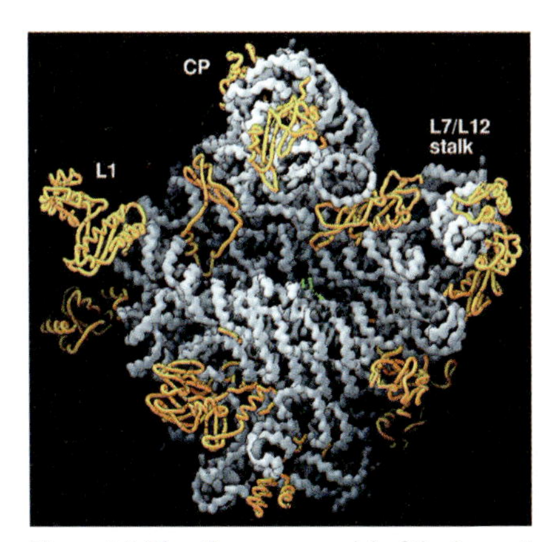

Figure 8.3 The ribosome: model of the large ribosomal subunit from *Haloarcula marismortui*. The protein is shown so that the surface of the subunit interacting with the small subunit faces the reader. RNA is shown in gray and the protein backbones in yellow. Reprinted from Ban *et al.* (2000) by permission from AAAS.

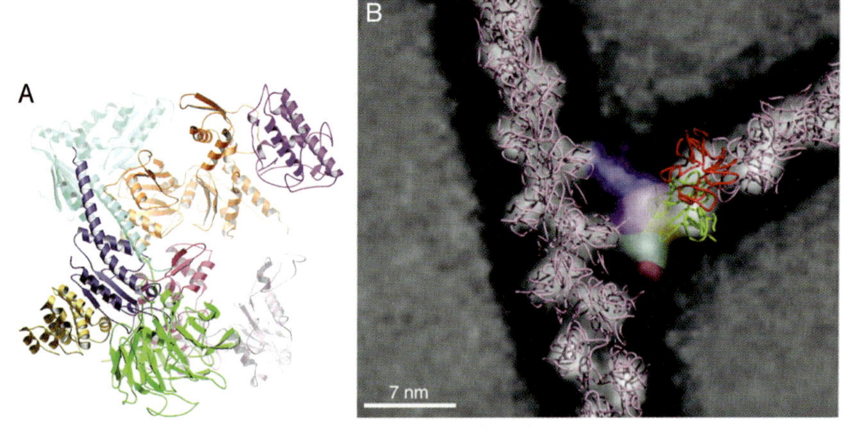

Figure 8.4 Arp 2/3 complex: the seven-subunit Arp2/3 complex choreographs the formation of branched actin networks at the leading edge of migrating cells. (A) crystallographic structure of the Arp2/3 complex. (B) Model of actin filament branches mediated by *Acanthamoeba* Arp2/3 complex. Reprinted from Robinson *et al.* (2001) and Volkmann *et al.* (2003) by permission from AAAS.

The **apoptosome** (Figure 8.5) is a key component of the process of induced cell death, termed apoptosis. It has an unusual 7-fold symmetry.

Figure 8.6 shows an example of a large multienzyme protein complex **pyruvate dehydrogenase**.

How can one best categorize these complexes? Some possible classifications would be categorizing them by their function, by their size or by their nonprotein components as protein complexes may also involve various other components involving nucleic acids, carbohydrates and lipids. One clearly functional classification scheme distinguishes **transient** complexes (enzyme inhibitor, signal transduction) from **stable/permanent** complexes with long lifetimes compared to those of typical

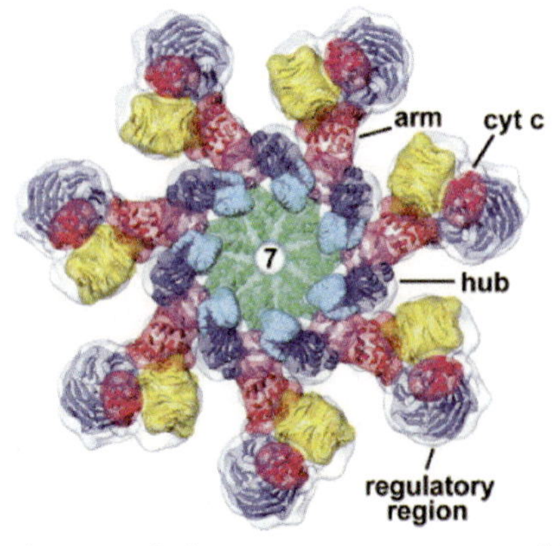

Figure 8.5 The human apoptosome. Reprinted from Yu *et al.* (2004) by permission from cell press.

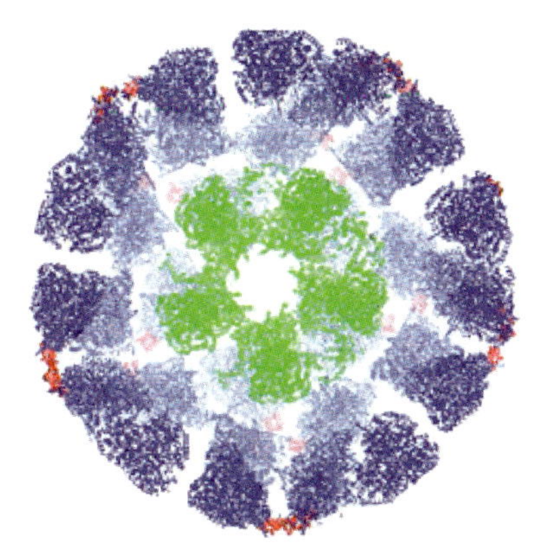

Figure 8.6 Pyruvate dehydrogenase is a huge multienzyme complex comprising 60 copies each of dihydrolipoyl acetyltransferase (E2), pyruvate decarboxylase (E1) and dihydrolipoyl dehydrogenase (E3), as well as binding proteins and regulatory kinases and phosphatases. The picture shows an atomic representation of the complete E1E2 complex involving the E2 catalytic and peripheral subunit-binding domains shown in green and in red as well as the E1 $\alpha_2\beta_2$ tetramers shown in purple (Milne *et al.*, 2002). The positions of the E1 $\alpha_2\beta_2$ tetramers and the peripheral subunit-binding domains were determined by core-weighted fitting to the density of the E1E2 complex. For visual clarity, only the back half of the model is presented. Reprinted from Milne *et al.* (2004) by permission from Macmillan Publishers Ltd.

biochemical processes. Obviously, obtaining structural information on transient complexes is much harder than for permanent complexes. One may also distinguish **obligate** and **nonobligate** complexes where the components of obligate complexes function only when in the bound state, whereas those of nonobligate complexes can also exist as monomers. Examples of the latter class are antibodies that exist in the free form in the cell until a suitable antigen target appears.

Similar to individual proteins, complexes do not need to be present in cells during the entire cell cycle, but only during phases when they are required. However, what is the "resting state" of a complex? Do the complexes exist as half-formed entities or are their components only expressed when needed for some job followed by their immediate assembly? The latter scenario would require a quite elaborate and efficient synthesis-and-assembly machinery. This question could be answered recently by analyzing gene expression data taken at various stages of the cell cycle which revealed which proteins oscillate with the frequency of the cell cycle. When overlapping this set with the set of high-quality complexes, all but two out of 200 complexes known at that time were not fully assembled during the entire cell cycle. Most of them remain half-formed in the cell once the fully assembled complex is not required (Section 10.1).

8.2 Complexeome of *S. cerevisiae*

In 2006, two research consortia published extensive surveys of the protein–protein interactions of the yeast *S. cerevisiae*.

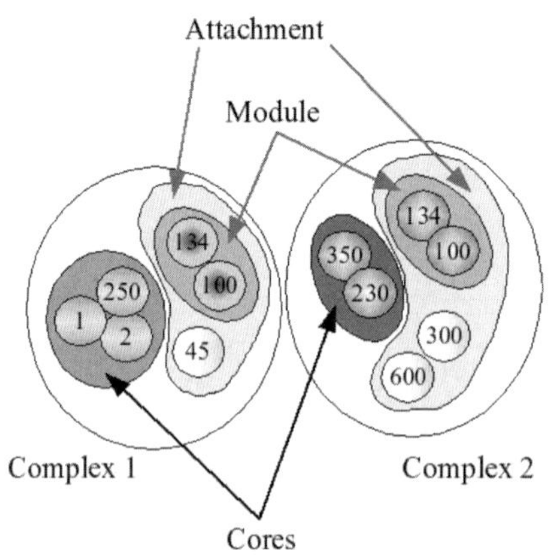

Figure 8.7 Definition and terminology used to define protein complex architecture (drawn after Gavin *et al.*, 2006). "Core" components are present in most isoforms, whereas "attachments" are present in only one of them. "Modules" are a subclass of "attachments" where two or more proteins are always together and present in multiple complexes.

Gavin *et al.* (2006) applied tandem affinity purification (TAP) (Section 3.2) to all 6466 open reading frames. In total, 1993 TAP fusion proteins could be purified, of which 88% were retrieved with at least one partner bound. A "socio-affinity" index was defined that quantifies the propensity of proteins to bind other proteins by measuring the log-odds ratio of the number of times two proteins are observed together, relative to what would be expected from their frequency in the data set. This analysis resulted in a list of 491 unique multiprotein complexes of two types – "core" components that are present in most isoforms and "attachments" present in only one of them (Figure 8.7). The cores of complexes contained 1–23 proteins with an average of 3.1 proteins. Among the attachments, in several instances two or more proteins were always found together and present in multiple complexes. These were termed "modules". On average, one such module is associated with 3.3 cores. The terminology "module" was already used in a different context in Chapter 4 where it described mostly functionally related proteins. Just think of "modules" as somehow tightly associated proteins.

Seventy-four known complexes could not be identified, possibly because they may not assemble under the growth conditions used in the experimental setup or because the purification tag interferes with complex assembly. By extrapolation based on the fraction of known complexes, the authors suggested that there may be an additional 300 core machines, leading to a total of 800 in yeast. When comparing the number of genes in yeast (6500) and humans (25 000), a simple extrapolation would then mean that there exist some 3000 core complexes in human cells. However, these would only be the stable, tightly bound complexes. The total number of complexes including transient and weakly binding ones may still be much higher.

A separate, parallel study (Krogan *et al.*, 2006) applied essentially the same methodology to all open reading frames of yeast and obtained 2357 purifications.

These were categorized into 547 distinct (nonoverlapping) heteromeric protein complexes. The average number of interactions per protein was 5.26 and the distribution of the number of interactions per protein followed an inverse power law (Section 4.1.1) indicating a scale-free topology. As expected, the authors found a significant amount of colocalization and semantic similarity (in terms of similarity of Genome Ontology annotation) among the members of particular complexes although the amount was lower than that in the list of curated MIPS complexes (Section 3.2.10).

It is quite encouraging that the two independent studies reported a similar number of successful purifications and an overall similar number of complexes. Due to the different definitions of complexes, the "core" complexes of the Gavin *et al.* (2006) screen ended up being smaller on average than those of Krogan *et al.* (2006). Future work will now compare the results of those two large-scale experiments and be complemented by additional data using different techniques. Thus, we can be quite hopeful that comprehensive lists of protein complexes will soon become available of other model organisms as well.

We keep in mind that the number of protein complex cores is small compared to the many cellular processes mediated by them. Shuffling functional modules between different complexes is an efficient way to multiply functionality and simplify temporal and spatial regulation.

8.3 Experimental Determination of Three-dimensional Structures of Protein Complexes

Section 3.1 already introduced several experimental methods that yield binary information whether a particular protein–protein interaction is present in the cell or not. This section will now briefly introduce several important experimental methods that are able to generate three-dimensional structural data on protein–protein interaction such as X-ray crystallography, nuclear magnetic resonance (NMR) spectroscopy, etc. Atomic protein structures are made available to the scientific community via the Protein Data Bank (www.rcsb.org).

8.3.1 X-ray Crystallography

X-ray crystallography is the most popular method for biomacromolecular structural determination. Successful structure determination yields very accurate structures of molecular complexes, from small molecules up to very large complexes such as the ribosome or viral capsids (Section 8.1). The basic principles of X-ray crystallography are shown in Figure 8.8.

In the early days of crystallography, reconstructing the molecular structure of the target molecule in the crystal was a huge numerical task. Modern software packages now allow processing the experimental data and structural modeling fairly automatically within a few hours of computing time. The two remaining bottlenecks of

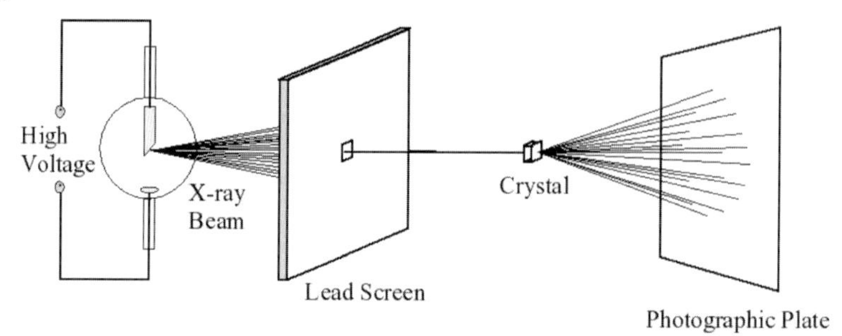

Figure 8.8 X-rays are electromagnetic waves in the ultra short ("hard") regime with wavelengths of the order of 0.1 nm. When they hit a sample, the electromagnetic X-rays undergo weak interactions with the electron clouds around the atomic nuclei which leads to partial diffraction of the incoming beam into different angles. As the interaction is quite weak, a noticeable diffraction intensity can only be detected in orientations where the diffracted beams from many molecules sum up in a constructive way. Here, we need to appreciate that electromagnetic waves are sinusoidal waves that may be described by an amplitude and phase. Therefore, intensities are only detected in those orientations where the path difference of waves originating from different molecules equals integer multiples of their phases. This requires, first of all, a very ordered orientation of all molecules like in a three-dimensional crystal. Still, in almost all orientations, the overlap of various waves will not be constructive. Images on the photographic plate (or charge coupled display detector) are recorded for various rotational orientations of the crystal. Structure determination involves reconstruction of the molecular structure of the target molecule that will give rise to the observed reflections. The numerical methods mostly involve Fourier transformation. A crystallographic structure determination ultimately reveals contours of the electron density. Atomic models are then refined using this electron density and information about typical bond lengths and bond angles of chemical bonds between atoms.

X-ray protein crystallography are (1) purification of milligram quantities of the protein(s) and (2) finding appropriate crystallization conditions where the proteins will assemble into three-dimensional crystals. In biological cells, proteins fortunately do not form crystals which would preclude their functioning. **Structural genomics** is the name for a series of large-scale research initiatives worldwide that tackle all these various steps in an automated manner. The crystallization trials are done by robots in parallel. Light diffraction may reveal which trials lead to microcrystals.

8.3.2
NMR

Some atomic nuclei (e.g. ^{1}H, ^{13}C, ^{15}N) possess nonzero magnetic moments. Under a strong external magnetic field, these nuclei orient themselves along this field. When an additional electromagnetic radio frequency field is applied, the nuclear spins will resonate with typical frequencies depending on their chemical environment. Interactions between different atomic spins via the nuclear Overhauser effect (NOE) then allow to determine the distances between ^{1}H nuclei that are less than 0.5 nm apart (cross-peaks). The measured distances can be used to compute three-dimensional structural models that agree with those distance restraints. One difficult task in protein NMR spectroscopy is assigning which resonance frequencies belong to which cross-interaction. Advantages of the methods are that no crystallization is required, the molecules are investigated under physiological conditions (in solution) and structural data is obtained at atomic resolution. Disadvantages are that the structural information is incomplete, NOE signals do not specify positions but distances and application of NMR is problematic for molecules larger than about 15 kDa due to overlapping signals. The last point is the main reason why NMR spectroscopy has not been frequently used in studies of large macromolecular assemblies.

8.3.3
Electron Crystallography/Electron Microscopy

The basic difference between electron crystallography and X-ray crystallography is that electron microscopy uses a fine beam of electrons pointed at the sample, instead of X-rays. Due to the much stronger interaction of an electron beam with the electrons of the molecular sample compared to that of the photons of an X-ray beam, electron microscopy can be applied to two-dimensional crystals (or even single molecules) instead of three-dimensional crystals that are required for X-ray crystallography. By combining data from two-dimensional images collected under varying angles or by averaging over thousands of images of single particles it is possible to reconstruct the three-dimensional structure of the diffracting molecule. The advantages of this method are that it does not require three-dimensional crystals, small amounts of protein are sufficient and the samples do not need to be as pure as in X-ray crystallography. Disadvantages of this method are that the electron beam is an ionizing radiation that destroys the probe over time, only small exposure times are possible, only medium resolution of 2 nm down to 0.4 nm can be achieved, and data collection and reconstruction may amount to several years for one molecular system. An ideal strategy is the combination of electron microscopy and X-ray crystallography where X-ray pictures of smaller subunits at high resolution are docked into medium-resolution electron microscopy maps of the full complex, see Section 8.4.

8.3.4
Immuno-electron Microscopy

Immuno-electron microscopy refers to imaging by electron microscopy of a target molecule together with an antibody that is labeled with a gold particle. The antibody is, for example, designed to bind specifically to one protein of a protein complex. Imaging will then detect the position of the gold particle within the complex by its increased contrast. This then allows to infer the position of the antibody bound to it and thus of the target protein bound to the antibody in the protein complex.

8.3.5
Fluorescence Resonance Energy Transfer

Fluorescence resonance energy transfer describes an energy transfer mechanism between two fluorescent molecules (Figure 8.9). A fluorescent donor is excited at its specific fluorescence excitation wavelength. By a long-range dipole–dipole coupling mechanism, this excitation energy is then nonradiatively transferred to a second molecule, the acceptor, while the donor returns to the electronic ground state. As the efficiency of this energy transfer decreases quickly with the sixth power of the inverse distance, the distance between donor and acceptor molecules can be deduced from observing the fluorescence of the acceptor and comparing it to a reference intensity, The described energy transfer mechanism is termed "Förster resonance energy transfer" (FRET), after the German scientist Theodor Förster. When both molecules are fluorescent, the term "fluorescence resonance energy transfer" is often used, although the energy is actually not transferred by fluorescence.

This method is very sensitive and specific and can even be applied between single molecules. Table 8.1 summarizes some key characteristics of the various experimental methods.

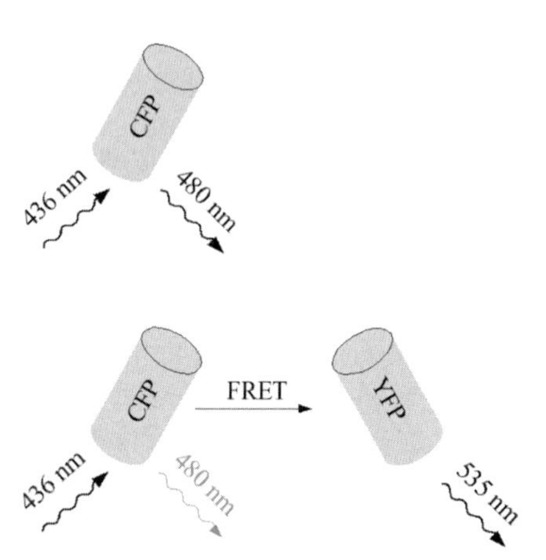

Figure 8.9 (Top) The chromophore of a Cyan Fluorescent Protein (CFP) absorbs light at 436 nm and emits light at 480 nm. (Bottom) This scenario involves the same CFP and a second protein, e.g. Yellow Fluorescent Protein (YFP), that absorbs light around 480 nm and emits light at 535 nm. If these two proteins are closer than 5 nm, the light emitted from CFP is partly absorbed by YFP due to fluorescence resonance energy transfer. In the upper scenario, illumination at 436 nm only leads to emission at one wavelength, in the bottom scenario one obtains two emission lines allowing one to conclude that the CFP and YFP molecules were closer than 5 nm. In the same way, additional proteins A and B may be fused to CFP and YFP to probe the interaction of A and B. A necessary condition for efficient fluorescence resonance energy transfer is that the emission spectrum of the first dye must overlap with the absorption spectrum of the second dye.

Table 8.1 Key data that can be obtained by various experimental techniques relevant to studying structural properties of protein complexes. After Sali *et al.* (2000).

	X-ray crystallography	NMR spectroscopy	Electron microscopy	Tomography	Immuno-electron microscopy	Fluorescence resonance energy transfer	Yeast two-hybrid	Tandem affinity purification	Mass spectrometry
Structure $\leq$ 3Å	×	×							
Structure $>$ 3Å	×	×	×	×					
Contacts	×	×	×	×		×	×	×	×
Proximity	×	×	×	×	×	×	×		
Stoichiometry	×	×			×			×	×
Complex symmetry	×	×	×	×	×				

8.4 Density Fitting

Over the past 15 years, many researchers have started using hybrid approaches for combining data from electron crystallography, showing the electron density of an entire complex at a somewhat lower resolution, and of high-resolution data from X-ray crystallography for individual components of the complex. The idea is that the constrained fitting of the atomic model would yield "pseudo-atomic precision" of the full complex or of parts of it with a 4–5-fold higher accuracy than the nominal resolution from electron microscopy. In the beginning, this task was completed manually by interactively manipulating the molecular objects on a graphics screen. However, manual procedures are very subjective and this task should better be done in an automated fashion. The problem can therefore be stated to perform an exhaustive or directed search for fitting a small piece of density into a larger density (Figure 8.10).

8.4.1 Correlation-based Fitting

The geometric match between two molecules A and B can be best measured when the shapes of the two molecules are discretized on a lattice. For this, a cubic lattice of sufficient size is placed around each molecule. A simple loop in a computer program then marches through all volume elements and assigns a value of 1 to them if any atom (or density) of the molecule is located in this volume element or 0 otherwise. The resulting shape functions of the two molecules on the

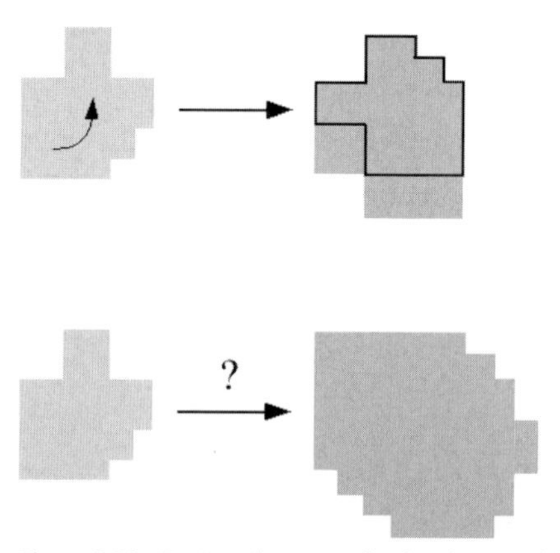

Figure 8.10 (Top) In this example, the shapes of the X-ray object (left) and the electron microscopy template (right) match fairly well. A simple 90° rotation will be sufficient to obtain a perfect match of the X-ray object in the lined area. (Bottom) Here, the X-ray object will fit into the much larger electron microscopy template in many ways.

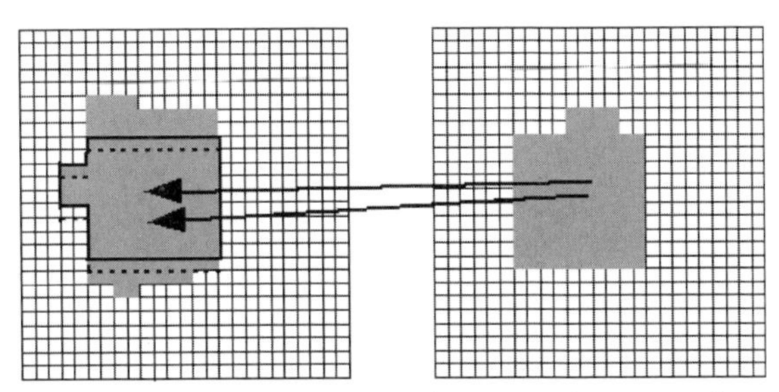

Figure 8.11 The steps involved in density matching. First, the shapes of the two molecules are discretized on a regular lattice. Then, the two lattices are overlaid at various relative positions and orientations, and the overlap of the two molecules is computed.

three-dimensional lattice with N^3 points and lattice indices l,m,n will be termed $a_{l,m,n}$ and $b_{l,m,n}$. Figure 8.11 shows a two-dimensional example of this discretization step.

Intuitively, we want to compute the overlap of the two densities after placing the two lattices on top of each other. However, what does "on top of each other" mean in mathematical terms? Orienting the two lattices can be done with respect to six degrees of freedom, three for translation along x, y and z, and three for rotation e.g. around each of these axes by angles α, β and γ. Among all these possibilities, one wishes to identify the relative orientation $x, y, z, \alpha, \beta, \gamma$ that minimizes the **sum of least squares**:

$$R(x,y,z,\alpha,\beta,\gamma) = (A - \mathbf{T}_{x,y,z}\mathbf{R}_{\alpha,\beta,\gamma}B)^2.$$

Here, $\mathbf{R}_{\alpha,\beta,\gamma}$ is a three-dimensional rotation matrix (Section 6.3.5) and $\mathbf{T}_{x,y,z}$ is a translation operator that translates molecule B to the position x, y, z. Minimizing the sum of squared errors is equivalent to maximizing the **linear cross-correlation** of A and B:

$$C_{x,y,z} = \sum_{l=1}^{N}\sum_{m=1}^{N}\sum_{n=1}^{N} a_{l,m,n} \cdot \mathbf{T}_{x,y,z}\mathbf{R}_{\alpha,\beta,\gamma} b_{l,m,n} \tag{8.1}$$

for a given translation vector (x,y,z) and rotation (α,β,γ) (Figure 8.11).

When fitting a high-resolution X-ray structure into an electron microscopy map of lower resolution, it has turned out to be beneficial to project the atomic structure B onto the cubic lattice of the electron microscopy data A by trilinear interpolation and convolute (i.e. "smear out") each lattice point $b_{l,m,n}$ with a Gaussian function g with a width corresponding to the lower resolution of molecule A. In this manner, the data sets will be compared at comparable resolution:

$$C_{x,y,z} = \sum_{l=1}^{N}\sum_{m=1}^{N}\sum_{n=1}^{N} a_{l,m,n} \cdot \mathbf{T}_{x,y,z}\mathbf{R}_{\alpha,\beta,\gamma}(g \otimes b_{l,m,n})$$

The complexity of computing this correlation for all translations in direct space is $O(N^6)$. The total effort is thus $O(N^6)$ times the number of rotations.

Researchers in the protein–protein docking field realized that this problem can be solved much more efficiently by applying fast Fourier transformation for the relative transformations of the shape functions $a_{l,m,n}$ and $b_{l,m,n}$ (see the Katchalski-Kazir algorithm in Section 8.7). We will therefore discuss in the next chapter the mathematical operation of Fourier transformation. Researchers working on the related problem of density fitting have borrowed that idea and the computation of the correlation coefficient is then formulated as:

$$C_{x,y,z} = FFT^{-1}\lfloor FFT(a_{l,m,n})^* \cdot FFT(b_{l,m,n})\rfloor.$$

A recent modification of this method only accelerates the three rotational degrees of freedom, while the three translational ones are simply scanned (Garzón *et al.*, 2007). By approximating the shapes of the molecules as linear combinations of spherical harmonics functions Y_{lm}, their Fourier transforms can be precomputed and tabulated. Using this modern variant, density fitting of objects as large as the ribosome (with 100^3 voxels) may be computed in CPU times of a few seconds to minutes.

8.5 Fourier Transformation

The **Fourier transform**, named after the French mathematician Joseph Fourier, is an integral transform that re-expresses a function in terms of sinusoidal basis functions, i.e. as a sum or integral of sinusoidal functions multiplied by some coefficients ("amplitudes"). Fourier transforms have many scientific applications in physics, signal processing, statistics, geometry and other areas. The discrete version of the Fourier transform (see below) can be evaluated quickly on computers using fast Fourier transform (FFT) algorithms. In this section we concentrate on those properties of Fourier transforms that are relevant for the topics of this textbook.

8.5.1 Fourier Series

A **Fourier series** is a representation of a periodic function $f(x)$ with period 2π as a sum of periodic functions of the form $x \rightarrow e^{inx}$ which are the **harmonics** (harmonics are functions e^{inx} with integer coefficients n; their name reflects their use to describe the vibrations of a string that is fixed at both ends such as in the violin) of e^{ix} using **Euler's formula**:

$$e^{ix} = \cos x + i \sin x \quad (8.2)$$

The Fourier series of a periodical function $f(x) = f(x + 2n\pi)$ is:

$$f(x) = \sum_{n=-\infty}^{\infty} F_n e^{inx},$$

where F_n are the (complex) amplitudes. The Fourier series for real-valued functions is often written using Eq. (8.2) as:

$$f(x) = \frac{1}{2}a_0 + \sum_{n=1}^{\infty}[a_n \cos(nx) + b_n \sin(nx)],$$

where a_n and b_n are the (real) Fourier series amplitudes:

$$a_n = \frac{1}{\pi}\int_{-\pi}^{\pi} f(x)\cos(nx)dx$$

$$b_n = \frac{1}{\pi}\int_{-\pi}^{\pi} f(x)\sin(nx)dx.$$

8.5.2 Continuous Fourier Transform

Most often, the term "Fourier transform" refers to the **continuous Fourier transform**, representing any square-integrable function $f(t)$ as a sum of complex valued exponentials with angular frequencies ω and complex valued amplitudes $F(\omega)$:

$$f(t) = F^{-1}(F)(t) = 1/\sqrt{2\pi}\int_{-\infty}^{\infty} F(\omega)e^{i\omega t}d\omega.$$

This is actually the *inverse* continuous Fourier transform, whereas the Fourier transform expresses $F(\omega)$ in terms of $f(t)$. The original function and its transform are called a *transform pair.* When $f(t)$ is an even or odd function, the sine or cosine terms disappear and one is left with the cosine transform or sine transform, respectively.

8.5.3 Discrete Fourier Transform

For use on computers, one must have functions f_k that are defined over *discrete* instead of continuous domains, again finite or periodic. In this case, one uses the **discrete Fourier transform**, which represents f_k as the sum of sinusoids:

$$f_k = \frac{1}{n}\sum_{j=0}^{n-1} F_j e^{2\pi ijk/n}, \quad k = 0, \ldots, n-1,$$

n is the number of grid points. Although applying this formula directly would require $O(n^2)$ operations, it can be computed in only $O(n \log n)$ operations using the FFT algorithm (see below) which makes Fourier transformation a practical and important operation on computers.

8.5.4 Convolution Theorem

One of the most useful properties of Fourier transformations is the immense facilitation in computing convolution integrals. In contrast to the product of two functions $f \cdot g$ – where the values of the two functions are simply multiplied at every point $f(t) \cdot g(t)$ – the **convolution** h of two functions f and g is defined as the integral over the products of the functions:

$$h(t) = \int_{-\infty}^{\infty} f(t')g(t-t')dt'.$$

Recall from Section 8.4.1 that we wanted to compute the correlation of the volumes of two molecules A and B for different relative translations and noted that this problem has a complexity of $O(N^6)$. We needed to compute:

$$C_{x,y,z} = \sum_{l=1}^{N}\sum_{m=1}^{N}\sum_{n=1}^{N} a_{l,m,n} \cdot \mathbf{T}_{x,y,z} b_{l,m,n}.$$

Taking the sum itself is not a convolution as the values of the functions A and $\mathbf{T} \cdot \mathbf{B}$ are simply multiplied at each grid point. This numerical task becomes a convolution when we want to evaluate $C_{x,y,z,\alpha,\beta,\gamma}$ for all possible translations.

Let us go back to the one-dimensional case. If $h(t)$ is the convolution of $f(t)$ and $g(t)$:

$$h(t) = \int_{-\infty}^{\infty} f(t')g(t-t')dt',$$

then the Fourier series transforms are related by:

$$H = 2\pi FG.$$

This means that once we have Fourier-transformed the functions f and g, computing their convolution becomes a simple multiplication!

8.5.5 Fast Fourier Transformation

Suppose that we have N consecutive sampled values of the function f at multiples of the sampling interval Δ:

$$f_k \equiv f(t_k),$$
$$t_k \equiv k\Delta$$
$$k = 0, 1, 2, \ldots, N-1.$$

Let us assume that N is even. The discrete Fourier transform of f is:

$$F_n \equiv \sum_{k=0}^{N-1} f_k e^{2\pi ikn/N}.$$

How much computation is involved in computing this sum? With W defined as the complex number:

$$W \equiv e^{2\pi i/N},$$

we can write:

$$F_n \equiv \sum_{k=0}^{N-1} W^{nk} f_k.$$

To evaluate this sum, the vector composed of f_ks needs to be multiplied by the matrix W^{nk} whose (n,k)th element is the constant W to the power $n \times k$. The matrix multiplication produces a vector result whose components are the F_ns. This matrix multiplication requires N^2 complex multiplications, plus a smaller number of operations to generate the required powers of W. So, the discrete Fourier transform appears to be an $O(N^2)$ process.

However, we will now see that the discrete Fourier transform (in 1 dimension) can be computed in $O(N \log_2 N)$ operations ($\log_2$ stands for the logarithm to the basis 2) by an algorithm called the **FFT**. For large N, the difference between $O(N^2)$ and $O(N \log_2 N)$ can mean a difference of CPU seconds versus CPU weeks! The FFT algorithm became generally known in the mid-1960s from the work of J. W. Cooley and J. W. Tukey. However, efficient methods to compute discrete Fourier transforms had in fact been independently discovered many times, starting with Gauss in 1805.

A discrete Fourier transform of length N can be rewritten as the sum of two discrete Fourier transforms, each of length $N/2$. One of the two is formed from the even-numbered points of the original N, the other from the odd-numbered points. Note that going from the first line to the second line, the counting index is changed from j to $2j$ (left sum) and to $2j+1$ (right sum), respectively. In this way, the even members of the sum in the first line are collected in the left sum of the second line and the odd ones in the right sum:

$$\begin{aligned}
F_k &= \sum_{j=0}^{N-1} e^{2\pi ijk/N} f_j \\
&= \sum_{j=0}^{N/2-1} e^{2\pi ik(2j)/N} f_{2j} + \sum_{j=0}^{N/2-1} e^{2\pi ik(2j+1)/N} f_{2j+1} \\
&= \sum_{j=0}^{N/2-1} e^{2\pi ikj/(N/2)} f_{2j} + W^k \sum_{j=0}^{N/2-1} e^{2\pi ikj/(N/2)} f_{2j+1} \\
&= F_k^e + W^k F_k^o.
\end{aligned}$$

Here, W is the same constant as before. F_k^e is the kth component of the Fourier transform of length $N/2$ formed from the even components of the original f_j values. F_k^o is the kth component of the Fourier transform of length $N/2$ formed from the odd components of the original f_js. The wonderful property of the *Danielson–Lanczos (DL) lemma* is that this decomposition can be used recursively. Having reduced the problem of computing F_k to that of computing F_k^e and F_k^o, we can again transform

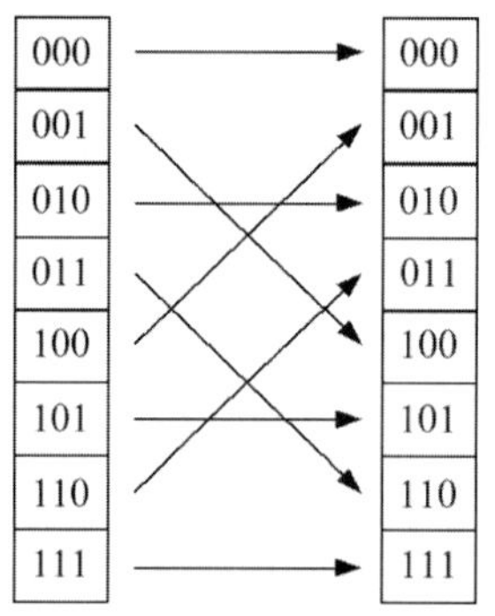

Figure 8.12 Reordering an array (here of length 8) by bit reversal. Bit reversal is a necessary part of the FFT algorithm.

the calculation of F_k^e to the problem of computing the transform of its *N*/4 even-numbered input data and *N*/4 odd-numbered data. We can continue applying the DL lemma until we have subdivided the data all the way down to transforms of length 1.

What is the Fourier transform of length 1? It is just the identity operation that copies one input number into one output slot. For every pattern of $\log_2 N$ e and o values, there is a one-point transform that equals one of the input numbers f_n for some *n*. The only missing element is now to figure out which value of *n* corresponds to which pattern of e and o values in:

$$F_k^{\text{eoeeoeo...oee}} = f_n.$$

The answer is to reverse the pattern of e and o values, then set e to 0 and o to 1, and we obtain, *in binary form*, the value of *n*. This works because the successive subdivisions of the data into even and odd are tests of successive low-order (least significant) bits of *n*. This idea of *bit reversal* can be exploited in a very clever way which, along with the DL lemma, makes FFT practical. Suppose we take the original vector of data f_j and rearrange it into bit-reversed order, so that the individual numbers are in the order not of *j*, but in the number obtained by bit-reversing *j*. The process is illustrated in Figure 8.12.

The given points are the one-point transforms. We combine adjacent pairs to get two-point transforms, then combine adjacent pairs of pairs to get four-point transforms, and so on, until the first and second halves of the whole data set are combined into the final transform. Each combination takes on the order of *N* operations, and there are $\log_2 N$ combinations of this type. This, then, is the structure of an FFT algorithm.

8.6 Advanced Density Fitting

The technique of density fitting based on correlation coefficients gives usually the best results when the surface edges of individual components in a complex are well defined and when there are only small regions of densities that cannot be assigned uniquely to a single component. However, as illustrated in Figure 8.10 (bottom), correlation mapping also faces situations where small fragments need to be placed into large templates. If the small fragments fit completely in the hull of the larger one, there may be a large number of ambiguous solutions where the smaller fragment is simply shifted around in the larger volume. In such cases, it turned out be helpful to emphasize the role of surface contours in the fitting procedure.

8.6.1 Laplacian Filter

A simple and computationally cheap filter for three-dimensional edge enhancement is the **Laplacian filter**:

$$\nabla^2 f = \frac{\partial^2 f}{\partial x^2} + \frac{\partial^2 f}{\partial y^2} + \frac{\partial^2 f}{\partial z^2},$$

that approximates the Laplace operator of the second derivative. (The full second derivative of f also involves the partial derivatives $\frac{\partial^2 f}{\partial x \partial y}, \frac{\partial^2 f}{\partial x \partial z}$ and so forth.) Applied to the density gradient on a grid, the Laplacian filtered density can be quickly computed by a finite difference scheme:

$$\begin{aligned}\nabla^2 a_{ijk} &= a_{i+1jk} - a_{ijk} + a_{i-1jk} - a_{ijk} + a_{ij+1k} - a_{ijk} + a_{ij-1k} - a_{ijk} \\ &\quad + a_{ijk+1} - a_{ijk} + a_{ijk-1} - a_{ijk} \\ &= a_{i+1jk} + a_{i-1jk} + a_{ij+1k} + a_{ij-1k} + a_{ijk+1} + a_{ijk-1} - 6a_{ijk}.\end{aligned}$$

where a_{ijk} and $\nabla^2 a_{ijk}$ represent the density and the Laplacian filtered density at grid point (i,j,k). The expression compares the values at grid points $+1$ and -1 along all three directions to the value of the grid point ijk (Figure 8.13).

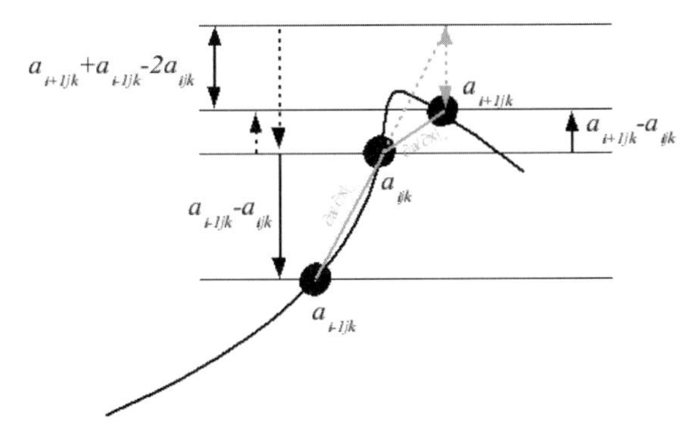

Figure 8.13 Schematic view of a Laplacian filter. a_{i-1jk}, a_{ijk} and a_{i+1jk} are the density values at three neighboring grid points in one direction. The grey lines denote the difference between the central point and the values to the left and to the right. These are finite difference approximations of the first derivative left and right of the grid point ijk. The dotted line and dotted arrow illustrate how the two first derivatives are combined to obtain an approximation of the second derivative at grid point ijk by finite difference as $a_{i+1jk} + a_{i-1jk} - 2a_{ijk}$.

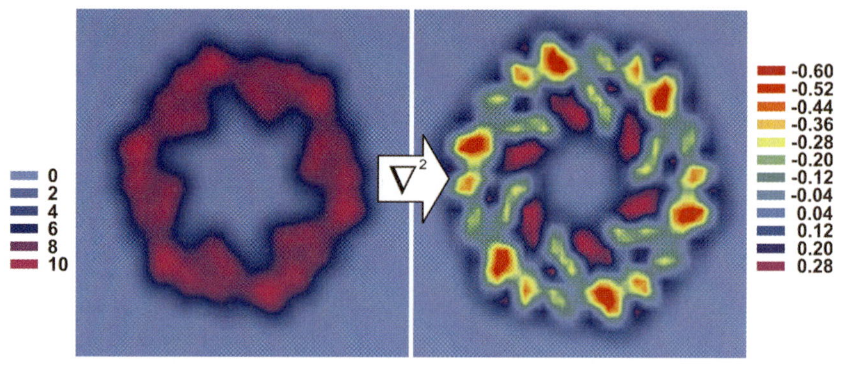

Figure 8.14 Example illustrating the effect of a Laplacian filter. The left picture shows a cross-section of 15 Å simulated density of the hexameric structure of the protein RecA. The right picture shows the same density after application of the Laplacian filter revealing much finer details of the electronic contour density. Reprinted from Chacón and Wriggers (2002) by permission from Elsevier.

The geometric match between two molecules A and B can then be measured by the Laplacian cross-correlation:

$$C = FFT^{-1}\left\lfloor FFT(\nabla^2 a_{l,m,n})^* \cdot FFT(\nabla^2 (g \otimes b_{l,m,n}))\right\rfloor$$

The effect of the Laplacian filter is demonstrated in Figure 8.14.

8.6.2
Fitting Using Core Downweighting

Core-weighted fitting is another technique to improve the sensitivity of standard correlation coefficients. It emphasizes the fitting of surface-exposed parts and reduces the importance of buried parts where the boundaries of individual components may not be clearly defined or may overlap significantly with each other (Wu *et al.*, 2003).

A **core index** f_{ijk} for grid point (i,j,k) describes the depth of the grid point with respect to the "core" of the object:

$$f_{ijk} = \begin{cases} 0 & a_{ijk} \leq \rho_c \text{ and } \min[f_{i\pm1jk}, f_{ij\pm1k}, f_{ijk\pm1}] = 0 \\ 0 & \nabla^2 a_{ijk} > 0 \text{ and } \min[f_{i\pm1jk}, f_{ij\pm1k}, f_{ijk\pm1}] = 0 \\ \min[f_{i\pm1jk}, f_{ij\pm1k}, f_{ijk\pm1}] + 1 & \text{otherwise} \end{cases} \tag{8.3}$$

where ρ_c is a cutoff density and $\min[f_{i\pm1jk}, f_{ij\pm1k}, f_{ijk\pm1}]$ represents the minimum core index of the neighboring grid points around grid point (i,j,k). The first condition is meant to suppress local density fluctuations that do not exceed a certain threshold. The second condition may appear counter-intuitive. It ensures that the core index rises at those points that are local elevations (hills) which are characterized by their negative second derivative.

This core index f_{ijk} is zero for grid points outside the core and increases progressively for grid points located deeper in the core (Figure 8.15). A grid point outside the core region must neighbor at least one other grid point that is also outside the core. On the other hand, a grid point within the core cannot neighbor a grid point outside the core unless it satisfies the condition $\nabla^2 a_{ijk} \leq 0$ and $a_{ijk} > \rho_c$. The core

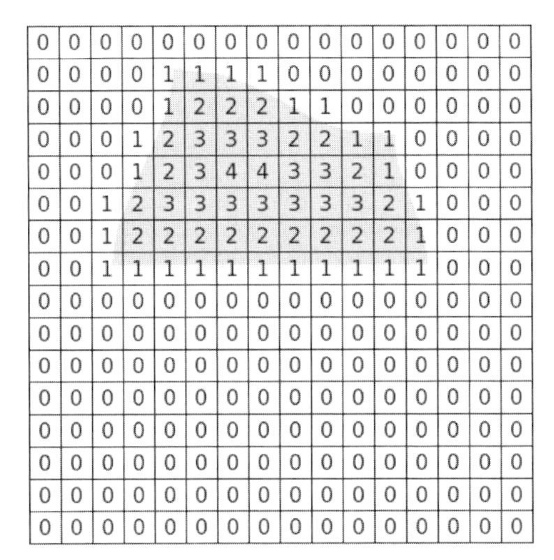

0	0	0	0	0	0	0	0	0	0	0	0	0	0	0	0
0	0	0	0	0	0	0	0	0	0	0	0	0	0	0	0
0	0	0	0	0	0	0	0	0	0	0	0	0	0	0	0
0	0	0	0	0	0	0	0	0	0	0	0	0	0	0	0
0	0	0	0	0	0	0	0	0	0	0	0	0	0	0	0
0	0	0	0	0	0	0	0	0	0	0	0	0	0	0	0
0	0	0	0	0	0	0	0	0	0	0	0	0	0	0	0
0	0	1	1	1	1	1	1	1	1	1	1	1	0	0	0
0	0	1	2	2	2	2	2	2	2	2	2	1	0	0	0
0	0	1	2	3	3	3	3	3	3	3	2	1	0	0	0
0	0	1	2	3	3	4	4	4	3	2	1	0	0	0	0
0	0	0	1	2	2	3	3	3	2	1	0	0	0	0	0
0	0	0	1	1	1	2	2	2	2	1	0	0	0	0	0
0	0	0	0	0	0	1	1	1	1	0	0	0	0	0	0
0	0	0	0	0	0	0	0	0	0	0	0	0	0	0	0
0	0	0	0	0	0	0	0	0	0	0	0	0	0	0	0

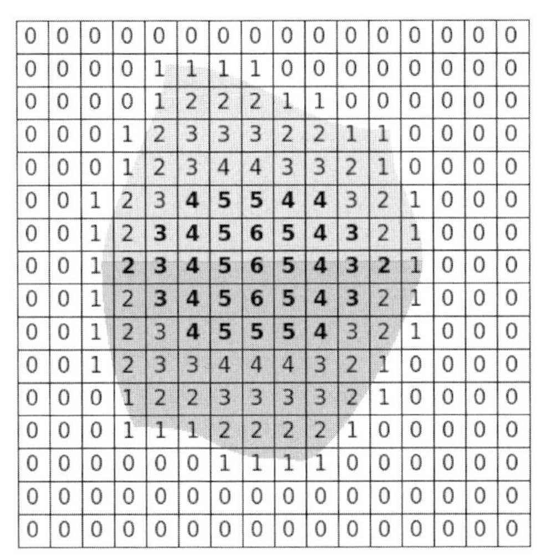

Figure 8.15 Example illustrating the effect of core weighting. Grid points outside and inside of protein A and B (top) and of the complex A:B (bottom) are labeled by the value of core index. For both proteins, the core index is 0 outside the domains, 1 at the outer edge and becomes larger inside the proteins. Bold labels are used for those grid points where the core indices of proteins A and B changes upon formation of the AB complex. Drawn after Wu *et al.* (2003).

index can be computed by an iterative procedure. First, all core indices are set to 1 except for the grid points at the boundary:

$$f_{ijk} = \begin{cases} 0 & i = 1 \text{ or } i = n_x \text{ or } j = 1 \text{ or } j = n_y \text{ or } k = 1 \text{ or } k = n_z \\ 1 & \text{otherwise.} \end{cases}$$

Then, a loop runs over all grid points and modifies their entries according to Eq. (8.3). This step is repeated until all grid points satisfy Eq. (8.3).

8.6.3
Core-weighted Correlation Function

Let us return to computing the correlation of two densities (Eq. 8.1):

$$\begin{aligned} C &= \rho_A \cdot \rho_{B,x,y,z,\alpha,\beta,\gamma} \\ &= \sum_{l=1}^{N}\sum_{m=1}^{N}\sum_{n=1}^{N} a_{l,m,n} \cdot \mathbf{T}_{x,y,z}\mathbf{R}_{\alpha,\beta,\gamma} b_{l,m,n}, \end{aligned}$$

and let us omit the translation and rotation operators from this expression for the moment. It is often preferable to use a normalized version of this expression in the form of a *Z*-score:

$$DC_{mn} = \frac{\overline{a_m a_n} - \overline{a_m}\,\overline{a_n}}{\delta(a_m)\delta(a_n)}, \tag{8.4}$$

where m and n refer to the two maps being compared:

$$\bar{a} = \frac{1}{n_x n_y n_z} \sum_{i}^{n_x} \sum_{i}^{n_y} \sum_{i}^{n_z} a_{ijk},$$

are the averages taken over the individual maps or over the product of the two maps computed at every grid point, and:

$$\delta(a) = \sqrt{\overline{a^2} - \bar{a}^2},$$

represents the average and fluctuation of the density. Considering Figure 8.15 and the objective to emphasize matching surface details over the interior of the complex, a weighting function is introduced to "downweight" the match between a region with low core index in the map of individual components and a region with high core index in the map of the complex:

$$w_{mn} = \frac{a_m^u}{a_m^u + v a_n^u + w},$$

where w_{mn} is the core-weighting function for the individual component, m, to the complex, n. The three parameters u, v and w control the dependence of the function on the core indices. (Wu *et al.* chose $u = 2, v = 1, w = 10^{-6}$.) Grid elements with equal core indices for the individual component and the complex ($a_n = a_m$) will be scaled down by a factor $1/(1+v+w/a_n^u)$. However, grid elements with a small core index in the individual component and a large core index in the complex will be scaled down more strongly. This is exactly what the authors planned.

Instead of computing the correlation from Eq. (8.4) above, the core-weighted correlation function is used:

$$DC_{mn} = \frac{\overline{(a_m a_n)_w} - \overline{(a_m)_w}\ \overline{(a_n)_w}}{\delta_w(a_m)\delta_w(a_n)},$$

where:

$$\overline{(X)_w} = \frac{\sum_{i,j,k} w_{mn}(i,j,k)X(i,j,k)}{\sum_{i,j,k} w_{mn}(i,j,k)},$$

and:

$$\delta_w(a) = \sqrt{\overline{(a^2)_w} - \overline{(a)_w}^2}.$$

Alternatively, one can compute the core-weighted Laplacian correlation (CWLC):

$$LC_{mn} = \frac{\overline{(\nabla^2 a_m \nabla^2 a_n)_w} - \overline{(\nabla^2 a_m)_w}\ \overline{(\nabla^2 a_n)_w}}{\delta_w(\nabla^2 a_m)\delta_w(\nabla^2 a_n)}.$$

These core-weighted correlation functions are designed to downweight the regions overlapping with other components, while emphasizing the regions with no overlap, which are typically the surface elements.

Core weighting was used to generate the docking model for the icosahedral multisubunit complex shown in Figure 8.6.

8.6.4
Surface Overlap Maximization (SOM)

Density correlation combined with suitable filters and core weighting are robust and reliable methods for detecting biological objects. When combined with FFT, density correlation also runs reasonably efficient. However, it is still desirable to look for more efficient alternatives. One such method, SOM (Ceulemans and Russell, 2004), exploits the fact that matching surface details is by far the most important element in density fitting. Recall from Section 8.2 that typical protein complexes have between five and 10 members. Think, for example, of a cube made from nine individual smaller cubes. The large cube only contains one small cube that is not at the surface. Therefore, we can expect that in small to medium sized complexes, many components will make contact to the surface of the complex. Instead of correlating volumes one may therefore try to correlate their surfaces. As the volume grows with r^3 and the surface only with r^2, r being the particle radius, this strategy appears promising in terms of computational efficiency.

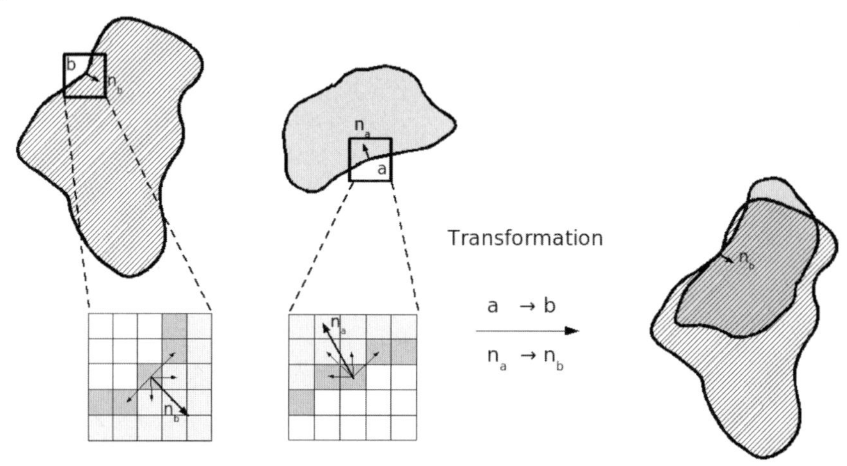

Figure 8.16 Principles of the SOM method.

The strategy of SOM is to score the goodness-of-fit by surface overlap as the fraction of surface voxels of the transformed target that are superimposed on the template surface. One needs to determine all combinations of translations and rotations (around the origin) that project at least one surface voxel of the target onto the template surface (Figure 8.16). This involves finding for every target surface voxel **a** and for every template surface voxel **b** a set of transformations that superimpose **a** onto **b**. Each such transformation can be decomposed into the unique translation of **a** to **b** and a rotation around **b**. If all rotations had to be searched exhaustively this process would be quite inefficient.

Interestingly, many rotations about **b** need not to be explored. If **a** really is the counterpart of **b**, the optimal transformation will superimpose the plane tangent to the target surface in **a** onto the plane tangent to the template surface in **b**. Therefore, only 1 rotational degree of freedom, around ν_b, has to be searched. In practice, the vectors ν_a and ν_b are approximated by considering all connecting vectors of **a** and those of their 26 spatial neighbors that contain the molecule.

The SOM method is a very fast method and will usually give many solutions. Unfortunately, it is usually difficult to distinguish the correct solutions from the incorrect ones.

8.7 FFT Protein–Protein Docking

Applications of the FFT method in structural bioinformatics involve image reconstruction, docking of X-ray structures into electron microscopy maps and protein–-protein docking. We have already encountered the problem of density matching in Sections 8.4 and 8.6. Here, we will now discuss the area of protein–protein docking.

As it is experimentally difficult and laborious to determine structures of protein complexes, and as many complexes may only exist transiently, it is very desirable to

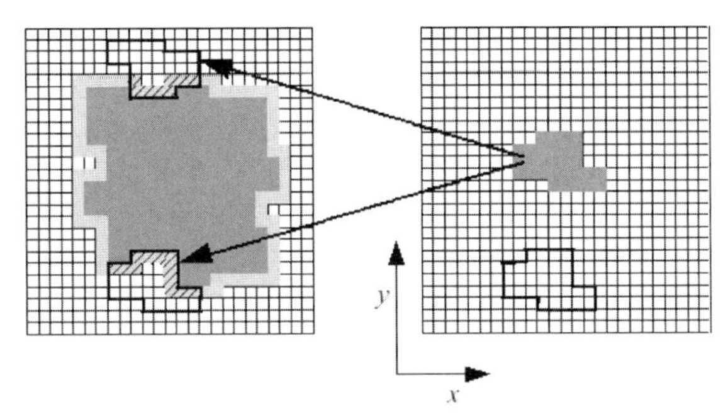

Figure 8.17 The left picture represents the shape of a protein when discretized on a regular grid. For simplicity, only two dimensions are shown. The darker area is the center of the protein, the lighter area the "surface". The right picture represents a smaller second protein. The surrounded area below the left protein shows the best possible fit of the second protein to the first protein that requires a translation in the x-direction by -1, a translation in the y-direction by -10 and no rotation. In this position, the shaded area marks the favorable overlap of the surface of protein 1 and the core of protein 2 (12 fields). The surrounded area above the left protein shows a second docking position that is less favorable (overlap only eight fields).

complement experimental studies by theoretical methods. The simplest approach to generate a putative model of a protein–protein complex is docking of two rigid protein structures. Here, we may exploit the experimental finding that **shape complementarity** is the main determinant of bound protein complexes. The task is therefore to find docked conformations without overlap which possess an optimal contact interface.

Maximization of shape complementarity is the basis for the most popular class of rigid body docking algorithms, the so-called correlation docking, that was introduced in 1992 (Katchalski-Katzir *et al.*, 1992). In this formulation, correlation-based docking algorithms use a purely geometric score to generate conformations featuring a large number of surface contacts and no significant overlap. As before in Section 8.4, the shapes of the two molecules will be discretized on a cubic lattice (Figure 8.17).

This time, the molecular geometries A and B are both represented by piecewise constant functions on two three-dimensional lattices with values a_{lmn} and b_{lmn}, that depend on whether this volume element belongs to the core of the protein, is at the surface of the protein, or outside:

$$a_{lmn} = \begin{cases} 1 \text{ on the surface of molecule A} \\ \rho \text{ inside molecule A}, \rho \ll 0 \\ 0 \text{ outside molecule A} \end{cases}$$

$$b_{lmn} = \begin{cases} 1 \text{ on the surface of molecule B} \\ 0 \text{ outside molecule B.} \end{cases}$$

The correlation of both grids is again computed as:

$$C_{x,y,z} = \sum_{l=1}^{N} \sum_{m=1}^{N} \sum_{n=1}^{N} a_{l,m,n} \cdot \mathbf{T}_{x,y,z} \mathbf{R}_{\alpha,\beta,\gamma} b_{l,m,n}.$$

This is the same formula that was used for fitting densities before. Note, however, that the $a_{l,m,n}$ and $b_{l,m,n}$ are defined differently so that overlap of the core regions of both molecules is heavily penalized. The best solutions are obtained when molecule B overlaps maximally with the grey-shaded surface of molecule A (Figure 8.17).

One important complication for this approach is that protein conformations are not rigid. Although backbone rearrangements are usually limited to within 0.2 nm, side-chains at the binding interface will adjust their positions upon binding to optimize the packing of the two surfaces in a process termed "induced fit". In particular, protein surfaces are covered with many long hydrophilic side-chains that may swing from one to another rotamer position with atomic coordinates changing by up to 0.5 nm. Therefore, the success of rigid-body docking has been limited. When the crystallographic structure of the complex is known, one can take the two or more proteins apart and use rigid-body docking to "redock" them into their bound conformation. The success rate for such exercises is quite high. However, as the correct answer was known beforehand, this is not a real biological problem to be solved. Much more interesting are predictions how two proteins bind of which we only know the conformations in the unbound state. In these cases, rigid-body docking based on shape complementarity alone often fails meaning that the correct solution only appears among the first few hundred solutions.

Fortunately, with additional experimental data at hand, it is often possible to employ distance filters to reduce the large number of solutions by requiring that certain pairs of amino acids have to be within a certain distance range. Also, one may score the docking solutions by additional properties such as electrostatic complementarity, sequence conservation or statistical potentials for amino acid propensities (Chapter 9).

8.8 Prediction of Assemblies from Pairwise Docking

As was just mentioned, the main difficulty in protein–protein pairwise docking is distinguishing the biologically correct from incorrectly docked conformations based on energetic or other criteria. This dilemma is mainly caused by conformational rearrangements at the binding interfaces due to induced fit effects upon association. Considering the relatively poor reliability of pairwise docking, there seems to be little hope to go beyond pairs of proteins and even attempt to predict the three-dimensional structures of oligomeric assemblies by docking methods. There, the combinatorial complexity seems even more problematic. Surprisingly, this task may actually be simpler than thought of. As will be shown it may be easier to correctly assemble a large macromolecular puzzle than any of its pairwise interactions.

In 2005, Nussinov and Wolfson introduced the first automated approach, termed CombDock, for predicting hetero multimolecular assemblies from structural models of their protein subunits (Inbar *et al.*, 2005). As the general solution of this problem is NP-hard, the main idea of the approach is to exploit additional geometric constraints

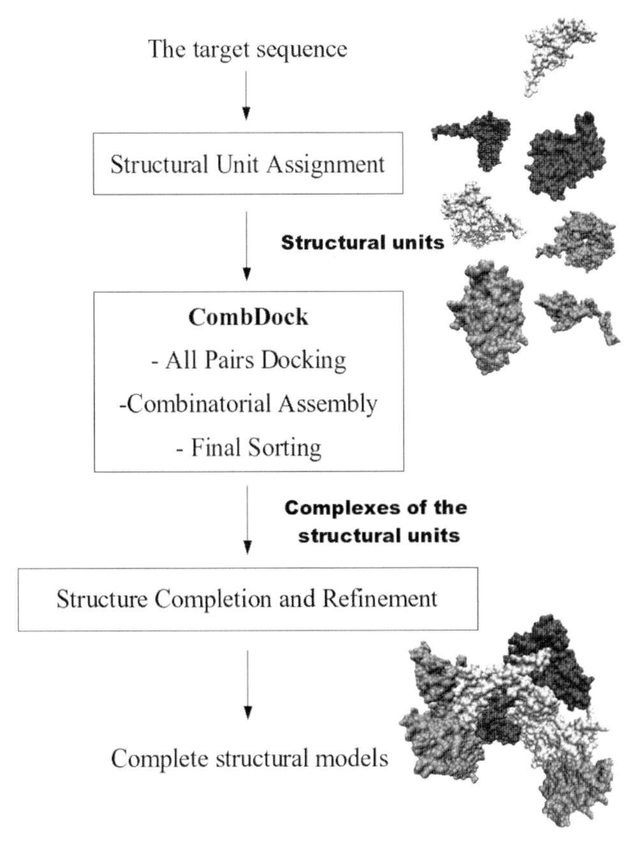

Figure 8.18 Flowchart of the CombDock algorithm. The protein subunits shown on the right are the seven subunits of the Arp2/3 complex. Drawn after Inbar *et al.* (2005).

during the combinatorial assembly process. Figure 8.18 gives an outline of the main steps of the algorithm.

The **All pairs docking module** performs pairwise protein–protein docking as discussed in Section 8.7 for all possible combinations of the N subunit structures given as input. Experience tells us that the correct docking solution may not appear at the very top of the list of solutions. Therefore, K best solutions are kept for each pair of proteins. In Inbar *et al.* (2005), K was varied from dozens to hundreds.

The **Combinatorial assembly module** receives as input the N subunits and the $N(N-1)/2$ sets of K scored transformations. These are the candidate interactions. The interesting algorithmic idea is now to view this problem as construction of a **spanning tree** (Section 2.5). A weighted graph is built representing the input where each structural unit corresponds to a vertex, each geometric transformation to match the surfaces of two subunits is an edge connecting the corresponding vertices and the edge weights are the scores for positioning the two vertices (subunits) in this particular orientation (i.e. the surface complementarity from the pairwise docking). Since the input contains K transformations for each pair of subunits, the complete graph has K parallel edges between each pair of vertices.

For two particular subunits, each candidate complex is represented by an edge and the two vertices. In the case of N structural units a candidate complex is represented by a spanning tree which is a subgraph of the input graph that connects all vertices and has no circles. Each possible spanning tree of the input graph represents an assembly of all the input structural units. The problem of finding complexes is therefore equivalent to finding spanning trees. The number of spanning trees in a complete graph with no parallel edges is N^{N-2} (Cayley's formula). Since the input graph has K parallel edges between each pair of vertices, the number of spanning trees is $N^{N-2}\ K^{N-1}$. Exhaustive searches of all these spanning trees are clearly infeasible.

To simplify the search step, the CombDock algorithm uses two basic principles – a hierarchical construction of the spanning tree and a greedy selection of subtrees. At the first stage, the algorithm constructs trees of size 1 where each tree contains a single vertex that represents a subunit. At stage i, the tree represents complexes consisting of exactly i vertices (subunits) that are generated by connecting two trees generated at a lower stage with an input edge transformation. Only tree complexes without penetrating subunits are kept for the next stages. Because it is impractical to search all valid spanning trees, the algorithm performs a greedy selection of subtrees. For each subset of vertices, the algorithm keeps only the D best-scoring valid trees that connect them.

At the end, the found solutions are clustered. To obtain a rough overview of what contacts are found in a particular spanning tree representing a structural model of the complex, a contact map of size $N(N-1)$ is computed between all subunits. If two subunits are in contact within this complex, the corresponding bit is set to one and to zero otherwise. Complexes having the same contact map are then superimposed and the root-mean-square deviation (RMSD) between their C_α atoms is computed. If this distance is smaller than a threshold, the complexes are considered as members of a cluster. For each cluster, only the complex with the highest score is kept.

The performance of CombDock was tested for five different targets involving between three and 10 subunits, the NF-κB p65 subunit, the Vhl/ElonginC/ElonginB complex, the Arp2/3 complex, RNA polymerase II and a major histocompatibility complex class II/T cell receptor/Sep3 complex. In each case, at least one near-native solution was obtained that was ranked in the top ten using both "bound" and "unbound" subunit conformations. Amazingly, the RMSDs of the best solutions were within 0.1–0.2 nm from the known crystallographic structures, even when using the structures of the unbound components. This is certainly a great success. It is unlikely, though, that this version of the algorithm using rigid protein conformations will be able to correctly assemble such complexes where the input subunits involve significant conformational changes. Future versions of the program are therefore planned to include hinge-bending movements of protein subunits.

8.9
Electron Tomography

The experimental techniques covered in this chapter so far have been able to generate three-dimensional structural representations of single proteins up to large protein complexes. At the end of this chapter, we introduce an experimental technique, electron tomography, that is able to provide structural information for objects as large as entire biological cells. Figure 8.19 shows the principle behind electron tomography.

The method allows recording noninvasive images of whole cells after they are instantaneously deep-frozen. In this manner one can obtain a momentary three-dimensional image of single protein complexes and even of the entire interactome of

a

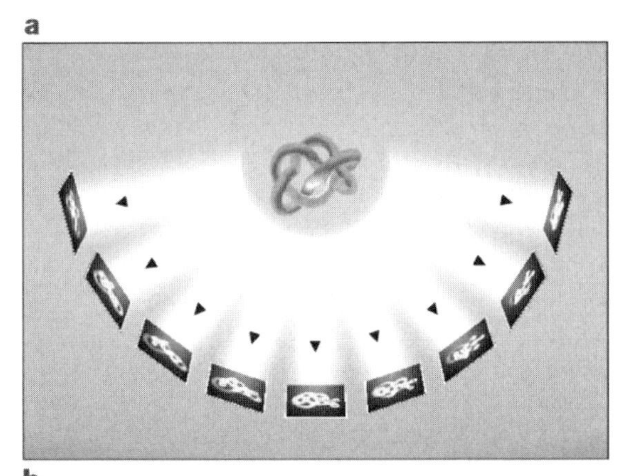

b

Figure 8.19 Principles of electron tomography. (a) The electron beam of an electron microscopy microscope is scattered by the central object and the scattered electrons are detected on the black plate. By tilting the object in small steps, electrons are scattered at different angles. (b) The reconstruction in the computer performs a back-projection (Fourier method) of the scatter information at different angles. The superposition generates a three-dimensional tomogram. Reprinted from Sali *et al.* (2003) by permission from Macmillan Publishers Ltd.

a cell. A current drawback is that the method is quite noisy compared to other imaging techniques. Also, due to a limited range of possible tilting angles, the resolution is limited to about 5–20 nm due to the missing data. This only allows visualization of very large complexes ($M_r > 400\,000$). As the cytoplasm is densely populated with about 30% of the cellular volume taken up by proteins, separation by eye or simple image detection is mostly impossible. Instead, objects have to be identified by sophisticated pattern matching methods, see Sections 8.4 and 8.6.

When imaging a cellular volume, it is in principle possible to scan the entire reconstructed volume by three-dimensional cross-correlation with a molecular template as was done in Section 8.4. When fitting individual components in the low-resolution map of the whole complex, a "brute-force" approach is computationally very expensive, because the orientation of the particles will be random and, consequently, the whole angular range has to be scanned by rotating the templates and calculating the cross-correlation coefficient for all independent combinations of Eulerian angles. As an alternative, a two-step approach was suggested.

In the first step, the tomographic volume is segmented by use of a nonlinear anisotropic diffusion procedure. This particular diffusion process equilibrates uncorrelated structures and highly curved features (e.g. small proteins, noise) faster with their environment than particles exhibiting surfaces with a lower curvature (e.g. macromolecules, cellular compartments). The appropriate adjustment of the number of iterations makes it possible to selectively detect the position of particles with a specific curvature, yielding subvolumes containing particles in the size range of interest with a high probability.

In a second step, the particles contained in the subvolumes are compared to known structures by calculating the three-dimensional cross-correlation of the segmented volumes with known protein templates. Compared with the brute-force approach, the number of necessary correlation functions is significantly reduced. Still, these scans have to be carried out for every segmented subvolume, for every template to be searched for and for every independent set of Eulerian angles. The maximum of a set of correlation peaks is assumed to yield the correct type of particle, as well as its precise position and orientation.

8.9.1
Reconstruction of a Phantom Cell

The potential of the method will be illustrated by results obtained for an artificial model system with well-defined properties (Frangakis *et al.*, 2002). So-called "phantom cells" were prepared that are liposomes of around 400 nm diameter containing a well-defined 1 : 1 mixture of two large protein complexes, the thermosome and the 20S proteasome. A thermosome has a size of 933 kDa, 16 nm diameter, 15 nm height and its subunits assemble into a toroidal structure with 8-fold symmetry. A 20S proteasome has a size of 721 kDa, 11.5 nm diameter, 15 nm height and its subunits assemble into a toroidal structure with 7-fold symmetry. Cryo-electron microscopy images were collected of such phantom cells for a tilt series from −70° to +70°. The aim was to identify and map the two types of proteins in the

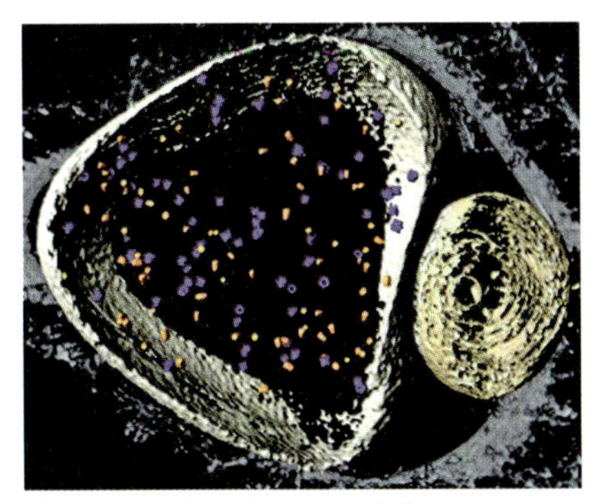

Figure 8.20 Three-dimensional density reconstruction from electron tomography recording of a "phantom cell" containing a 50 : 50% mixture of two different proteins – the thermosome and the proteasome. The identities of each density spot were assigned to either of the two proteins on the basis of a density correlation. Reprinted from Frangakis *et al.* (2002) by permission from the National Academy of Sciences, USA.

phantom cell by density fitting. In good agreement with the experimentally controlled 1 : 1 ratio of both proteins in the phantom cells, the algorithm identified 52% as thermosomes and 48% as 20S proteasomes. Figure 8.20 shows a volume-rendered representation of a reconstructed ice-embedded phantom cell containing a mixture of thermosomes and 20S proteasomes. After applying the template-matching algorithm, the protein species were identified according to the maximal correlation coefficient. The molecules are represented by their averages; thermosomes are shown in blue, the 20S proteasomes in yellow.

Recently, in 2007, researchers have even managed to apply this technique to map the interior of a biological cell (Figure 8.21).

In summary, the method of electron tomography can collect still images of very large assemblies up to entire cells, allowing the identification of single proteins and protein complexes. Accurate identification of all particles involves extensive computations. Problems for real cells arise from molecular crowding where identification of spots

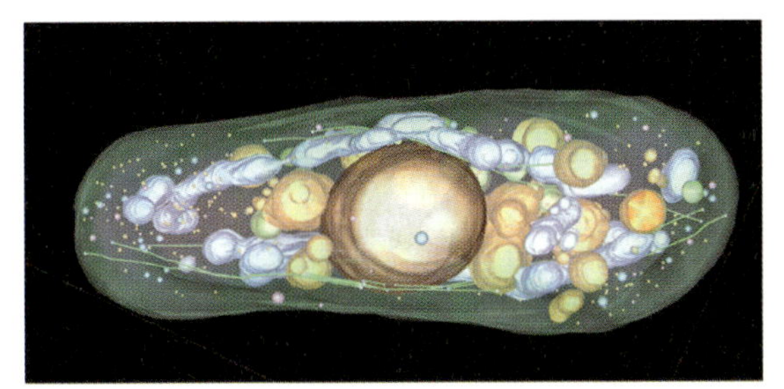

Figure 8.21 Electron tomography reconstruction of the full cell volume of *Schizosaccharomyces pombe* with all microtubule bundles (green), mitochondria (blue), vacuoles (yellow) and trafficking vesicles. Picture courtesy of Dr. Claude Antony (EMBL).

becomes a problem. In order to allow detection of smaller complexes with higher precision, the spatial resolution of the tomograms needs to be further increased.

Summary

Our current atomistic understanding of the functioning of many large macromolecular machines such as the ribosome or RNA polymerase is based on remarkable experimental breakthroughs mainly in the area of protein crystallography during the last 10–15 years. These discoveries have been rewarded with several Nobel Prizes in Chemistry and Medicine. In the future, the structural characterization of large multiprotein complexes and the resolution of cellular architectures will likely be achieved by a combination of methods in structural biology: X-ray crystallography and NMR for high-resolution structures of single proteins and pieces of protein complexes, (cryo) electron microscopy to determine medium-resolution structures of entire protein complexes, stained electron microscopy for still pictures at medium resolution of cellular organelles, and (cryo) electron tomography for three-dimensional reconstructions of biological cells and for identification of the individual components.

When aiming at integrating the results from different methods, all the approaches based on density fitting and the incorporation of additional biochemical or bioinformatics data as restraints during structural modelling require important contributions from computational methods. The coming years will lead to a partial structural understanding of cellular processes because the resolutions accessible to crystallographic and molecular modeling approaches and those of light microscopy and its modern variants with resolution down to 10 nm are slowly starting to converge.

Problems

Mapping of crystal structures into electron microscopy maps

For some protein complexes one has both a low-resolution image, e.g. an atomic force microscopy image or an electron microscopy density map of the whole complex, and the atomic structures of the individual constituents. This allows asking where these constituents are located in the cluster. The task is consequently to fit the position and orientation of a given structure with atomic resolution into a blurred density map such that the correlation is maximized. To achieve a maximal overlap, the high-resolution structure has to be blurred, i.e. convoluted with the experimental resolution.

An efficient way to calculate the convolution of the atomic structure data with the experimental resolution is via the Fourier theorem. Therefore, you will look at various properties of the Fourier transform in the first exercise. However, you will not perform a full three-dimensional reconstruction of multiple fragments into a blurred complex but try to fit a two-dimensional structure into a smeared image of itself.

(1) Properties of the Fourier transform

The (complex) Fourier transform of a function $f(x)$ is defined as:

$$FT[f(x)] = F(k) = \int dxe^{-ikx} f(x).$$

Note the change of the variable from x to its conjugate variable k. Its inverse is, consequently:

$$f(x) = FT^{-1}[F(k)] = \frac{1}{2\pi}\int dke^{ikx} F(k).$$

Remember that the complex exponential is defined as $\exp[\pm ix] := \cos(x) \pm i\sin(x)$ and that the complex integral can be split up into real and imaginary parts, which can be evaluated independently.

(a) The Fourier transform and its inverse
With the δ distribution defined as:

$$\delta(x_1 - x_2) = \frac{1}{2\pi}\int dke^{ik(x_1 - x_2)},$$

show that $FT^{-1}\,[FT\,[f(x)]] = f(x)$, i.e. that the definitions of the Fourier transform and its inverse given above do match.

> *Hint: Be careful not to mix up the different (integration) variables of the Fourier transform and its inverse. Use, e.g.* x_1, x_2, *etc.*

(b) Shift of the argument
Show that a shift of the argument of $f(x) \to f(x + \Delta x)$ shows up as a phase factor exp $[ik\Delta x]$ in the Fourier transform of f. Consequently, the Fourier transform can be used to shift the image of, e.g. a protein:

$$T(\Delta x)b = FT^{-1}[e^{ik\Delta x} FT(b)].$$

(c) Linearity
Show that $FT[f(x) + g(x)] = F(k) + G(k)$.

(d) The convolution theorem
The convolution of two functions $f(x)$ and $g(x)$ is defined as:

$$(f * g)(x) = \int dyg(y)f(x - y).$$

Show that $FT[(f\ {}^*g)](k) = F(k)G(k)$, i.e. that the Fourier transform of the convolution of f and g equals the product of the individual Fourier transforms of f and g.

(2) Blurring the structure

(a) Calculate the convolution of a model molecule with an experimental uncertainty, which is described by a Gaussian distribution:

$$g(x; x_0) = \frac{1}{\sigma\sqrt{2\pi}} \exp\left[-\frac{(x - x_0)^2}{2\sigma^2}\right]$$

of width σ, centered around x_0. The density $\rho(x)$ of the model molecule is given by a sum of delta peaks with masses m_i at the atom positions x_i:

$$\rho(x) = \sum m_i \delta(x - x_i).$$

Hint: Note that $\int dx \delta(x - x_0) f(x) = f(x_0)$: *you do not need the Fourier transform to evaluate* $g^* \rho$.

Hint: The result should be a sum of displaced Gaussians.

Now combine the convolution of the molecular structure and the Gaussian uncertainty with a displacement by Δx [cf. problem (1b)].

Hint: The result should be a sum of displaced Gaussians.

(3) Reconstruction of low-resolution images

For this two-dimensional fit you are given a file hello.dat with the atomic "structure" of the hypothetical HLO (nicknamed "hello") protein and various smeared images, where this structure was shifted, rotated and blurred. Implement a reconstruction program with which you perform the tasks given below.

The objective is to minimize the difference between the given "experimental" maps and the blurred map from the structure. For this, use the sum of the squared differences between the two maps at each grid point.

Note that the difference between this fitting and docking is that here the whole structure has to match the map, whereas in docking only the interface regions needs to match.

Hint: For creating the blurred map, see problem (2) above.
Hint: In the structure file each line contains, in this order, the x- and y-positions of an atom and its mass. The mass determines how much a given atom contributes to the image, i.e. how visible this atom is to the imaging technique.
Hint: You can start from the supplied Python script construct_example.py, which was used to generate the blurred maps. This should explain you how to read and interpret the structure file and how to create a blurred image on a grid. Note that the shifting and rotation parameters saved in this script are not the ones used to generate the given density maps!

(a) Resolution calibration

To calibrate the resolution to be used for the reconstruction, minimize the difference between the given map hello_map.dat and the map generated from the "atomic" structure of hello.dat by varying the width σ for the Gaussian used to smear the high resolution structure. To generate hello_map.dat, the structure was not displaced nor rotated. The only parameter you have to change is σ. Plot the sum of the squared differences against the width σ.

Keep this optimal σ for the subsequent reconstructions.

> *Hint: the best σ is somewhere in the range 0.1–0.2. Choose enough values from this interval to create a plot of the σ-dependent difference. Do not forget to give the optimal value of σ that you find.*

Create a two-dimensional plot of the smoothed image with the optimal σ. Try to include the atom positions, too.

> *Hint: Most spreadsheet programs give you an option to plot a colored "height map" or a contour plot of the gridded two-dimensional data.*

(b) Angular correlation

In the next "experimental" map, hello_map_rot.dat, the HLO protein is rotated, but not displaced. Calculate the difference between the given map and the blurred known structure for rotation angles between 0 and 2π in at least 100 angular steps. Plot this difference vs. the rotation angle and determine the best fit rotation angle. Plot the reconstructed image.

(c) Angular correlation displaced

Repeat the above angular fit *without* displacement with the map of hello_map_rot_dx-dy.dat. Here, the protein is rotated and displaced. Again, plot the difference versus the angle; this time plot both the given "experimental" data and your best fit reconstruction. What do you observe? Can you determine the optimum rotation angle?

(d) Displacement alone

Now try to fit the HLO protein into hello_map_dxdy.dat. Here, the protein is displaced in both the x- and y-direction, but not rotated. For the fit, determine the centre of mass for both the given "atomic" structure and for the density map. Then, displace the atomic structure such that both centers coincide. When you blur the displaced structure, the difference with the density map should become comparably small to the minimal difference in (b). Give the necessary displacement for best overlap and the corresponding squared distance.

(e) Rotation plus displacement

Go back to hello_map_rot_dxdy.dat, and determine both the optimal rotation and the displacement in the x- and y-direction. To do so, first rotate the atomic structure and then shift it as in (d). Give the values of the rotation and the displacement for best overlap (minimal difference). Create two plots which give (i) the squared difference vs. the rotation angle and (ii) the displacements in the x- and y-direction versus the rotation angle.

Further Reading

Protein Complexes

Azubel M, Wolf SG, Sperling J, Sperling R (2004) Three-dimensional structure of the native spliceosome by cryo-electron microscopy, *Molecular Cell*, **15**, 833–839.

Ban N, Nissen P, Hansen J, Moore PB, Steitz TA (2000) The complete atomic structure of the large ribosomal subunit at 2.4 Å resolution, *Science*, **289**, 905–920.

Cramer P, Bushnell DA, Fu J, Gnatt AL, Maier-Davis B, Thompson NE, Burgess RR, Edwards AM, David PR, Kornberg R (2000) Architecture of RNA polymerase II and implications for the transcription mechanism, *Science*, **288**, 640–649.

Cramer P (2002) Multisubunit RNA polymerases, *Current Opinion in Structural Biology*, **12**, 89–97.

Robinson RC, Turbedsky K, Kaiser DA, Marchand JB, Higgs HN, Choe S, Pollard TD (2001) Crystal structure of Arp2/3 complex, *Science*, **294**, 1679–1684.

Volkmann N, Amann KJ, Stoilova-McPhie S, Eglie C, Winter DC, Hazelwood L, Heuser JE, Li R, Pollard TD, Hanein D (2001) Structure of Arp2/3 complex in its activated state and in actin filament branch junctions, *Science*, **293**, 2456–2459.

Yu X, Acehan D, Ménétret JF, Booth CR, Ludtke SJ, Riedl SJ, Shi Y, Wang X, Akey CW (2005) A structure of the human apoptosome at 12.8 Å resolution provides insights into this cell death platform, *Structure*, **13**, 1725–1735.

Milne JLS, Shi D, Rosenthal PB, Sunshine JS, Domingo GJ, Wu X, Brooks BR, Perham RN, Henderson R, Subramaniam S (2002) Molecular architecture and mechanism of an icosahedral pyruvate dehydrogenase complex: a multifunctional catalytic machine, *EMBO Journal*, **21**, 5587–5598.

Yeast Complexeome

Gavin AC, Aloy P, Grandi P, Krause R, Boesche M, Marzioch M, Rau C, Jensen LJ, Bastuck S, Dümpelfeld B, Edelmann A, Heurtier MA, Hoffman V, Hoefert C, Klein K, Hudak M, Michon AM, Schelder M, Schirle M, Remor M, Rudi T, Hooper S, Bauer A, Bouwmeester T, Casari G, Drewes G, Neubauer G, Rick JM, Kuster B, Bork P, Russell RB, Superti-Furga G (2006) Proteome survey reveals modularity of the yeast cell machinery. *Nature*, **440**, 631–636.

Krogan NJ *et al.* (2006) Global Landscape of Protein complexes in the yeast Saccharomyces cerevisiae, *Nature*, **440**, 637–643.

Correlation-based Density Fitting

Wriggers W, Chacón P (2001) Modeling tricks and fitting techniques for multiresolution structures, *Structure*, **9**, 779–788.

Chacón P, Wriggers W (2002) Multi-resolution contour-based fitting of macromolecular structures, *Journal of Molecular Biology*, **317**, 375–384.

Fabiola F, Chapman MS (2006) Fitting of high-resolution structures into electron microscopy reconstruction images, *Structure*, **13**, 389–400.

Garzón JI, Kovacs J, Abagyan R, Chacón P (2007) ADP_EM: fast exhaustive multi-resolution docking for high-throughput coverage, *Bioinformatics*, **23**, 427–433.

Milne JLS, Wu X, Borgnia MJ, Lengyel JS, Brooks BR, Shi D, Perham RN, Subramaniam S (2006) Molecular structure of a 9-MDa icosahedral pyruvate dehydrogenase subcomplex containing the E2 and E3 enzymes using cryoelectron microscopy, *Journal of Biological Chemistry*, **281**, 4364–4370.

Wu X, Milne JLS, Borgnia MJ, Rostapshov AV, Subramaniam S, Brooks BR (2003) A core-weighted fitting method for docking atomic structures into low-resolution maps: application to cryo-electron microscopy, *Journal of Structural Biology*, **141**, 63–76.

SOM Method

Ceulemans H, Russell RB (2004) Fast fitting of atomic structures to low-resolution electron density maps by surface overlap maximization, *Journal of Molecular Biology*, **338**, 783–793.

CombDock

Inbar Y, Benyamini H, Nussinov R, Wolfson HJ (2005) Prediction of multimolecular assemblies by multiple docking, *Journal of Molecular Biology*, **349**, 435–447.

Protein–Protein Docking

Katchalski-Katzir E, Shariv I, Eisenstein M, Friesem AA, Aflalo C, Vakser IA (1992) Molecular surface recognition: determination of geometric fit between proteins and their ligands by correlation techniques. *Proceedings of the National Academy of Sciences USA*, **89**, 2195–2199.

Electron Tomography

Frangakis AS, Böhm J, Förster F, Nickell S, Nicastro D, Typke D, Hegerl R, Baumeister W (2002) Identification of macromolecular complexes in cryoelectron tomograms of phantom cells, *Proceedings of the National Academy of Sciences USA*, **99**, 14153–14158.

Sali A, Glaeser R, Earnest T, Baumeister W (2003) From words to literature in structural proteomics, *Nature*, **422**, 216–225.

9
Biomolecular Association and Binding

In this chapter, we will discuss computational approaches complementary to those discussed in the last chapter that were mostly based on fitting and matching geometric objects. Here, we will try to exploit statistical data on the composition of protein–protein interfaces, data on sequence evolution and energetic considerations to predict the structures and assembly pathways of protein complexes. The specificity of protein–protein interactions requires a strong complementarity of the two interfaces in terms of shape and physicochemical properties. As a result, one expects a certain amount of evolutionary conservation in the interaction patterns between similar proteins and domains. Indeed, it has been found that close homologs are likely to interact in the same way, and protein–protein interfaces place certain evolutionary constraints on protein sequence evolution and structural divergence. However, we note upfront that the biophysics governing protein–protein interactions is extremely complex and currently not well understood. As in other chapters, we will focus on presenting concepts and computational approaches rather than providing a comprehensive review of the current understanding of protein–protein interactions.

9.1
Modeling by Homology

In biology, "homology" of two biomolecules describes the expectation that they should have certain similar properties (e.g. their three-dimensional structure, function, etc.) because they are thought to have descended from a common ancestor. In the area of single protein structure, a very important connection was found between the level of sequence identity of two sequences and the resulting similarity of their three-dimensional structures. Figure 9.1 illustrates the so-called "twilight zone".

It was therefore tempting to test whether a similar relation also holds for pairs of protein complexes. The open question was whether given the known crystal structure of the protein–protein complex A:B, and the facts that the sequence of a protein A is similar to that of A′ and the sequence of B is similar to that of another protein B′, will the two proteins A′ and B′ interact in a structurally similar way as A:B? This was tested extensively and the results are shown in Figure 9.2.

Principles of Computational Cell Biology – From Protein Complexes to Cellular Networks. Volkhard Helms

ISBN: 978-3-527-31555-0

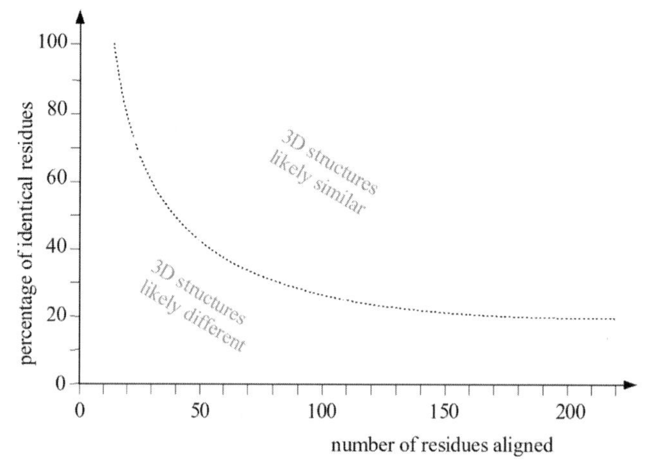

Figure 9.1 Plot illustrates the connection between the sequence similarity of two protein sequences and the structural similarity of the two proteins. The dotted line represents the "twilight zone". Sequences with lower identity than this threshold are likely to have different structures. Note the influence of the length of the sequence stretch aligned. Short stretches need to be more similar than long stretches to have a similar structure. Drawn after Rost (1999).

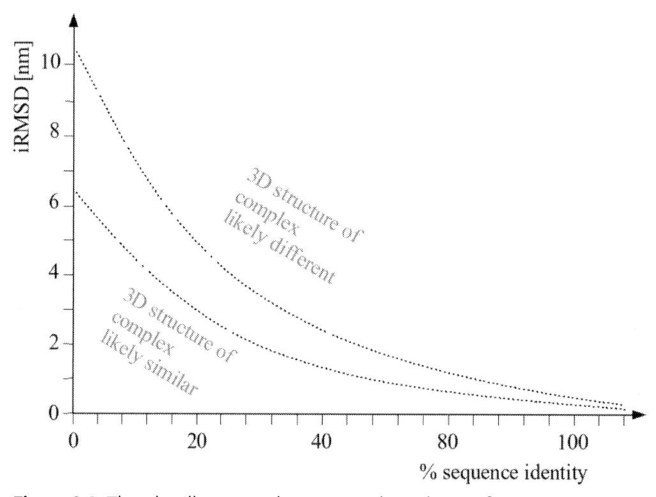

Figure 9.2 The plot illustrates the structural similarity of complexes A–B to A′–B′ measured by their interface root mean square deviation (iRMSD) as a function of the average sequence similarity of A with A′ and of B with B′. The two lines drawn are 80 and 90% percentile lines meaning that 80 or 90% of the solutions are found below the lines. Drawn after Aloy *et al.* (2003).

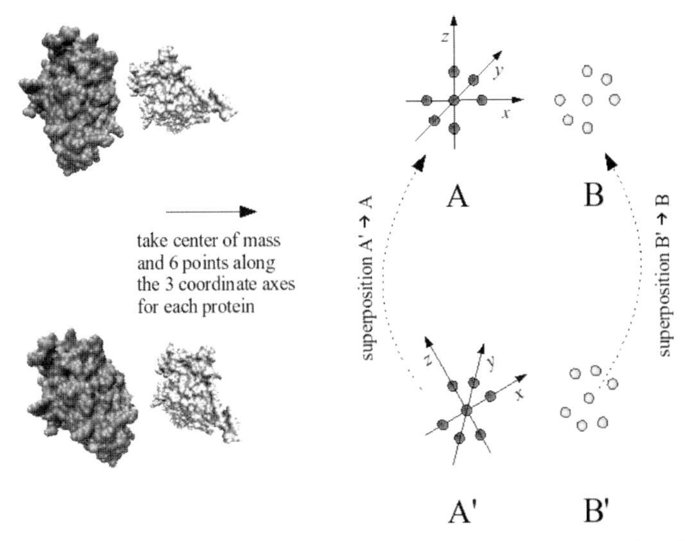

Figure 9.3 The plot illustrates the computation of the iRMSD between the complex A–B and A′–B′. Around each center of mass of any of the four proteins, six points are added along the axes of the corresponding coordinate system. Then, either the seven points representing A′ are optimally superimposed on those of A and the RMSD between the points of B′ with those of B is measured or vice versa for superimposing B′ on B and measuring the RMSD of the A domains. The iRMSD is taken as the smaller one of both values.

The values shown on the *y*-axis refer to the **iRMSD** (Aloy *et al.*, 2003). The calculation of iRMSD is explained in Figure 9.3.

It was suggested that interacting pairs with an iRMSD below 0.5 nm should be considered similar, whereas an iRMSD between 0.5 and 1 nm could indicate a similar positioning of domains, but with one domain being rotated relative to the other one.

As shown in Figure 9.2, a 40% level of sequence identity usually means that the binding mode of interaction is conserved. Various online servers are available that allow users to (a) test whether a suitable template complex exists to model the binding mode of two protein sequences and (b) to generate such a model. The generated model can be scored by energy terms scoring side-chain interactions, and additional terms scoring the residue conservation and residue interface propensity.

9.2 Structural Properties of Protein–Protein Interfaces

9.2.1 Size and Shape

When small molecules interact with proteins, the small molecules usually bind to the deepest cleft on the protein surface. The situation for protein–protein interaction sites is, however, more complicated because the surface area involved is rather large and the binding surfaces are relatively flat. The size of a protein–protein interface is

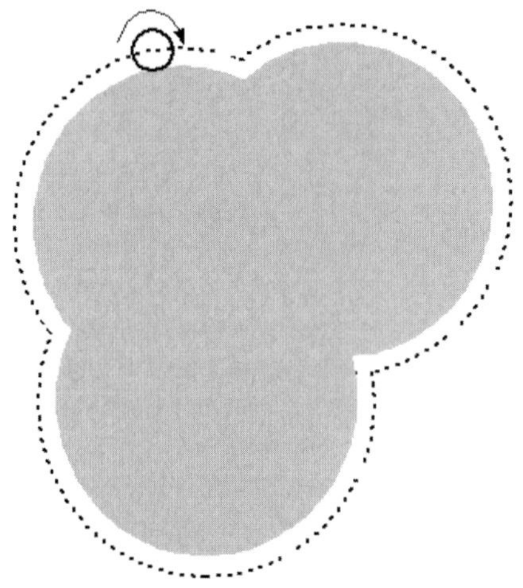

Figure 9.4 Computation of the SASA. A small probe is rolled over the complete surface of the large molecule shown in grey. The dashed line connects the positions of the center of the probe. In three dimensions, it is a surface. Its area is the SASA.

commonly computed from the **solvent-accessible surface area** (SASA) of the protein complex and of the individual proteins (Figure 9.4). A small sphere of a water molecule's radius ($r = 0.14$ nm) is rolled over the van der Waals surface of each protein and over that of the complex. A second surface is generated that contains all the center points of the rolling sphere. This surface area is termed the SASA. The magnitude of the interface area ΔSASA is defined as:

$$\Delta\text{SASA} = \text{SASA}_\text{A} + \text{SASA}_\text{B} - \text{SASA}_\text{AB}.$$

When aiming at computing binding affinities (Section 9.5), one has to define ΔSASA as the negative of the right side.

When analyzing protein interfaces, we first need to specify what is meant by an interface. Often, the interface is defined to contain all those residues that are in contact with the other protein. Two residues from different proteins are considered to be in contact when the distance between any of their heavy atoms is smaller than a threshold value (e.g. often taken as 0.5 nm or as the sum of the two atoms' van der Waals radii plus 0.1 nm). Alternatively, we may specify an interface to consist of those residues (atoms) whose SASA changes upon binding. For example, weak transient homodimers have been characterized with reasonably flat (RMSD 0.025 nm^2) and small (less than 10 nm^2) interfaces. Obligate complexes, such as homodimers, have about twice as large interface areas than nonobligate complexes.

Due to mixed results reported in different studies, it is unclear whether protein interfaces are generally more hydrophobic than the remainder of the protein surface. At the moment, this depends largely on the data set that was used for analysis. Some interfaces have a significant number of polar residues, usually when interactions are less permanent. Protein–protein complexes and biological homodimers form about one hydrogen bond per 1.7–2.1 nm^2 interface. Biological interfaces are enriched in aromatic residues. Charged side-chains are often excluded from protein–protein interfaces with the exception of arginine, which is one of the most abundant interface residues regardless of interaction types. The highly variable polarity of the interfaces may be related to the control mechanism of oligomerization. Small interfaces are probably easier to control and can accommodate more readily to changes in the cell environment. To change the oligomeric state in such interfaces, only a limited number of interactions would need to be broken/made and no large structural rearrangements would be required.

9.2.2 Hot Spots

For the two proteins barnase and the human growth hormone (hGH) about three-quarters of the interface residues can be substituted by alanine with no or little effect on the affinity of binding of their interaction partners barstar and hGH receptor (hGHR). The remaining residues that make the largest contributions to the binding affinity are termed "hot spots". The vast majority of interfaces consists of smaller, evenly distributed hydrophobic patches, interspersed with interacting, polar residues and buried water molecules.

Table 9.1 presents a classification schema of protein–protein interfaces and typical properties collected for them from analyzing crystal structures of protein–protein complexes in the Protein Data Bank (PDB; www.rcsb.org). A "standard size" of about 16 nm^2 has been identified as the interface area that needs to be buried for the formation of a stable complex involving two components that can exist independently

Table 9.1 Classification of protein–protein interfaces (the four classes of hetero-permanent complexes, hetero-transient complexes, homo-permanent complexes and homo-transient complexes possess similar residue interface propensities).

	Shape of interface	Size of interface (nm^2)	Physicochemical properties	Comments
Weak transient homodimers	flat	<10	most hydrophilic and polar	
Strong transient complexes			average polarity	are often induced by allosteric or cooperative effector binding or phosphorylation; often involve large conformational changes and allow a change of affinity
Hetero-permanent interfaces and homo-permanent interfaces		>16	more hydrophobic than transient complexes	

and a lesser area would make the association between the two molecules rather transient.

Keskin *et al.* (2004) derived a structurally unique data set of two-chain interfaces derived from the PDB. The interfaces were clustered based on their spatial structural similarities into three types. Type I clusters consist of similar interfaces whose parent chains are also similar. In Type II clusters, the interfaces are similar; however, the overall structures of the parent proteins from which the interfaces derive are different. In all Type II cases, the clustered proteins belong to different Structural Classification of Protein (SCOP) families, with different functions. The Type III category introduces clusters of interfaces where only one side of the interface is similar, but the other side differs. Type III clusters illustrate that a binding site can interact with more than one chain, with different geometries, sizes, and compositions. This suggests an extension of the paradigm of protein science which states that similar global structures may have similar functions: similar interface architectures may have different functions. As in protein structures, evolution has reused "good" favorable interface structural scaffolds and adapted them to diverse functions.

9.2.3
An Experimental Model System: Human Growth Hormone and its Receptor

In gross contract to the previous approximation used in rigid protein–protein docking, protein surfaces have fluid-like properties that give rise to an astounding structural plasticity, allowing contact points to adapt to conformational changes and multiple amino acid substitutions. Important questions to be asked are which types of amino acids are the most versatile in forming productive contacts in different packing environments and which types are the most specific? What is the relative mutational tolerance of the hot spot residues compared with that of neighboring residues? Does sequence conservation among species provide insights into the steric and chemical restrictions imposed on side-chains at interfaces? Are there functionally homologous hydrophilic side-chains, and if so, in which circumstances can they be exchanged?

Answering these questions requires a quantitative understanding of the structure–function relationship of protein–protein binding interfaces and should ideally be based on a full sampling of all the available structural and chemical diversity generated by the 20 amino acids that can be genetically encoded at each position in the binding interface. Using conventional site-directed mutagenesis techniques would require construction and purifying of many hundreds of mutants for a single protein–protein pair and this has never been done so far. However, modern molecular biology provides different means of achieving this goal. We will discuss results from a recent study (Pál *et al.*, 2006) which investigated the binding interface of hGH with its receptor hGHR and with a specific antibody using phage display. This study provides probably the most comprehensive picture of adaptability in a large protein–protein interface to date.

As mentioned before, the use of alanine scanning has been extremely powerful for determining the binding contributions of individual residues. This has resulted in

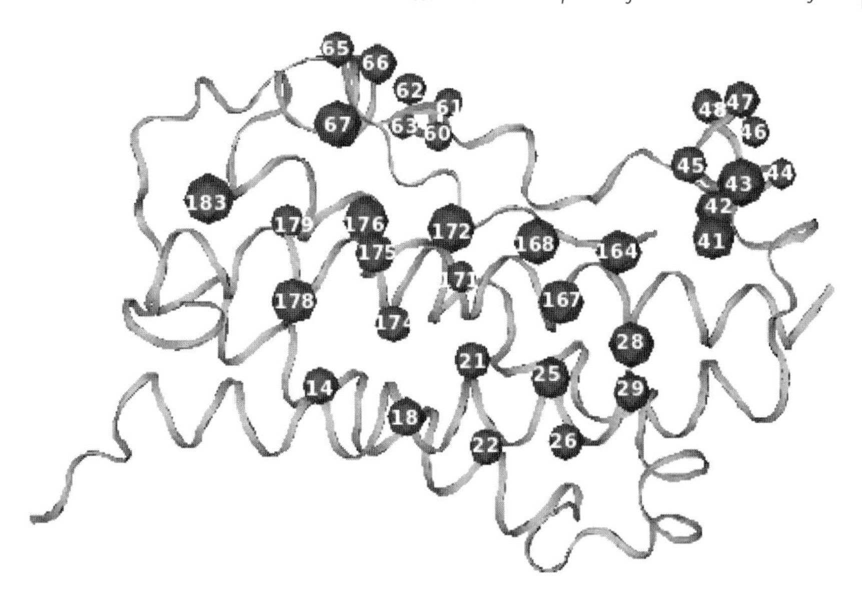

Figure 9.5 The high-affinity receptor-binding site on hGH for binding to the hGHR contains 35 residues distributed across four regions: helices 1 and 4 of the four-helix bundle (residues 14–29 and 164–183) and two connecting loops (residues 41–48 and 60–67). The site was broken down into six libraries covering five to six residues each. Each phage-displayed hGH library on M13 bacteriophage contained about 2×10^{10} unique members. Strong binders to hGHR and to an anti-hGH monoclonal antibody were selected in two rounds of binding selectivity. An average of 180 enzyme-linked assay-positive binding clones from each selection were sequenced and subjected to statistical analysis. Drawn after Pál *et al.* (2006).

the paradigm that binding energy is generally concentrated in a limited region of the interface, termed the "hot spot" of binding energy. Figure 9.5 shows the binding interface of hGH to hGHR.

Table 9.2 shows the amino acid frequencies p_i observed at each randomized position of the binding interface of those clones selected via the binding assays. The results of the phage display experiment revealed that the hGH binding interface is extremely tolerant to mutations. In many cases, even substitutions were tolerated that were expected to introduce steric and physicochemical characteristics incompatible with the wild-type residue. To quantify these results, the Shannon entropy for each protein site was computed by:

$$H_i = -\sum_{k=1}^{20} p_k \ln p_k,$$

where p_k is the fraction of the 20 naturally occurring amino acids at the site that are of type k. The transformed Shannon entropy (TH in Table 9.3) was computed as e^{H_i}. Also, a specificity index (SI) for each position was calculated as the difference between the TH values for the antibody binding and receptor binding selections.

For a frequency distribution of 20 amino acids, the TH value varies between values of 1 or 20 for positions that are completely conserved or completely random,

Table 9.2 At each position (wild-type), following selection for binding to the hGHR (top row) or an anti-hGHR antibody (bottom row), the percent occurrence of each amino acid was calculated after correction for codon bias. Shown are data for 8 out of 35 positions.

WT	Consensus	TH	SI	W	F	Y	M	L	I	V	A	G	S	T	R	K	H	N	Q	D	E	P	C
K41	FV	13	1	1	**20**	9	1	8	3	**15**	5	9	7	2	5	4	4	1	1	3	2	2	0
		14		7	**21**	4	7	**11**	5	**10**	4	6	5	2	3	1	4	1	2	3	3	4	0
Y42	V	17	-1	4	8	9	5	6	7	10	6	3	7	4	2	8	6	3	3	7	3	0	0
		16		2	**15**	9	7	8	5	7	8	7	7	2	7	2	2	3	2	4	1	3	1
S43	RWF	12	4	**16**	**10**	1	1	8	5	3	9	2	7	7	**16**	4	8	1	1	0	0	2	0
		16		9	**15**	8	4	4	4	7	5	5	4	4	5	1	5	2	**10**	4	0	5	1
F44	FVL	10	4	**10**	**33**	8	7	**10**	3	5	4	2	3	4	4	1	3	0	0	1	2	1	0
		14		5	**21**	4	8	9	5	7	4	5	9	1	3	0	4	0	2	2	6	4	2
L45	FL	10	6	7	**30**	6	9	**16**	3	4	3	6	2	0	3	1	1	3	3	0	0	3	0
		16		2	**16**	6	3	**10**	3	9	6	8	5	2	7	1	6	3	4	2	2	3	3
Q46	FYRM	13	3	7	**12**	**11**	**10**	7	3	8	7	3	8	1	**11**	9	1	3	0	0	0	0	0
		16		4	**14**	6	3	**12**	7	7	6	6	6	2	5	6	1	2	2	5	1	4	2
N47	FSV	13	5	1	**18**	6	2	6	2	**11**	5	4	**15**	9	5	0	1	4	1	4	3	1	2
		18		8	6	3	7	5	5	9	3	5	9	4	8	7	2	4	3	6	1	4	3
P48	FVL	13	4	7	**22**	0	3	**11**	4	**12**	3	8	5	3	3	3	5	1	0	3	2	6	1
		17		7	**11**	6	5	7	4	5	**10**	5	6	2	2	1	4	3	2	3	3	6	9

To facilitate the analysis, the wild-type occurrences are in bold when the frequencies are equal to or larger than 10% and printed in light grey when they are equal to or smaller than 2%. The column labeled "consensus" lists the over-represented amino acids in the hGHR-binding assay. Hydrophilic groups can generally be replaced by small hydrophobic groups, but are resistant to replacement by larger ones.

respectively. Following selection for the native fold, the average TH value for the 35 positions is quite high (14 ± 2). This indicates that the hGH fold is highly tolerant to mutations at these solvent-exposed positions. The low standard deviation suggests that the mutational tolerance of the individual positions is very similar. Overall, the distributions are not biased in favor of the wild-type, which shows average abundance at 19 positions and above or below average abundance at seven or nine positions, respectively. While the isolated chemical properties of each of the 20 amino acids can be described by well-established physicochemical criteria, it is understood that these properties are highly context dependent, and will certainly be influenced by the local dielectric and packing environments within protein interfaces.

The picture emerging from this data is that of an energy landscape composed of three different regions. The first is the high-specificity region that correlates reasonably well with the binding hot spot paradigm. The second is the low-specificity region where many residues types can be inserted without noticeable effects on the binding affinity. The third consists of positions that are sensitive to subtle changes and are not optimized for binding function. It is within this last region that the detailed nature of the binding energy surface is poorly characterized and not well understood. Based on these results, the authors concluded that the concept of evolution being a dynamic process that is ever fine-tuning an interaction may not apply. Conservation across species does not necessarily mean that a particular residue is important for structure or function but rather may reflect other constraints imposed by the requirements of the complex biological systems.

9.3 Bioinformatic Prediction of Protein–Protein Interfaces

Protein-binding sites on protein surfaces can be identified by the best current methods with an accuracy of about 70%. The problem seems not to devise the prediction methods themselves, but it appears that we need to understand better what characteristics of protein-binding interfaces should be considered. In the light of the increasing number of experimental structures of protein complexes, some established ideas such as the typical composition of interfaces and the role of sequence conservation have been criticized lately. Therefore, we will take a cautious approach here. We will discuss (i) methods that are solely based on geometrical, physicochemical and statistical properties of the surfaces, and (ii) methods incorporating evolutionary information in the form of certain conservation measures derived from multiple sequence alignments (MSAs) that are projected on the protein surface.

9.3.1 Amino acid Composition of Protein Interfaces

Statistical analysis of protein–protein interfaces in crystal structures of protein–protein complexes (Chapter 1 and Section 4.1) shows that interfaces have a somehow different **amino acid composition** than the rest of the protein surface. We previously analyzed a data set of 170 nonobligate protein complexes, and found that interfaces contained 30.4% hydrophobic residues, 32.8% polar residues and 36.8% charged residues. This finding may reflect that transient complexes need to bind quickly and specifically, but need not be stable for a long period of time, which would require a higher rate of hydrophobic residues. On the other hand, it is not expected for permanent protein complexes to have a stable unbound state which would require a rather hydrophilic interface.

Tyrosine and arginine are overrepresented in hot spots. The enrichment of tyrosine as an aromatic residue can be explained by its ability to form favorable contacts via hydrophobic interactions without a large entropic penalty since tyrosine has few rotatable bonds. Furthermore, tyrosine is capable of forming multiple types of interactions in the lowered effective dielectric environment of hot spots. A preference is also found for arginine, which may contribute to binding through electrostatic steering and is capable of forming multiple types of interactions. It can form salt bridges with its positively charged guanidinium motif, its guanidinium π-electron system allowing for a delocalization of the electron leads to an aromatic character, and it can form hydrogen-bond networks with up to five hydrogen bonds. The high preferences for arginine could also be explained with the ability of arginine to "guide away" water molecules from the interface during complex formation, or, conversely, upon dissociation.

9.3.2
Pairing Propensities

Given the set of interface residues on both proteins, one may analyze what contacts each of them forms with residues on the other protein. A typical distance threshold between pairs of atoms is 0.5 nm. The computed statistics are given in a 20×20 matrix shown in Figure 9.6. These scores are normalized against the fractional abundance of each residue at the interface.

9.3.3
Interface Statistical Potentials

From the observed statistics in Figure 9.6, interfacial pair potentials $P(i,j)$ ($i = 1, \ldots, 20$, $j = 1, \ldots, 20$) may be calculated.

$$P(i,j) = -\log\left(\frac{N_{\text{obs}}(i,j)}{N_{\text{exp}}(i,j)}\right). \tag{9.1}$$

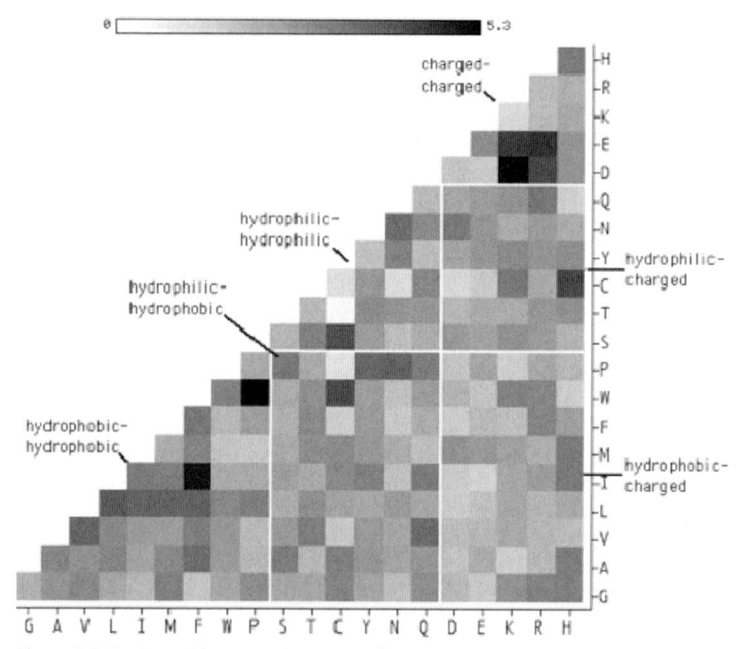

Figure 9.6 Amino acid propensity matrix of transient protein–protein interfaces. Scores are normalized pairing frequencies of two residues that occur on the protein–protein interfaces of transient complexes.

Here, $N_{obs}(i,j)$ is the observed number of contacting pairs of i,j between two chains, and $N_{exp}(i,j)$ is the expected number of contacting pairs of i,j between two chains derived from their frequencies in the proteins chains and assuming that there are no preferential interactions among them. $N_{exp}(i,j)$ is computed as:

$$N_{exp}(i,j) = X_i \times X_j \times X_{total}.$$

where X_i is the mole fraction of residue i among the total surface residues and X_{total} is the total number of contacting pairs.

Figure 9.7(a–d) highlights a few representative rows from this matrix for specific residues. Figure 9.7(a) shows the contacts formed by the hydrophobic amino acid leucine reflecting that hydrophobic residues prefer to interact with other hydrophobic residues. Figure 9.7(b) shows the contacts of the polar amino acid asparagine that show a slight preference for contacts with other polar and charged residues over those with hydrophobic residues. Figure 9.7(c and d) shows the contacts formed by the negatively charged aspartic acid and the positively charged lysine. As expected, they preferentially contact residues of the opposite charge whereas contacts to hydrophobic residues are only about half as frequent as in Figure 9.7(a).

We will also comment on some of the most frequent contacts found in Figure 9.6. Unexpectedly, contacts between tryptophan and proline (W–P) are very frequent. Such contacts are often found at the binding interfaces for proline-rich peptides on adapter domains such as SH3 domains. Another high score was observed for the interactions between phenylalanine and isoleucine. This is not surprising since both hydrophobic amino acids have rather flat and elliptic side-chains leading to good geometric matches. As expected, one of the highest interaction peaks of Figure 9.6 is found between arginine and glutamic acid that carry opposite electrostatic charges. Whereas the relative orientation of the charged groups of both residues simply suggests electrostatic attraction such as formation of salt bridges between both groups, a closer look reveals a broad range of residue–residue side-chain distances and angles reflecting an interesting mixture of electrostatic interactions, including salt bridges and hydrogen bonding.

9.3.4 Conservation at Protein Interfaces

As mentioned above, functional constraints are expected to limit the amino acid substitution rates, resulting in a higher conservation of functional sites such as binding interfaces with respect to the rest of the protein surface. However, evolutionary conservation of particular residues across species, also termed "purifying selection", can also be due to geometrical constraints on the folding of the protein into its three-dimensional structure, constraints at amino acid sites involved in enzymatic activity or in ligand binding. For example, the family of G-proteins has been crystallized with more than 90 interactors. At least for this particular family of proteins, there are virtually no surface residues that can be considered as noninteracting.

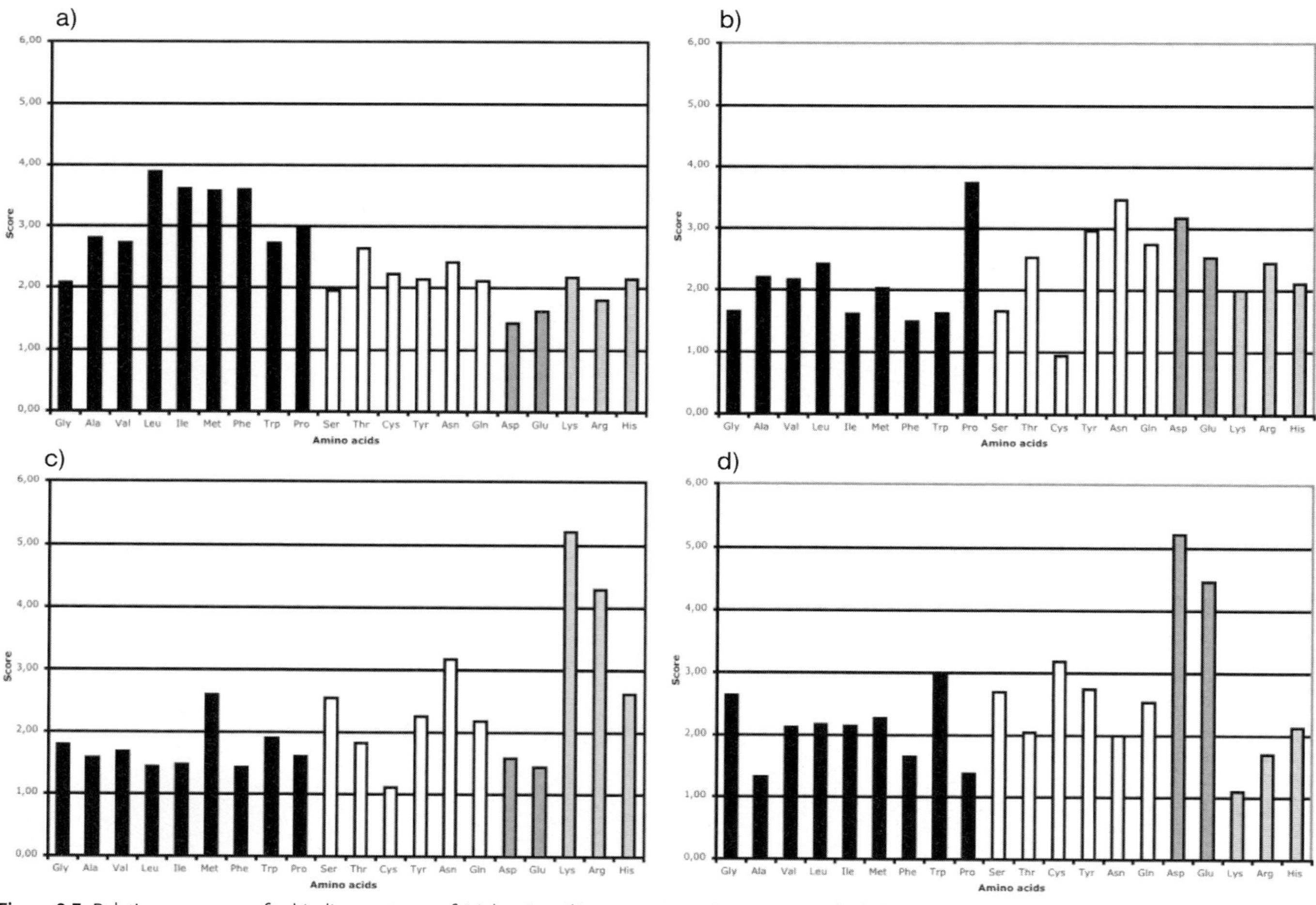

Figure 9.7 Relative occurrence for binding partners of (a) leucine, (b) asparagine, (c) aspartate and (d) lysine. Black bars indicate hydrophobic residues, white bars indicate hydrophilic residues and grey bars indicate charged residues. The higher the score, the more frequently such pairs occurred in the dataset.

Early analysis of crystal structures of protein–protein complexes combined with MSAs of the respective protein families suggested that interface regions are **evolutionarily** more **conserved** than other regions on the protein surface. In general, surface residues show much lower conservation than amino acid sites in the protein interior. This is understandable because surface residues generally do not form specific protein contacts, but their side-chains interact with solvent. Residues at interfaces, on the other hand, should be more conserved because family members typically interact with each other in the same manner (Section 9.1).

There exist various approaches that analyze evolutionary conservation at surfaces. For example, the **evolutionary trace** method generates a phylogenetic tree that is split into evenly distributed partitions. For each partition, the sequences connected by a common vertex are clustered together. Next, a consensus sequence is generated for each cluster. Then, the consensus sequences for all clusters are compared. A position is defined as "conserved" if all consensus sequences have an invariant residue at that position. A position is "class specific" if it is invariant for each cluster, and varies between them. A position is "neutral" if it is variable in at least one cluster.

The popular online tool Consurf (consurf.tau.ac.il) calculates residue conservation scores and projects these scores on a three-dimensional protein structure. First, the program collects a number of homologous sequences and performs a MSA among these homologous sequences. Based on specific rules for scoring amino acid exchanges and gap penalties for insertions or deletions, the program calculates an average score for each position in the query sequence and applies normalization for each score. A residue that is detected by Consurf as the most conserved is considered as conserved as a residue that is buried in the core of the protein.

However, the reliability of evolutionary conservation scores derived from MSA as determinants of protein interfaces has recently been questioned by several authors. In spite of the broad evidence that the interfaces generally mutate at slower rates than the rest of the protein surface, it was argued that a conservation score alone is not sufficient for accurate discrimination. For example, alignments can be easily contaminated with paralogs that do not share the same interfaces. Moreover, many protein interfaces are not expected to be better conserved at all, either because of their function (e.g. the adaptable binding surfaces of the immune system proteins) or because they were formed late in evolution. Such interfaces are undetectable with alignment-dependent methods.

9.3.5
Correlated Mutations at Protein Interfaces

Noting that three-dimensional complexation patterns are often conserved in protein families (Section 9.1), one may expect that a mutation changing the nature of an amino acid position at the interface could be compensated by a corresponding change at the surface of the binding partner. Identifying such **correlated mutations** in MSAs may be a very sensitive pattern to identify interacting residues and binding proteins,

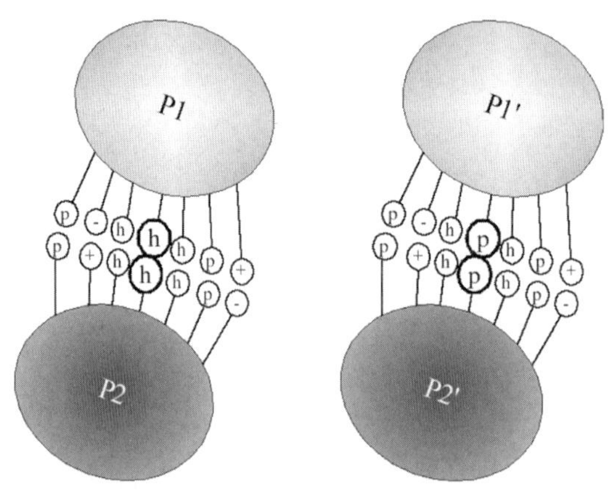

Figure 9.8 (Left) Schematic drawing of a protein–protein interface involving contacts between oppositely charged amino acids, between hydrophobic amino acids (labeled "h") and between polar amino acids (labeled "p"). (Right) In a related organism, one hydrophobic residue on P1′ is changed into a polar residue. A compensating, correlated mutation is observed on the contacting residue on P2′.

respectively. This method may even be applied on a large scale without knowledge of the three-dimensional structures of the proteins (Figure 9.8).

Assuming that correlated pairs of mutations in both monomers tend to accumulate at the contact interface, analysis of correlated protein mutations may therefore help in bioinformatics prediction of protein–protein interfaces. To identify such correlated mutations, MSAs need to be created for both protein families supposed to interact. Then, one has to assess the similarity between all combinations of positions in a MSA. Such similarities may be detected by computing a correlation score CM_{ij} (cf. Section 8.6.3) weighted by the residue complementarities for each pair of positions i and j in the sequence as:

$$CM_{ij} = \frac{1}{N^2}\frac{\sum_{k,l}(S_{ikl} - \langle S_i \rangle)(S_{jkl} - \langle S_j \rangle)}{\sigma_i \sigma_j} C_{ik,jk} C_{il,jl}$$

Here, the summations run over every possible pair of proteins k and l from different species in the MSA, S_{ikl} is the ranked similarity between residue i in protein k and residue i in protein l, S_{jkl} is the same for residue j, and $C_{ik,jk}$ is the complementarity of residue ik and residue jk (e.g. according to an amino acid substitution matrix such as PAM or BLOSUM) (Figure 9.9).

Residues in the interfaces of obligate complexes tend to evolve at a relatively slower rate, allowing them to coevolve with their interacting partners. In contrast, the plasticity inherent in transient interactions leads to an increased rate of substitutions for the interface residues and leaves little or no evidence of correlated mutations across the interface (Mintseris and Weng, 2005). A recent study concluded that

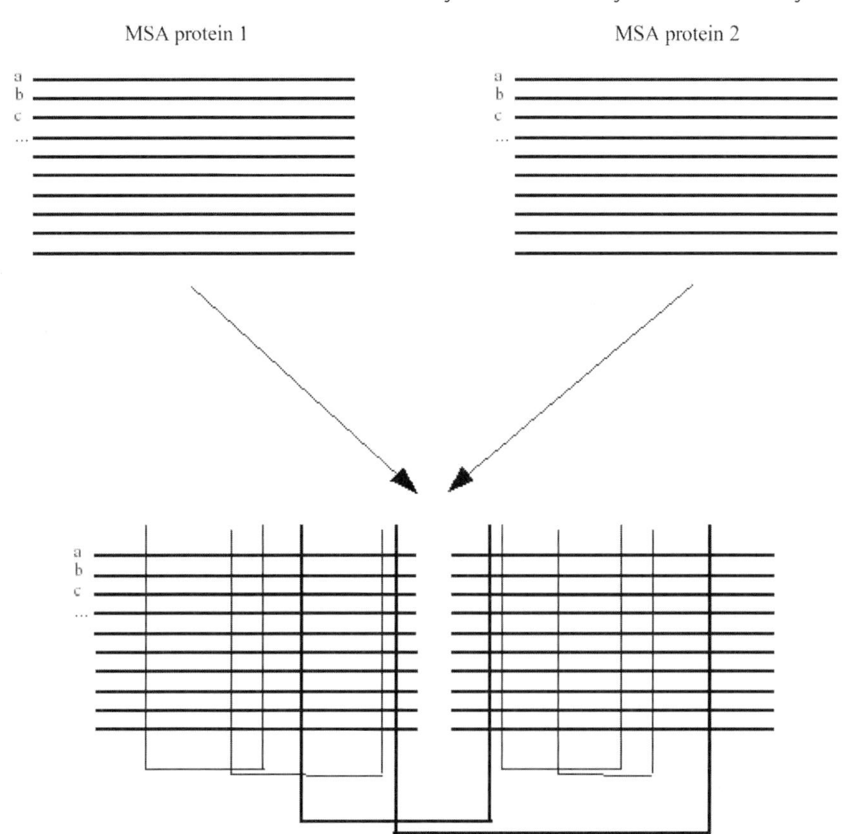

Figure 9.9 Identification of correlated mutations. (Top) Family alignments are collected for two different proteins, 1 and 2, including corresponding sequences from different species (a, b, c, . . .). (Bottom) A virtual alignment is constructed, concatenating the sequences of the probable orthologous sequences of the two proteins to identify correlated mutations of residues *i* and *j*. Drawn after Pazos and Valencia (2002).

including correlated mutations does not improve the identification of protein-binding sites (Halperin *et al.*, 2006).

9.3.6
Classification of Protein Interfaces

Interfaces of obligate complexes that are mostly formed by homodimers are larger and more hydrophobic than nonobligate associations. (See Section 8.1 for a definition of obligate and nonobligate.) The stable association results from the cofolded and coexpressed protomers and the large hydrophobic surface patches that cause strong and tight interactions. In contrast, nonobligate interactions exhibit a more polar interface ensuring the stable unbound state of the monomers. Once the interface area exceeds 10 nm^2, conformational changes may be noticed that lead to an induced fit with an increased lifetime of the interaction. However, there exists a continuum between nonobligate/obligate or transient/permanent interactions and structural characterization alone is inadequate to distinguish between their different affinities or specificities.

In the same manner as presented above for the amino acid pairing preferences (Eq. 9.1), interfaces can be classified according to features such as amino acid pairing composition, propensity of secondary structure elements, pairing preferences of secondary structure elements, interface size, polarity and tightness of fit.

Given a three-dimensional structure of a protein–protein complex, these properties can be quickly computed using automated scripts. Can one distinguish separate classes of interfaces based on their characteristics? One important classification of protein–protein complexes is that into obligate and non-obligate interactions. Obligate complexes (e.g. the ribosome) always exist in the oligomeric, fully assembled state, whereas components of non-obligate complexes (e.g. electron transfer partners or elements of signal transduction chains) may also exist in their monomeric uncomplexed forms. As expected, the interface areas of obligate complexes are larger than those of non-obligate complexes. Using support vector machines or a decision tree, one may successfully separate both classes with more than 95% reliability using only three interface properties.

9.4 Forces Important for Biomolecular Association

Many of the cellular processes covered in previous chapters can be very well modeled using continuum descriptions, of, for example, the ionic strength or of protein concentrations. This allows studying the temporal evolution of such quantities using ordinary differential equations or the coupled temporal and spatial evolution by partial differential equations (Chapter 7).

Quite a few cellular processes, however, require modeling at molecular scale (see Section 7.7) rather than describing the concentrations by particle densities. In doing so, we need to define what level of detail should be included, whether a protein should be modeled as one sphere of certain radius, as a collection of a few spheres, as a collection of spheres that each correspond to protein residues or in atomic detail? Table 9.3 gives a quick overview over what cellular processes require what detail.

The electrostatic interaction between two charged point particles is described by the well-known Coulomb law:

$$U_{ij}(r) = \frac{1}{4\pi\varepsilon_0\varepsilon_r}\frac{q_i q_j}{r_{ij}}, \tag{9.2}$$

where q_i and q_j are the net electrostatic charges of both particles, r_{ij} is their distance, ε_0 is the (constant) dielectric permittivity of vacuum, and ε_r is the relative permittivity of the medium (here, the solution) between both particles. The net charge on a protein at

Table 9.3 Modeling different cellular processes at different detail.

Process	Molecular detail required	Internal flexibility	Electrostatic interactions	Excluded volume
Protein diffusion	1 protein = 1 or more spheres	no	no	yes
Protein association	1 protein = collection of beads each corresponding to a residue	desirable	yes	yes
Conformational dynamics of a protein	residue or atomic level	required	(yes)	yes
Protein–ligand binding	atomic level required	required (ligand); desirable (protein)	yes	yes

a given pH is determined by the pK_a values of its ionizable groups. The net charge on a protein is zero at the isoelectric point (pI), positive at pHs below the pI and negative at pHs above the pI. Typical protein charges are on the order of between 0 and 10 times the charge of an electron. For example, ribonuclease Sa has seven Asp, five Glu, two His, zero Lys and five Arg residues. Consequently, it has a net charge of $-7e$ at pH 7. Equation 9.2 can be rewritten in a more convenient form:

$$U_{ij}(r) = 139\frac{q_i q_j}{r_{ij}},$$

where the atomic charges are to be inserted in multiples of an electron charge and the distance r_{ij} is given in nanometers. Figure 9.10 plots the strength of the electrostatic interaction energy given by Eq. 9.2 as a function of the distance between the two proteins.

In a water environment with a typical concentration of salt ions of 150 mM (with an according ionic strength κ), one often uses a screened Debye–Hückel formulation:

$$U_{ij}(r) = \frac{1}{4\pi\varepsilon_0\varepsilon_r}\frac{q_i q_j}{r_{ij}}e^{-\kappa r_{ij}}, \qquad (9.3)$$

that accounts for the additional ion screening of the charge-charge interaction. Interactions modeled in this way will consequently decay much more quickly than the right side of Figure 9.10.

The typical diameter of one protein is 3–5 nm. Typical energetic interactions of protein–protein pairs decay quickly over such distances to very small magnitudes of a fraction of the thermal fluctuations k_BT. Therefore, we often need not be concerned with modeling fine details such as electrostatic interactions between single residues when working at supramolecular levels. Several cellular systems, on the other hand, have very strong electrostatic interactions such as DNA, the ribosome, and the interior of virus capsids. Figure 9.11 shows electrostatic potentials mapped onto the surface of the two proteins barnase and barstar that will be later used as a model

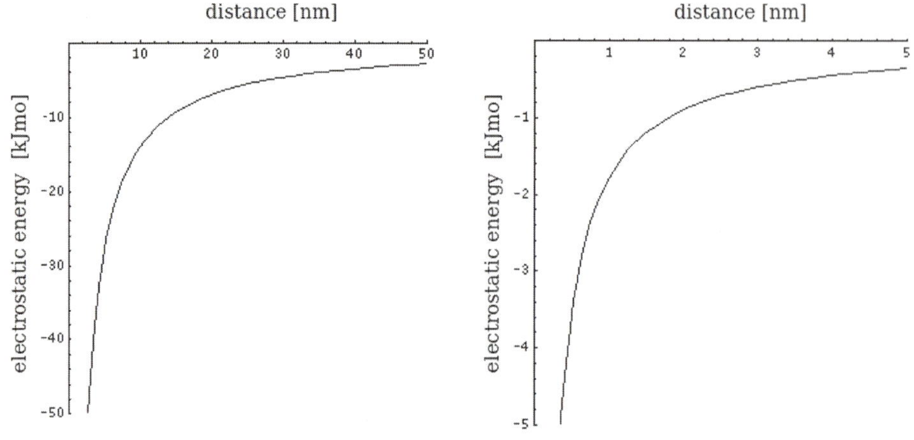

Figure 9.10 Electrostatic interaction energy of two oppositely charged particles with charges $-e$ and $+e$. (Left) Here, the interaction is computed in vacuum with $\varepsilon_r = 1$. (Right) The same interaction is computed in aqueous environment using $\varepsilon_r = 78$. This leads to a large dampening of range where electrostatic interactions are strongly felt. Note that the plotted distance range is very different in the two plots.

system for protein–protein association. The electrostatic potential is generated by the partial atomic charges of the protein(s) and taking into account the different relative dielectric permittivities of protein ($\varepsilon_r \approx 4$) and water ($\varepsilon_r = 78$). The potential can be computed by numerically solving the Poisson–Boltzmann equation using a finite difference scheme.

Another important class of interactions relevant for macromolecular associations are the hydrophobic interactions. These describe the finding that hydrophobic particles do not mix well with water and rather prefer to bind to each other. An intuitive explanation was given by Chandler (2007). At the boundary between liquid

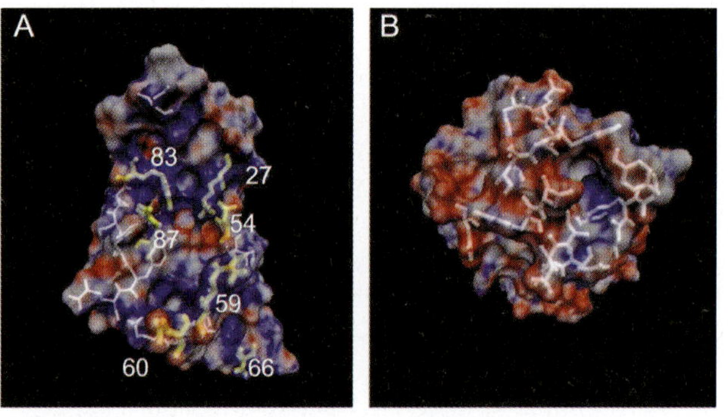

Figure 9.11 (A) Surface representation of the RNase barnase colored according to its electrostatic potential mapped to the surface from -7 (red) to $+7\ kT/e$ (blue). Interface residues are drawn as white sticks. (B) The same for its inhibitor, barstar.

water and water vapor, an interface forms that is characterized by an area with a density lower than the average density of bulk water. The same sort of "depletion layer" also occurs when water comes into contact with a sufficiently large hydrophobic surface. This happens because hydrophobic surfaces provide no opportunity for water molecules to establish their usual hydrogen bonds. Without this adhesive force, the molecules move away from the surface to seek such bonds in the bulk of the liquid.

The nonpolar term is typically modeled as proportional to the change of the SASA:

$$\Delta G_{\mathrm{np}} = \gamma \cdot \Delta\mathrm{SASA},$$

with the microscopic surface tension γ as proportionality coefficient. As SASA decreases upon binding, this contribution favors association.

9.5 Protein–Protein Association

Often, protein diffusion in cellular media can simply be modeled as random diffusion using an effective diffusion constant (see Section 7.3 for a discussion of the diffusion equation). For example, experiments on Green Fluorescent Protein showed that its diffusion in cells is only reduced by a factor of 5 compared to diffusion in pure water. The Einstein relation:

$$D = \mu\, k_{\mathrm{B}} T,$$

relates the diffusion constant D of a particle to its mobility μ times the Boltzmann constant k_{B} and the temperature T. In the limit of low Reynolds number, the mobility μ is the inverse of the drag coefficient γ. For spherical particles of radius r, Stokes' law gives:

$$\gamma = 6\pi\eta r,$$

with the viscosity η of the medium. The Einstein relation becomes:

$$D = \frac{k_{\mathrm{B}} T}{6\pi\eta r},$$

which is also known as the **Stokes–Einstein relationship**. Using this equation, we can estimate the diffusion coefficient of a globular protein. The mean squared displacement *msd* in a time interval of length Δt is:

$$msd = 6D\,\Delta t,$$

in three dimensions.

A typical diffusion constant for a small protein like barnase is $0.15\,\mu\mathrm{m}^2/\mathrm{ms}$ (corresponding to $1.5 \times 10^{-6}\,\mathrm{cm}^2/\mathrm{s}$ in the typical units). Typical cellular dimensions are $1\,\mu\mathrm{m}$ for the cell diameter or 100 nm for the diameter of a vesicle. This means that such a protein explores the volume of a cell roughly in one millisecond.

In Chapter 8 we have already touched on the methods of protein–protein docking. Here, we will now address the question of how the two proteins approach each other. As was just discussed, biomolecules move in the cell by undirected diffusion, also termed Brownian motion after its discoverer, the Scottish botanist Robert Brown. The association of two proteins has therefore been studied using the Brownian dynamics method that is the name of an algorithm to generate diffusional trajectories of particle motion.

9.5.1 Brownian Dynamics Simulations

The Ermak–McCammon algorithm Ermak *et al.* (1978) is an iterative algorithm to compute trajectories for the diffusional motion of a particle using the translational/rotational diffusion coefficients D and D_R. The algorithm computes the forces **F** and torques **T** acting between diffusing particles and with the rest of the system that may be kept fixed. From these forces and torques, it computes translational and rotational displacements from the particles' current positions during the next time step. Importantly, the algorithm accounts for the solvent influence on the particle dynamics by a second stochastic term **R**. This noise term will generate the characteristic flickering motion of a Brownian particle:

$$\Delta \mathbf{r} = (kT)^{-1} D\,\mathbf{F}\,\Delta t + \mathbf{R} \text{ with } \langle \mathbf{R} \rangle = 0 \quad \text{and} \quad \langle \mathbf{R}^2 \rangle = 6\,\mathrm{D}\,\Delta t$$

and

$$\Delta \mathbf{w} = (kT)^{-1} D_R\,\mathbf{T}\,\Delta t + \mathbf{W} \quad \text{with} \quad \langle \mathbf{W} \rangle = 0 \quad \text{and} \quad \langle \mathbf{W}^2 \rangle = 6\,D_R\,\Delta t.$$

When talking about protein–protein association, we first need to choose a good coordinate system or reaction coordinate. Possible choices are the center–center distance d_{1-2} or the average distance between contact pairs cd_{avg} (Figure 9.12).

Figure 9.13 shows a sketch of the interaction free energy versus the separation between proteins. At large distances (region (1) in the scheme), e.g. beyond a few times the Debye length, the electrostatic interactions of both proteins are screened from each other by the ionic cloud of the solution (Eq. 9.8). Both proteins will undergo free diffusion.

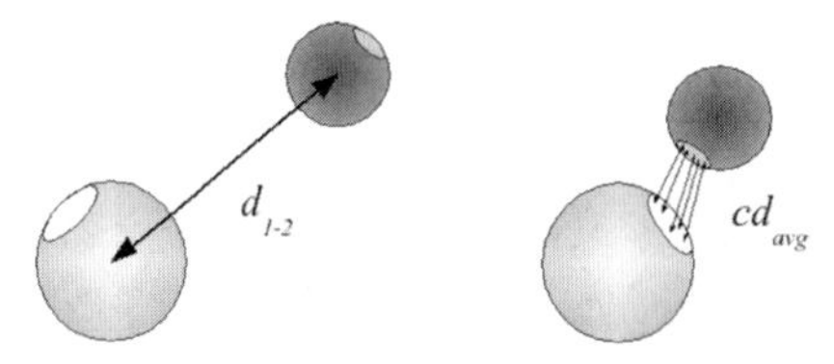

Figure 9.12 Definition of different criteria to describe the relative orientation of the binding interfaces of two proteins. (Left) d_{1-2} is the distance between the centers of both proteins. This criterion is particularly suitable at large particle distances. Here, the orientation of the particles is not accounted for. (Right) cd_{avg} is the average distance between interaction pairs on the two interfaces (atoms or residues). This criterion is most useful at short distances.

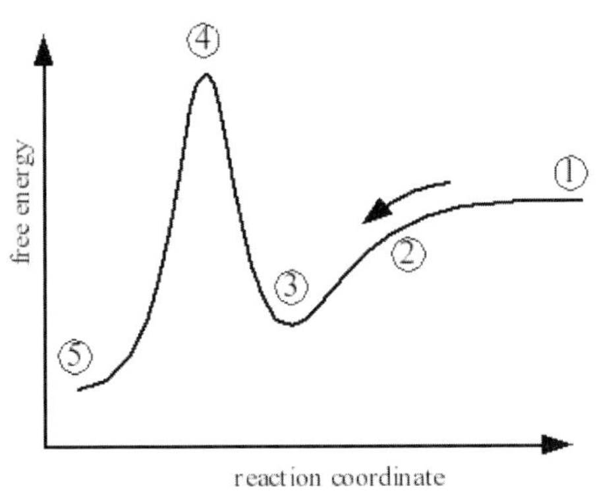

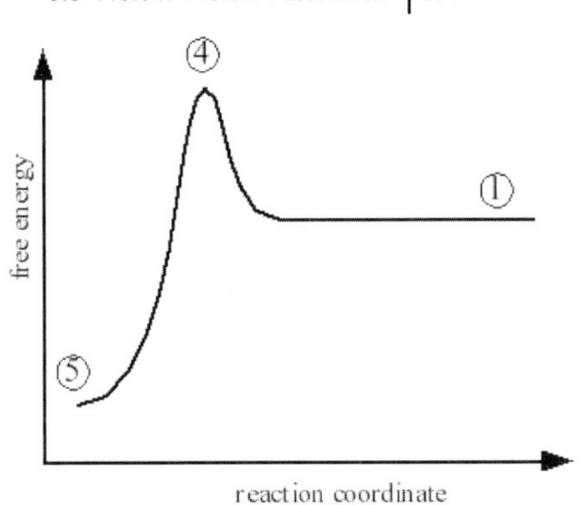

Figure 9.13 Schematic representation of the free energy for protein–protein interaction along an idealized one-dimensional reaction coordinate such as the distance of their centers of masses. (Left) Example of a protein–protein pair attracted by long-range steering in region (2) (Right) Example of a protein–protein pair that undergoes random diffusion until the two proteins eventually collide and bind to each other via hydrophobic contacts.

At closer distance (region 2), the proteins will start to experience the electrostatic field of the other protein. First, they will feel the effect of its net charge (monopole field), and at closer distances this field will be modulated by its dipolar character and other fine details.

Eventually, the two proteins will be so close to each other (region 3) that they have favorably oriented themselves in the electrostatic field of their binding partner. Before they can approach each other even more closely, they need to fully or partially strip off their solvation shells and to fine-tune the orientation of side-chains at the binding interface. The former process is energetically costly, the latter one simply takes some time. Therefore, we imagine the appearance of an energy barrier (region 4) that needs to be overcome before both proteins can finally bind (region 5). This scenario applies to electrostatically attracted proteins. However, by far not all pairs of proteins are electrostatically attracted. Those that are not simply do not have the steering region (2) and the encounter region (3), see right part of Figure 9.13.

Our group has studied the association pathway for the protein–protein pair of barnase and barstar using Brownian dynamics simulations (Spaar *et al.*, 2006). As mentioned before, the surfaces of these proteins show strong electrostatic complementarity (Figure 9.11), leading to one of the fastest known association rates due to strong electrostatic steering. Also, the interaction is one of the tightest known protein–protein interactions characterized by a very high k_D. In these simulations, one protein is kept fixed at the origin, and the other is started at a defined distance at a random angle and randomly oriented. We have mapped the association trajectories from 200 000 such simulations onto a cubic lattice and computed the occupancy of the diffusing particle in each voxel. The idea here is that very favorable locations should be highly populated and unfavorable regions for the approach should have low populations. Figure 9.14 shows the observed occupancy plots. Figure 9.14 shows that the orientation (B) is less constrained than the position (A).

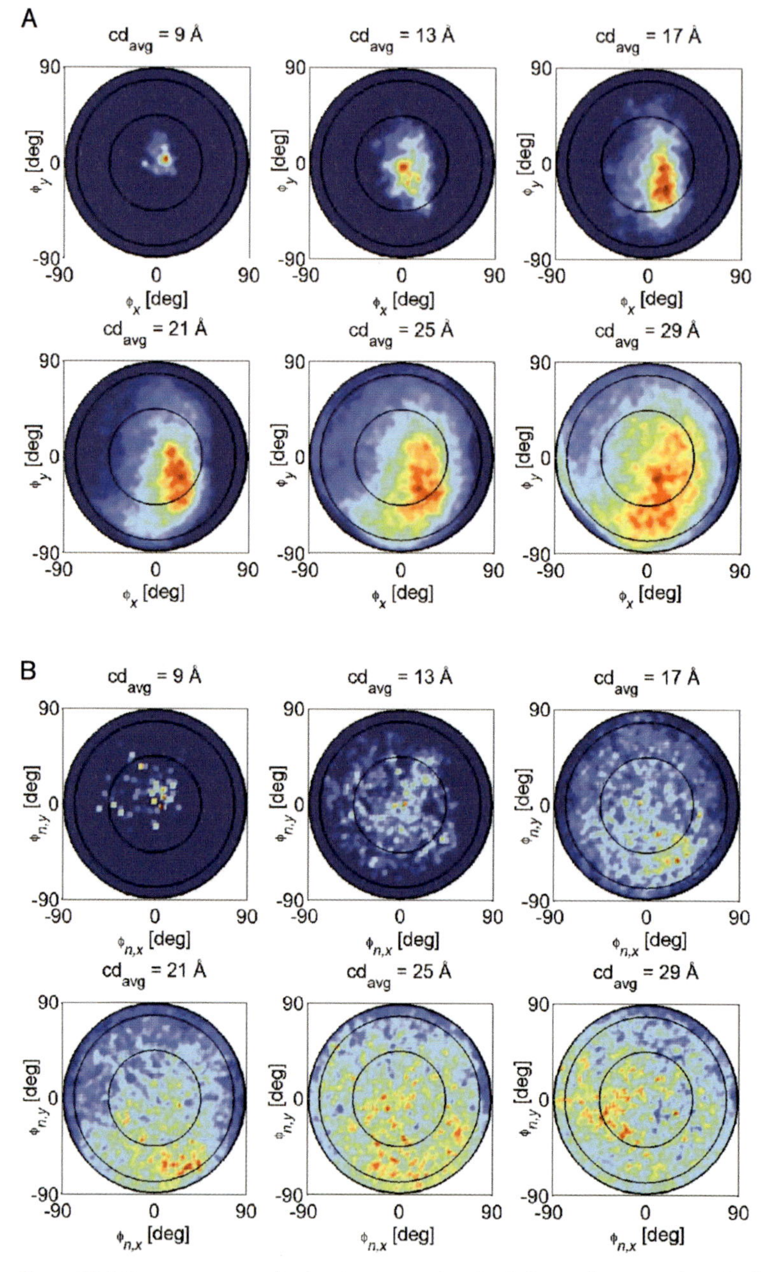

Figure 9.14 Occupancy maps for barstar at various distances cd_{avg} from barnase: (A) positions of barstar and (B) orientations of barstar. Preferred orientations for barstar at $cd_{\mathrm{avg}} = 2.9$–0.9 nm from barnase. To understand these plots you should imagine yourself standing on the binding interface of protein 1 and looking perpendicularly in the sky. The maps show the projected positions where protein 2 is found in the sky. At larger distances, there is almost no orientational preference. Once the two proteins approach each other more closely, there is a clearly preferred maximum of occupancy perpendicular to the binding patch. This means that the two proteins approach each other in a very carefully carried out maneuver that may be compared to docking a space ship to the international space station.

Occupancy maps can be interpreted as probability distributions for the computation of an entropy landscape. The entropy of a system with N states is:

$$S = -k_B \sum_{n=1}^{N} P_n \ln P_n,$$

where P_n is the probability for each state. Using this relationship, we computed local entropy differences for distribution within spheres around grid points, separately for the translational and for the orientational maps.

The free energy difference for protein–protein encounter was then computed as:

$$\Delta G(cd_{avg}) = \Delta E_{Coul}(cd_{avg}) + E_{desolv}(cd_{avg}) - T\Delta S_{transl}(cd_{avg}) - T\Delta S_{rot}(cd_{avg}),$$

involving the Coulombic interaction energy ΔE_{Coul} computed by solving the Poisson–Boltzmann equation, a desolvation term ΔE_{desolv} that accounts for the unfavorable change in solvation energy of the two proteins, and the contributions of the translational and rotational entropy loss. The results for the association of barnase and barstar are shown in (Figure 9.15).

Interestingly, the position of the encounter state for this protein–protein system is found as a well-localized minimum where the two biological interfaces are about 1 nm apart from each other.

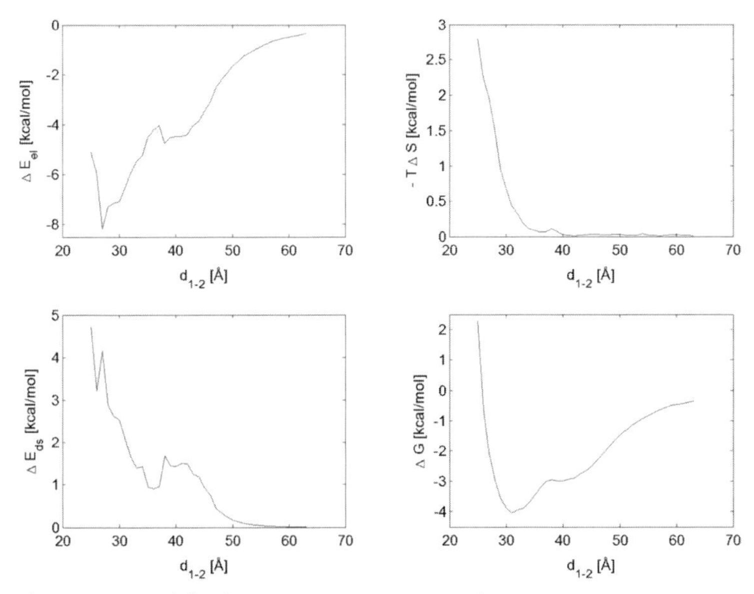

Figure 9.15 (Top left) Electrostatic interaction energy between barnase and barstar at various centre to centre distances d_{1-2}. (Bottom left) Desolvation energy of the two proteins. (Top right) Entropy loss at close distances due to barstar adopting a preferred orientation. (Bottom right) Free energy of interaction between barnase and barstar obtained as the sum of the other three panels.

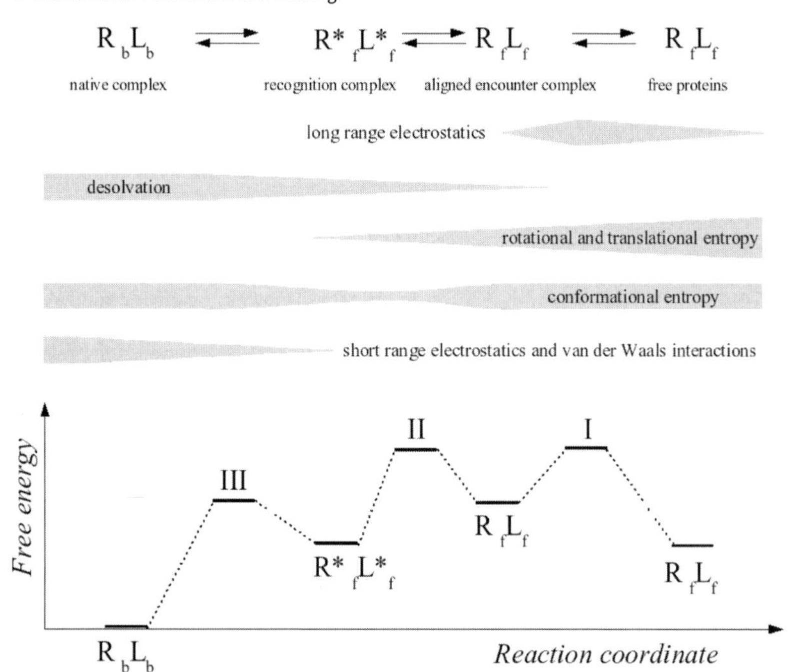

Figure 9.16 Protein–protein association may be governed by diffusion, selection of matching conformers, and refolding. R_f and L_f are the free structure ensembles of receptor and ligand, respectively. R^*_f and L^*_f are subsets of the free receptor and ligand ensembles (recognition conformers). The middle and lower sections of the figure suggest, schematically, the forces involved at the different stages and the resulting free energy profile. Drawn after Grünberg *et al.* (2004).

This model is so far incomplete as association is not followed through to the final, bound complex. Based on molecular dynamics simulations that include explicit solvent molecules, a modified free energy schema was suggested as shown in Figure 9.16.

9.6
Assembly of Macromolecular Complexes: the Ribosome

The way how macromolecular complexes assemble is still a different issue from just asking for their final structures as in Chapter 8. We have seen in Chapters 4 and 8 that the lifetimes of complexes may be fairly limited, and they may be competing for the same proteins as some complexes share common components. Therefore, the issue of dynamic assembly of complexes may have true biological implications. Unfortunately, not much is known so far about these assembly processes. Some of the best studied examples are the virus hulls of icosahedral viruses that are composed of a large number of identical protein copies and the assembly of the 30S subunit of the ribosome. We will use the latter example to discuss both an experimental and a computational way of deriving the potential energy landscape for assembly that ultimately determines the assembly pathway.

Ribosomes are ribonucleoprotein assemblies that are responsible for the translation of the genetic code. They are composed of small and large asymmetric subunits, named according to their sedimentation coefficients – 30S and 50S for bacterial ribosomes. The subunits are held together by a number of intermolecular non-covalent interactions. The 30S subunit consists of the 16S ribosomal RNA and 21 proteins (named S1, S2, etc.) that contain globular protein domains with extended loops that contact the ribosomal RNA. The small ribosomal subunit from *Escherichia coli* can reassemble *in vitro* from the 16S RNA and a mixture of the 30S proteins and form an active particle. Therefore, the pathway and the mechanism of the assembly have been of significant interest. The assembly of the 30S ribosomal subunit resembles a complex dance of macromolecules folding and binding in which 20 proteins bind to rRNA as it folds, creating a complete particle that is competent to participate in the translation of mRNA.

The complexation of the small subunit *in vitro* was found to proceed in a sequential and ordered manner that shows a cooperative character meaning that all the information needed for the small subunit assembly is present in the ribosomal RNA and protein components.

A recent experimental study (Talkington *et al.*, 2005) initiated the assembly of 30S subunits by incubating the *E. coli* 16S rRNA with a mixture of 30S proteins (S2–S21) that were uniformly labeled with the ^{15}N nitrogen isotope instead of the normal ^{14}N isotope. At various time points, binding of the ^{15}N-labeled proteins was competitively blocked with an excess of unlabeled ^{14}N proteins. Completely formed 30S subunits were then purified and the ^{15}N/^{14}N ratio for each protein was determined by mass spectroscopy. In this way, the binding kinetics of all ribosomal proteins could be determined by a single series of experiments. The experiment showed that protein binding rates were weakly affected by concentration indicating that RNA folding and protein binding occur at similar rates. Figure 9.17 shows the binding rates of individual subunits (top), their roles in the overall assembly map (middle) and their positions in the crystal structure of the ribosome (bottom).

Just as macromolecular folding pathways are nowadays pictured as folding landscapes that can be traversed by any of various parallel pathways, so too can the assembly of a multi-component protein:RNA complex, the 30S subunit, now be represented by a free energy landscape (Figure 9.18).

In this landscape representation, all possible conformations of the 16S rRNA map onto a free energy surface, but in the absence of proteins the native 30S conformation is energetically unfavorable. Folding can proceed along many possible pathways to the native state because the landscape is composed of many local and modes barriers. A unique feature of the 30S landscape, as compared with unimolecular folding landscapes or those for biomolecular association (see Figure 9.16) is the intermolecular protein binding, which alters the shape of the free-energy surface during the assembly process. Once RNA folding produces a new binding site, protein binding creates new downhill directions by which further RNA folding can proceed. Each protein binding event further stabilizes the native 30S conformation until all assembly pathways converge at this stage.

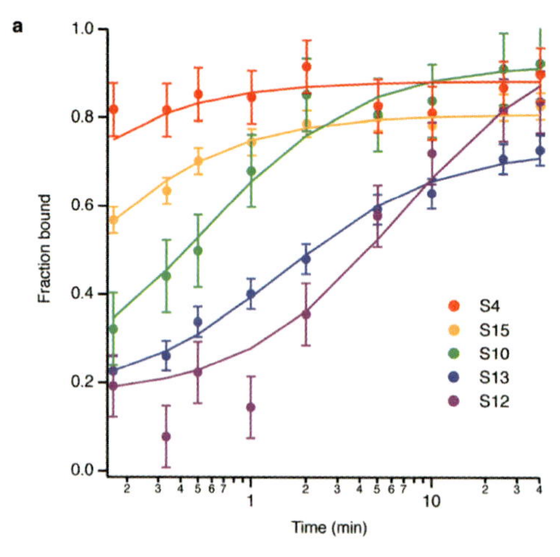
a
Fraction bound
Time (min)
S4
S15
S10
S13
S12

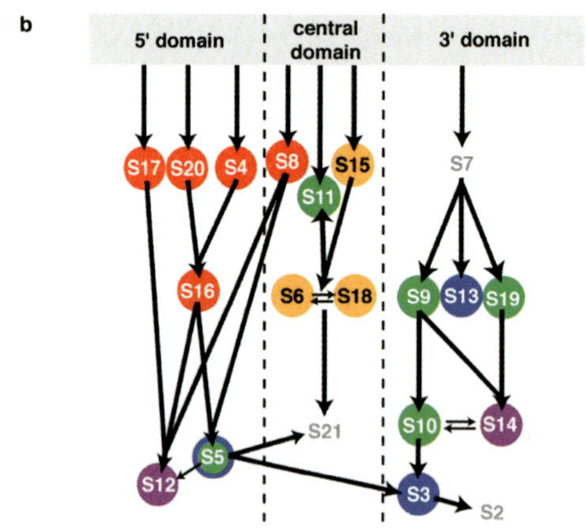
b
5' domain
central domain
3' domain
S17
S20
S4
S8
S15
S11
S7
S16
S6
S18
S9
S13
S19
S10
S14
S21
S5
S12
S3
S2

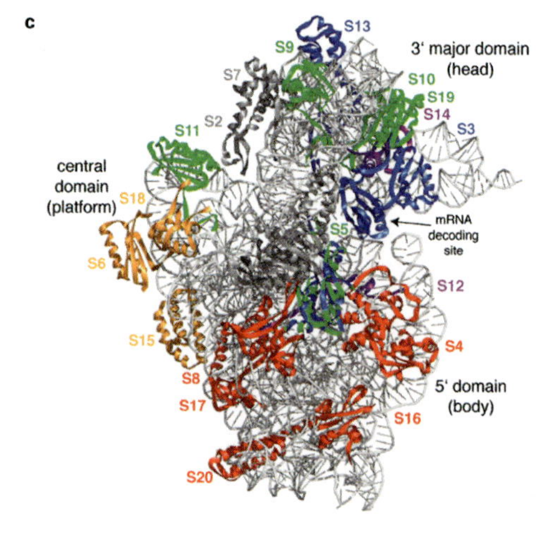
c
S13
S9
3' major domain (head)
S7
S10
S19
S14
S3
S11
S2
central domain (platform)
S18
S6
S5
mRNA decoding site
S12
S15
S4
S8
5' domain (body)
S17
S16
S20

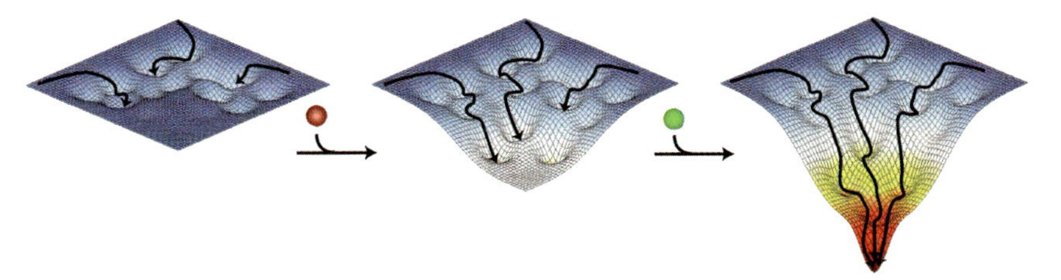

Figure 9.18 Assembly landscape for 30S assembly. The horizontal axes of the surface correspond to 16S rRNA conformational space and the vertical axis is free energy. The native conformation of the 16S rRNA adopted in the 30S subunit is located at the bottom corner. In the absence of proteins, this is not the lowest-energy conformation of the RNA. Parallel folding pathways are indicated by the arrows on the energy surface. Local folding creates protein binding sites, and large changes in the landscape accompany protein binding (colored spheres). Sequential protein binding eventually stabilizes the native 30S conformation. All pathways converge at this point, and there is no bottleneck through which all folding trajectories must pass. Reprinted from Talkington *et al.* (2005) by permission from Macmillan Publishers Ltd.

The same question of ribosome assembly was also addressed in a recent computational study (Trylska *et al.*, 2005). Generally, the free energy of association of two molecules may be written as:

$$\Delta G_{\mathrm{calc}}^{\mathrm{bind}} = \Delta G_{\mathrm{elec}} + \Delta G_{\mathrm{np}} + \Delta G_{\mathrm{strain}} - T\Delta S_{\mathrm{conf}} - T\Delta S_{\mathrm{trans+rot}}.$$

where ΔG_{elec} is the electrostatic contribution to binding, ΔG_{np} is the nonpolar interaction free energy, $-T\Delta S_{\mathrm{conf}}$ describes the loss of configurational main-chain and side-chain entropy upon complexation, $-T\Delta S_{\mathrm{trans+rot}}$ represents the loss of rigid body translational and rotational degrees of freedom and a possible change in vibrational motions, and $\Delta G_{\mathrm{strain}}$ represents the reorganizational cost and distortions upon complexation of the binding species.

The electrostatic ΔG_{elec} interaction free energy includes the electrostatic interactions between both molecules and the surrounding model (that was modeled as a continuum of high dielectric) computed from the Poisson–Boltzmann equation and the direct Coulombic interaction between the atomic charges of both molecules. As mentioned in Section 9.4, the nonpolar term is typically modeled as:

$$\Delta G_{\mathrm{np}} = \gamma \cdot \Delta \mathrm{SASA},$$

with the microscopic surface tension γ as proportionality coefficient. Unfortunately, it is currently unfeasible to compute the entropic terms and the strain term in an accurate fashion. Instead, one may use an empirical relationship for the loss of protein

Figure 9.17 Measured binding kinetics of 30S proteins. (Top) Progress curves for protein binding of five representative subunits. (Middle) Proteins in the assembly map are colored by their binding rates. The fastest rates from 20 to $\geq$30 min^{-1} are shown in red. Green proteins bind at rates from 1.2 to 2.2 min^{-1}, blue from 0.38 to 0.73 min^{-1} and the slowest rates are found for the purple proteins with 0.18–0.26 min^{-1}. (Bottom) Positions of the subunits in the X-ray structure of the 30S ribosome from *Thermus thermophilus*. Reprinted from Talkington *et al.* (2005) by permission from Macmillan Publishers Ltd.

side-chain conformational entropy, which takes on values between 2 and 8 kJ/mol per side-chain buried at the interface. The final estimate of the binding free energy was:

$$\Delta G_{\text{calc}}^{\text{bind}} \approx \Delta G_{\text{elec}} + \gamma \cdot \Delta\text{SASA} + \Delta G_{\text{strain}} + B \cdot NR_{\text{buried residues}} + C,$$

where NR_{buried} $\text{RLR}_{\text{residues}}$ is the number of residues buried upon binding for each protein, and B and C are constants that were fitted to experimental data available for four of the 30S proteins by a least square fit.

It turned out that, in most cases, the calculated ΔG_{elec} interactions opposed binding because it is more favorable for proteins to interact with the high dielectric solvent, which has a strong ability to reorganize, than with RNA which is more constrained and cannot adjust that much to a protein. The nonpolar contribution depends on the amount of SASA buried upon complexation of the molecules. Interestingly, a roughly linear relationship was found between the SASA buried upon binding to the naked 16S RNA and to the 30S complex and the charge of the protein. This means the higher the charge of the 30S protein, the more buried it gets upon binding. This is a counter-intuitive behavior as the surface is typically the most polar part of a biomolecule. Here, it may reflect a unique property of the ribosome that largely consists of highly charged ribonucleic acid. The proteins binding to the 5′ domain show an overall higher change in SASA when binding to 16S RNA. The

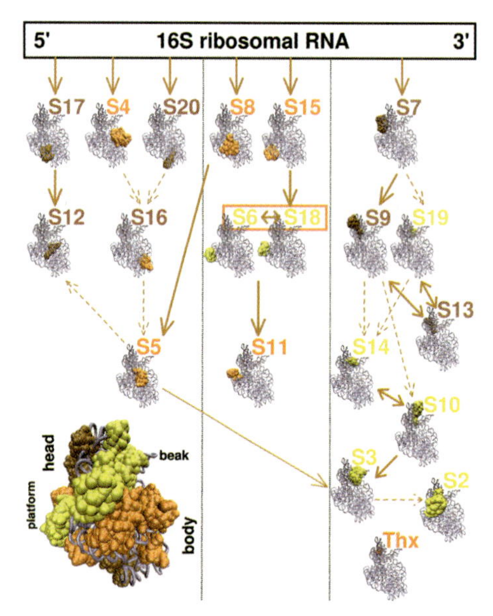

Figure 9.19 Binding affinity assembly map of the *T. thermophilus* 30S ribosomal subunit based on the results from calculation of binding free energies. Strong binders = dark brown, middle binders = orange, weak binders = yellow (colors based on their binding free energies to the naked 16S RNA). Insets denote the positions of appropriate proteins in the 30S subunit looking from the solvent side. Filled brown arrows indicate detected increase in the binding free energy due to the presence of the protein the arrow originates from. Brown dashed arrows indicate no increase in the binding free energy. Courtesy of Dr. Joanna Trylska. Reprinted with permission from the American Chemical Society.

lowest change in SASA upon binding is associated with proteins attaching to the central domain. Remarkably, all proteins get buried to some extent upon binding to 16S RNA, meaning that the tertiary proteins do not bind on the top of the primary or secondary ones but all of them interact with RNA. Figure 9.19 shows the assembly map based on the computed association energies.

We note an overall similarity of Figures 9.19 and 9.17(middle) with some differences in the positioning of individual subunits such as S5 and S11, for example. One has to take into account, however, that the computational study relied on the folded conformation of 16S rRNA as seen in the crystal structure, whereas the experimental study argued that rRNA folding proceeds simultaneously with complex assembly.

The area of characterizing the assembly of protein complexes is still a rather young research area. Further progress can be expected by combining energetic analyses with approaches such as CombDock (Section 8.8).

Summary

It is very hard to give a general answer as to how much detail we currently understand the interaction and association of particular protein–protein pairs. Certainly, much progress has been made for certain model systems such as the barnase:barstar complex. Here, a thorough level of understanding can be attested due to the combination of a large number of experimental and computational studies. The situation is much worse if we were asked to predict the modes of interaction and of association for an arbitrary protein:protein pair if we were only given the information that they do interact. As was shown for the example of human growth hormone and its receptor, even for such a well characterized system, there are still many surprises that we cannot explain today. This area will certainly require some patience and continuous efforts in the near and mid future.

Further Reading

Twilight Zone

Rost B (1999) Twilight zone of protein sequence alignments, *Protein Engineering*, **12**, 85–94.

Modeling Complexes by Homology

Aloy P, Ceulemans H, Stark A, Russell RB (2003) The relationship between sequence and interaction divergence in proteins, *Journal of Molecular Biology*, **332**, 989–998.

Lu L, Arakaki AK, Lu H, Skonick J (2003) Multimeric threading-based prediction of protein–protein interactions on a genomic scale: application to the *Saccharomyces cerevisiae* proteome, *Genome Research*, **13**, 1146–1154.

Phage Display Study for hGH

Pál G, Kouadio J-LK, Artis DR, Kossiakoff AA, Sidhu SS (2006) Comprehensive and quantitative mapping of energy landscapes for protein–protein interactions by rapid combinatorial scanning, *Journal of Biological Chemistry*, **281**, 22378–22385.

Protein–Protein Interactions

Grünberg R, Leckner J, Nilges M (2004) Complementarity of Structure Ensembles in Protein-Protein Binding, *Structure*, **12**, 2125–2136.

Reichmann D, Rahat O, Cohen M, Neuvirth H, Schreiber G (2007) The molecular architecture of protein–protein binding sites, *Current Opinion in Structural Biology*, **17**, 67–76.

Conservation at Protein–Protein Interfaces

Mintseris J, Weng Z (2005) Structure, function, and evolution of transient and obligate protein–protein interactions, *Proceedings of the National Academy of Sciences USA*, **102**, 10930–10935.

Correlated Mutations

Pazos F, Valencia A (2002) *In silico* two-hybrid system for the selection of physically interacting protein pairs, *Proteins*, **47**, 219–227.

Halperin I, Wolfson H, Nussinov R (2006) Correlated mutations: advances and limitations. A study on fusion proteins and on the Cohesin–Dockerin families, *Proteins*, **63**, 832–845.

Forces in Biomolecular Interactions

Chandler D (2007) Oil on troubled waters, *Nature*, **445**, 831–832.

Brownian Dynamics Simulations

Ermak DL, McCammon JA (1978) Brownian Dynamics with Hydrodynamic Interactions, *Journal of Chemical Physics*, **69**, 1352–1360.

Spaar A, Dammer C, Gabdoulline RR, Wade RC, Helms V (2006) Diffusional encounter of barnase and barstar, *Biophysical Journal*, **90**, 1913–1924.

Assembly of the Ribosome

Talkington MWT, Siuzdak G, Williamson JR (2005) An assembly landscape for the 30S ribosomal subunit, *Nature*, **438**, 628–632.

Trylska J, McCammon JA, Brooks III CL (2005) Exploring assembly energetics of the 30S ribosomal subunit using an implicit solvent approach, *Journal of the American Chemical Society*, **127**, 11125–11133.

10
Integrated Networks

At various places throughout the text, it was mentioned that separate studies of the protein–protein interaction network, the gene regulatory network or the metabolic network are only revealing pieces of the full network – ideally, we should be investigating the different networks all at once. Many researchers in this field have of course realized this, and more and more studies are forthcoming where different aspects of cellular networks are integrated. Here, we will look at several of those studies published in the last few years. We will not be concerned with the technical details so much as most of these studies integrate methods that we already encountered throughout this textbook.

10.1
Correlating Interactome and Gene Regulation

Most research on biological networks in the past has been focused on static topological properties, describing networks as collections of vertices and edges rather than as dynamic structural entities. In this first example, protein–protein interaction data on yeast proteins (Chapter 3) was integrated with information on the timing of the transcription of specific genes during the yeast cell cycle, obtained from DNA microarray timeseries (De Lichtenberg *et al.*, 2005). The expression studies identified 600 proteins whose expression levels change periodically during the cell cycle. Out of these, 416 were not involved in physical interactions of significant confidence. These are shown in Figure 10.1 outside of the circle.

The remaining 184 of these dynamic proteins were involved in interactions with 116 static proteins whose expression levels throughout the cell cycle are about constant. These interaction pairs are shown in the interior of the cycle of Figure 10.1. Their positions reflect the expression of the dynamic components. Apart from binary interactions, the derived cell-cycle network contains 29 heavily interconnected modules. The general design principle appears to be that only some subunits of each complex are transcriptionally regulated in order to control the timing of final assembly. This just-in-time assembly would have an advantage over just-in-time

Principles of Computational Cell Biology – From Protein Complexes to Cellular Networks. Volkhard Helms

ISBN: 978-3-527-31555-0

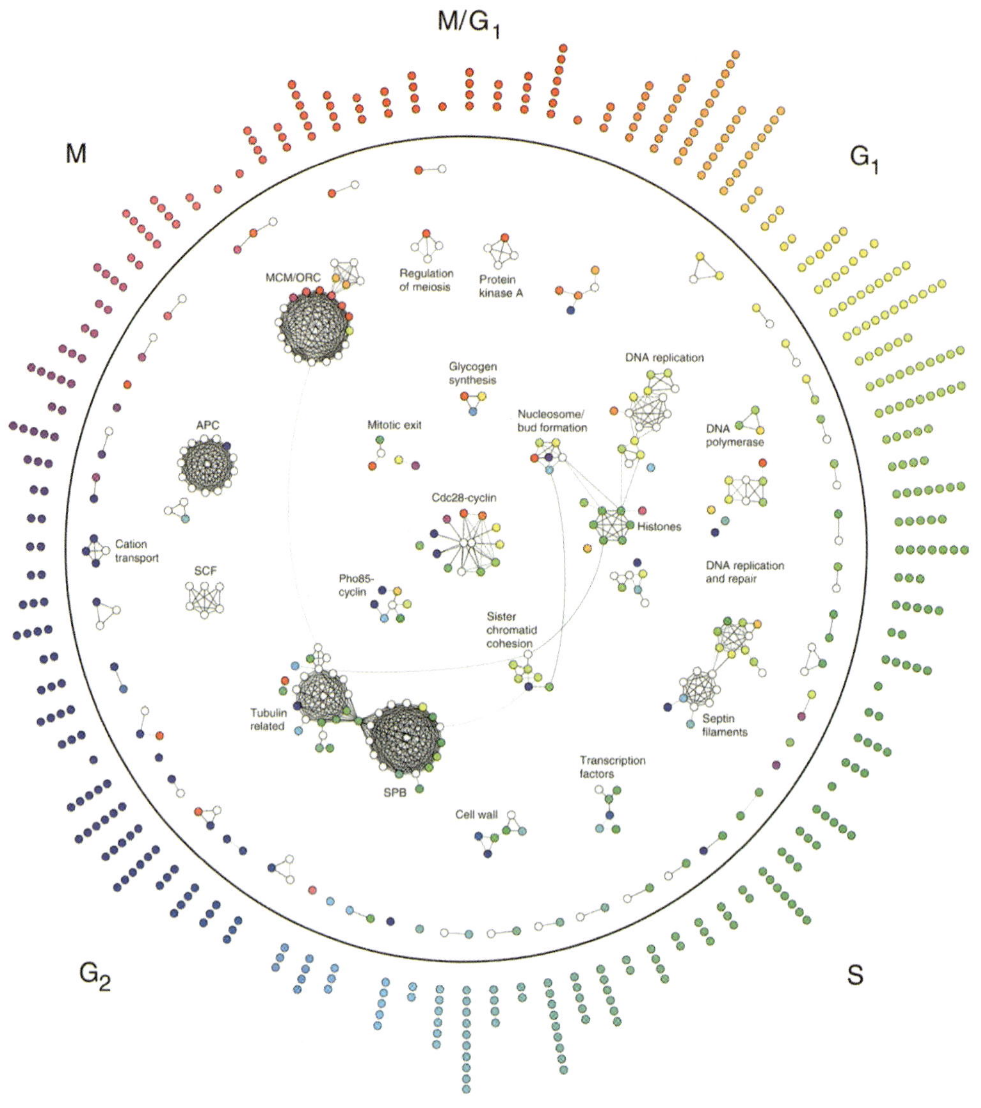

Figure 10.1 Temporal protein interaction network of the yeast mitotic cell cycle. All yeast proteins that are parts of complexes or are involved in other physical interactions are represented inside the cycle. For the dynamic proteins, the time of peak expression is indicated by the vertex color. Static proteins are shown as white nodes. Placed outside the circle are those dynamic proteins without (known) interactions. They are also colored according to their peak maximum. Reprinted from De Lichtenberg *et al.* (2005) by permission from AAAS.

synthesis of entire complexes in that only a few components need to be tightly regulated in order to control the timing of final complex assembly.

Dynamic complex assembly also functions as a mechanism for temporal regulation of substrate specificity, as exemplified by the association of the cyclin-dependent

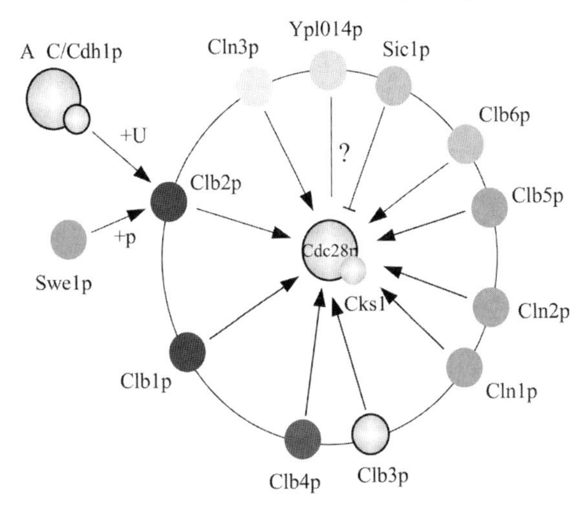

Figure 10.2 Cdc28p module, with the different cyclins and interactors placed at their time of synthesis. At the end of mitosis, the cyclins are ubiquitinated and targeted for destruction by the anaphase-promoting complex (APC) and Skp1/Cullin/F-box (SCF) complexes, reflected in the network by the interaction between Cdh1p and Clb2p. The latter also interacts with Swe1p, which inhibits entry into mitosis by phosphorylating Cdc28p in complex with Clb-type cyclins. Drawn after De Lichtenberg *et al.* (2005).

kinase Cdc28p with its various transcriptionally regulated cyclins and inhibitors (Figure 10.2).

The approach taken accurately reproduces this key regulatory system and its temporal dynamics, correctly placing each of the Cdc28p interactions' partners at their time of function and capturing even very transient interactions such as phosphorylation and ubiquitination.

10.2 Response of Gene Regulatory Network to Outside Stimuli

The second example is a study that integrated transcriptional regulatory information and gene expression data for multiple conditions in yeast (Luscombe *et al.*, 2004). Figure 10.3 (top left) shows the integrated static gene expression network from 240 microarray experiments for five conditions. Altogether, the network contains 7042 regulatory interactions between 142 transcription factors and 3420 target genes. Topological analysis of the underlying network characterized properties such as the in-degree of genes (by how many transcription factors is a gene regulated?), the out-degree of transcription factors (how many genes are regulated on average by the binding of a particular transcription factor?), the path length and the clustering coefficients, see (Figure 10.3, bottom). We have already encountered these topological descriptors in Chapter 4 on protein–protein interactions. Figure 10.3 (top right) presents the subnetworks active under different cellular conditions, and gross changes are apparent in the distinct sections of the network.

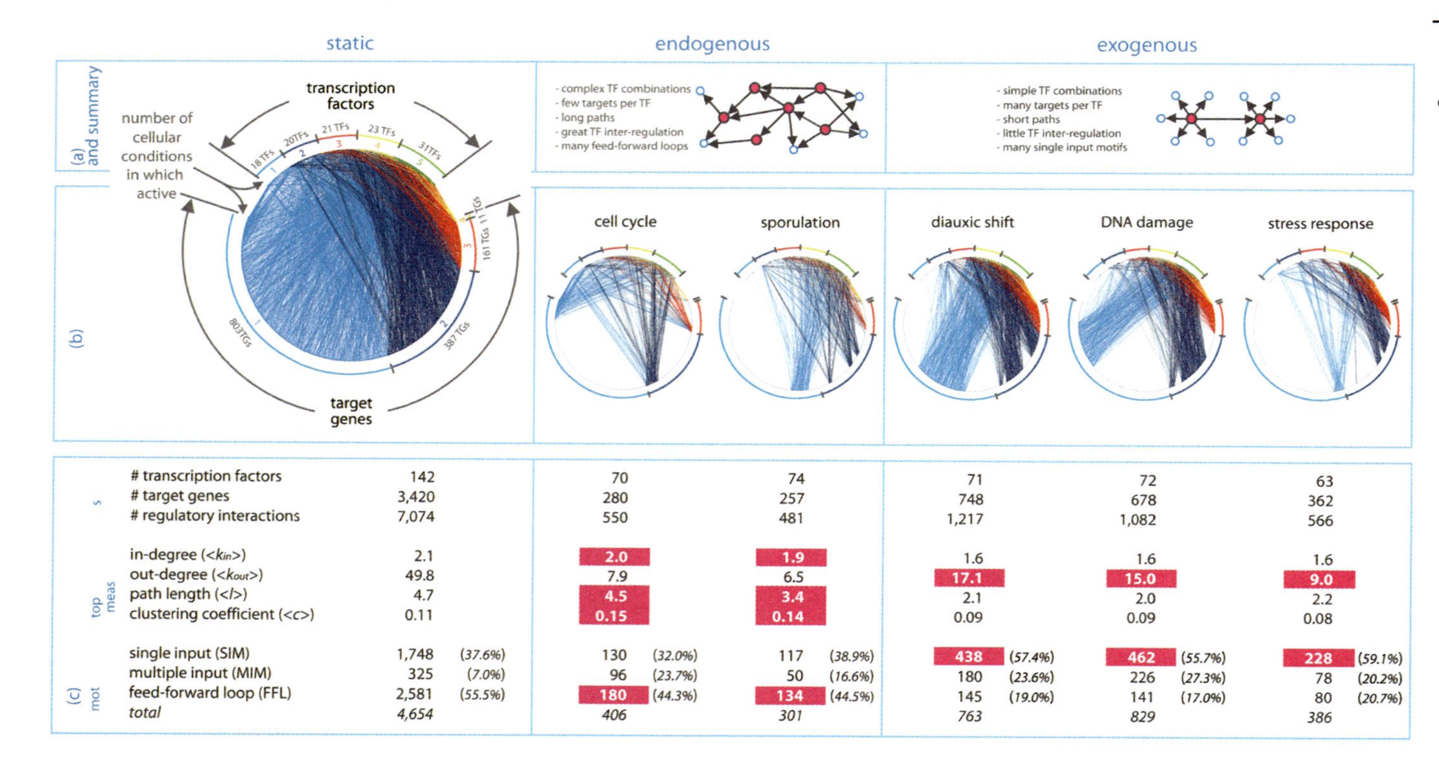

	static		cell cycle		sporulation		diauxic shift		DNA damage		stress response	
# transcription factors	142		70		74		71		72		63	
# target genes	3,420		280		257		748		678		362	
# regulatory interactions	7,074		550		481		1,217		1,082		566	
in-degree ($<k_{in}>$)	2.1		2.0		1.9		1.6		1.6		1.6	
out-degree ($<k_{out}>$)	49.8		7.9		6.5		17.1		15.0		9.0	
path length ($<l>$)	4.7		4.5		3.4		2.1		2.0		2.2	
clustering coefficient ($<c>$)	0.11		0.15		0.14		0.09		0.09		0.08	
single input (SIM)	1,748	*(37.6%)*	130	*(32.0%)*	117	*(38.9%)*	438	*(57.4%)*	462	*(55.7%)*	228	*(59.1%)*
multiple input (MIM)	325	*(7.0%)*	96	*(23.7%)*	50	*(16.6%)*	180	*(23.6%)*	226	*(27.3%)*	78	*(20.2%)*
feed-forward loop (FFL)	2,581	*(55.5%)*	180	*(44.3%)*	134	*(44.5%)*	145	*(19.0%)*	141	*(17.0%)*	80	*(20.7%)*
total	*4,654*		*406*		*301*		*763*		*829*		*386*	

Half of the target genes are uniquely expressed in only one condition. In contrast, most transcription factors are used across multiple processes. The active subnetworks maintain or rewire regulatory interactions and over half of the active interactions are completely supplanted by new ones between conditions. Only 66 interactions are retained across four or more conditions. These comprise "hot links" that are always on (compared with the rest of the network) and mostly regulated housekeeping functions.

Overall, the calculations divided the five condition-specific subnetworks into two categories of endogenous and exogenous processes. This separation allows rationalizing the different subnetwork structures in terms of the biological requirements of each condition. Endogenous processes (cell cycle and sporulation) are multistage and operate with an internal transcriptional programme, whereas exogenous states (diauxic shift, DNA damage and stress response) constitute binary events that react to external stimuli with a rapid turnover of expressed genes.

In biological terms, the small in-degrees for target genes in exogenous conditions indicate that transcription factors are regulating in simpler conditions and the large out-degrees reflect that each transcription factor has greater regulatory influence by targeting more genes simultaneously. The short paths imply faster propagation of the regulatory signal. Conversely, long paths in the multistage endogenous conditions suggest slower action arising from the formation of regulatory chains to control intermediate phases. Finally, high clustering coefficients in endogenous conditions signify greater inter-regulation between transcription factors. In summary, subnetworks have evolved to produce rapid, large-scale responses in exogenous states and carefully coordinated processes in endogenous conditions.

The authors also performed a motif search as introduced in Chapter 5. Most of the common regulatory motifs introduced in Section 5.5, i.e. feed-forward loops, single-input multiple-output(SIM) data and multiple-input multiple-output (MIM) data were observed in the experimental data set analyzed here. This is not surprising because the basic motifs were discovered on a similar data set on the expression of yeast genes. Interestingly, the relative occurrence of motifs varies considerably between endogenous and exogenous conditions. SIMs are favored in exogenous networks (more than 55% of regulatory interactions in motifs). In contrast, endogenous processes favor feed-forward loops (44%) over SIMs (35%). In previous studies, SIMs and MIMs were implicated in conferring similar regulation over large groups of genes, so they are ideal for directing the large-scale gene activation found in

Figure 10.3 Dynamic representation of the transcriptional network and standard statistics. (a) Schematics and summary of properties for the endogenous and exogenous subnetworks. (b) Graphs of the static and condition-specific networks. Transcription factors and target genes are shown as vertices in the upper and lower sections of each graph, respectively, and regulatory interactions are shown as edges; they are colored by the number of conditions in which they are active. Different conditions use distinct sections of the network. (c) Global topological measures and local network motifs describing network structures. These vary between endogenous and exogenous conditions. Those that are high compared with other conditions are shaded. Reprinted from Luscombe *et al.* (2005) by permission from Macmillan Publishers Ltd.

exogenous conditions. Feed-forward loops are buffers that respond only to persistent input signals. They are suited for endogenous conditions, as cells cannot initiate a new stage until the previous one has stabilized.

10.3
Integrated Analysis of Metabolic and Regulatory Networks

The picture emerging from studying the connectivities of large-scale cellular networks showed a densely woven web where almost everything is connected to everything. In the cell's metabolic network, hundreds of substrates are interconnected through biochemical reactions. Although this could in principle lead to the simultaneous flow of substrates in numerous directions, in practice metabolic fluxes pass through specific pathways (high flux backbone, see Section 6.8). Topological studies so far did not consider how the modulation of this connectivity might also determine network properties.

The third example discussed in this chapter is a study that correlated the experimental expression data for yeast with the structure of traditional metabolic pathways (Ihmels *et al.*, 2003). To identify relevant experimental conditions from a data set of over 1000 conditions, all conditions in the data set were scored by their average expression over the input set of genes. Only those conditions were selected whose absolute score exceeds a given condition threshold. Then, all genes were scored by the weighted average for this set of selected conditions. Again, only those genes were selected having a score greater than a gene threshold. This two-stage analysis was termed a "signature algorithm" because it reveals the "signature" of each experimental setup and the "signature" of each gene across various experiments. Its output is a set of genes together with their association scores. Using this technique showed that genes belonging to the traditional Kyoto Encyclopedia of Genes and Genomes (KEGG) pathways show a significantly higher correlation than random pairs of genes. However, typically only subsets of the genes assigned to a given pathway were in fact coregulated. For example, of the 46 genes assigned to the glycolysis pathway in the KEGG database, only 24 showed a correlated expression pattern.

Based on this reassuring finding, the authors went on to search for additional associated genes showing similar expression profiles with a particular pathway using the signature algorithm. Interestingly, but not unexpected, this analysis revealed numerous genes that are not directly involved in the enzymatic steps of metabolic pathways. Examples of such coexpressed genes are genes coding for membrane transporters that provide a particular metabolic pathways with its metabolites (Figure 10.4).

Other genes found to be coregulated are transcription factors that are often coregulated with their regulated pathways. One may wonder whether transcriptional regulation also defines a higher-order metabolic organization, by coordinated expression of distinct metabolic pathways? Indeed, feeder pathways (which synthesize metabolites) are frequently coexpressed with pathways using the synthesized

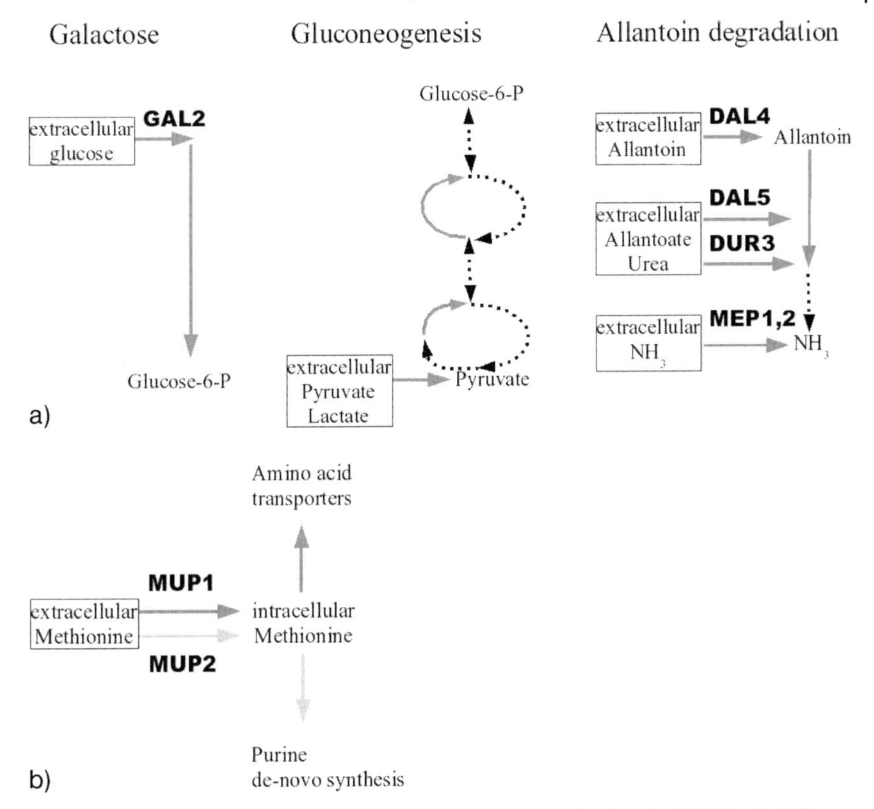

Figure 10.4 (Top) Transporter genes are coexpressed with the relevant metabolic pathways providing the pathway with its metabolites (coexpression is marked by grey arrows). (Bottom) Here, differential coexpression of either the MUP1 or MUP2 transporter (that transport the amino acid methionine into the cell) with the genes belonging to either the upper or lower pathway. Drawn after Ihmels *et al.* (2003).

metabolites (Figure 10.5). These results can be interpreted in that the organism will produce those enzymes that are needed.

Finally, the authors addressed the question whether gene regulation affects the topological properties of the metabolic network discussed in Chapter 6. Figure 10.6(left) shows the hierarchical organization of the metabolic network that was derived by Jeong *et al.* (2000) by analyzing the connectivity between substrates considering all potential connections. When analyzing the connectivity between substrates, considering all potential connections, the structural connectivity between metabolites was shown to impose a hierarchical organization of the metabolic network (Jeong *et al.*, 2000) (Fig. 10.6, left). Importantly, when including expression data the connectivity pattern of metabolites changes from a power-law dependence to an exponential one, corresponding to a network structure with a defined scale of connectivity (Fig. 10.6, right). This finding reflects that transcription regulation reduces the complexity of the metabolic network of *Saccharomyces cerevisiae*. Transcription of genes leads the metabolic flow toward linearity.

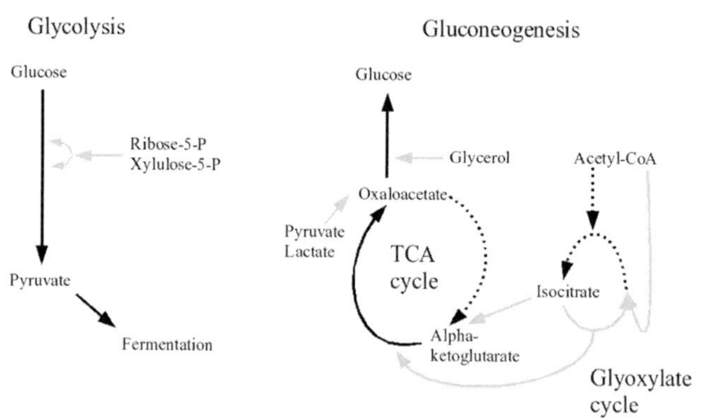

Figure 10.5 Feeder pathways or genes coexpressed with the pathways they fuel. The feeder pathways (light grey line) provide the main pathway (dark line) with metabolites in order to assist the main pathway, indicating that coexpression extends beyond the level of individual pathways. Drawn after Ihmels *et al.* (2003).

Summary

In daily life, looking at many properties simultaneously is generally considered more difficult than looking at them one by one. We argue that this is likely not the case for cellular networks. From the few examples collected in this chapter, one can feel that solving the secrets of cellular networks is going to be achieved by some of these integrated approaches. As simple as they may still be, they have often revealed a far richer behavior than those studies that focused on isolated features. Therefore, we can be quite hopeful to see fast progress in this research area.

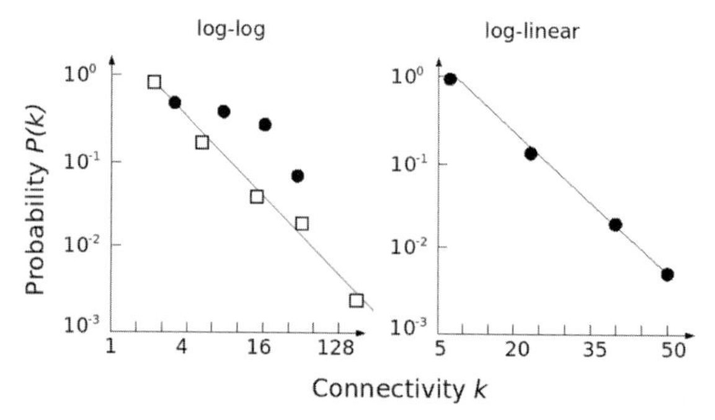

Figure 10.6 The connectivity of a metabolite is defined as the number of reactions connecting it to other metabolites. Shown are the distributions of connectivity between metabolites in an unrestricted network (open squares) and in a network where only correlated reactions are considered (filled circles). In accordance with previous results (Jeong *et al.*, 2000), the connectivity distribution between metabolites follows a power law (log-log plot on the left). In contrast, when coexpression is used as a criterion to distinguish functional links, the connectivity distribution becomes exponential (log-linear plot on the right).

Further Reading

Integrating Gene expression with Protein–Protein Interactions

De Lichtenberg U, Jensen LJ, Brunak S, Bork P (2005) Dynamic complex formation during the yeast cell cycle, *Science* **307**, 724–727.

Dynamics and Rewiring of Gene Regulatory Networks

Luscombe NM, Babu MM, Yu H, Snyder M, Teichmann SA, Gerstein M (2004) Genomic analysis of regulatory network dynamics reveals large topological changes, *Nature* **431**, 308–312.

Correlating Metabolic and Regulatory Networks

Ihmels J, Levy R, Barkai N (2003) Principles of transcriptional control in the metabolic network of *Saccharomyces cerevisiae, Nature Biotechnology* **22**, 86–92.
Jeong H, Tombor B, Albert R, Oltvai ZN, Barabási AL (2000) The large-scale organization of metabolic networks, *Nature* **407**, 651–654.

11
Outlook

Our understanding of cellular networks has been greatly advanced over the past few years due to the advent of modern proteomics techniques. Much to be discovered about these networks and their components still lies ahead of us. Therefore, this textbook has tried to put an emphasis on computational methodologies used in analyzing cellular networks instead of presenting the latest details about the networks of particular organisms. Here, one can expect rapid progress and any monograph will be quickly outdated.

We feel that the connections between network modeling and the structural details of the molecules involved need to be strengthened in the forthcoming years. When it comes to engineering cellular networks, one has to deal with molecules, not with fluxes. Therefore, the structural details of interaction patches should certainly be part of the systemic description as well.

Several robust methodologies have been developed that allow studying individual levels of the entire cellular network with, apparently, sufficient levels of success. Here, we have placed more emphasis on bottom-up approaches instead of data-driven approaches as we feel that results from bottom-up approaches are easier to interpret. However, particularly when it comes to integrating several layers of networks, we expect data-driven methods or hybrid approaches to be of great importance.

Challenges that lie ahead involve data, software and model accessibility. Much of the software and code is not made readily available, and needs to be developed into formats that are accessible and accepted by experimental biologists. Here, the "virtual cell" initiative is truly at the forefront of such initiatives. One problem is certainly that Advanced Calculus is not typically part of biology curricula, at least not in Germany. Much closer to the topics covered in this textbook seem Bioengineering and Biotechnology programs.

Principles of Computational Cell Biology – From Protein Complexes to Cellular Networks. Volkhard Helms

ISBN: 978-3-527-31555-0

Index

Principles of Computational Cell Biology – From Protein Complexes to Cellular Networks. Volkhard Helms

ISBN: 978-3-527-31555-0